FUNDAMENTALS OF GENERALIZED RECURSION THEORY

STUDIES IN LOGIC

AND

THE FOUNDATIONS OF MATHEMATICS

VOLUME 105

NORTH-HOLLAND PUBLISHING COMPANY
AMSTERDAM · NEW YORK · OXFORD

FUNDAMENTALS OF GENERALIZED RECURSION THEORY

MELVIN FITTING
Herbert H. Lehman College
City University of New York

1931 NHPC 1981

NORTH-HOLLAND PUBLISHING COMPANY
AMSTERDAM · NEW YORK · OXFORD

ISBN:0 444 86171 8

Published by:

North-Holland Publishing Company — Amsterdam · New York · Oxford

Sole distributors for the U.S.A. and Canada:

Elsevier North-Holland, Inc.
52 Vanderbilt Avenue
New York, N.Y. 10017

Library of Congress Cataloging in Publication Data

Fitting, Melvin Chris.
Fundamentals of generalized recursion theory.

(Studies in logic and the foundations of mathematics ; v. 105)
Bibliography: p.
1. Recursion theory. I. Title. II. Series.
QA9.6.F57 511.3 81-2145
ISBN 0-444-86171-8 AACR2

PRINTED IN THE NETHERLANDS

To my wife Greer,
who heard this book,
patiently,
many times,
and never failed to make it better.

"That's another thing we've learned from *your* Nation," said Mein Herr, "map-making. But we've carried it much further than *you.* What do you consider the *largest* map that would be really useful?"

"About six inches to the mile."

"Only *six inches*!" exclaimed Mein Herr. "We very soon got to six *yards* to the mile. Then we tried a *hundred* yards to the mile. And then came the grandest idea of all! We actually made a map of the country, on the scale of *a mile to the mile*!"

"Have you used it much?" I enquired.

"It has never been spread out, yet," said Mein Herr: "the farmers objected; they said it would cover the whole country, and shut out the sunlight! So we now use the country itself, as its own map, and I assure you it does nearly as well. ..."

Sylvie and Bruno Concluded
Chapter XI, The Man in the Moon
Lewis Carroll

INTRODUCTION

I. We describe, briefly, the contents of the book, and the approach to the material that we have elected to follow.

We treat *ordinary recursion theory* and one of its natural generalizations to arbitrary structures, *search computability*; and we treat *hyperarithmetic theory* and one of its abstractions to arbitrary structures, *hyperelementary theory*. Both are dealt with simultaneously using *generalized elementary formal systems*, which are simple axiomatic systems of derivation, and which can be thought of as being rules for the generation of a set or relation.

Elementary formal systems were announced in Smullyan [1956A], and given in detail in Smullyan [1961]. As presented there, they constitute a mechanism for developing ordinary recursion theory only, thought of as a theory of words over a finite alphabet. Traced further back, elementary formal systems themselves have their origin in Post's canonical systems (Post [1943]), and can be looked at as a special case, albeit a very elegant and simple special case. The various generalizations of elementary formal systems presented here (to arbitrary structures, allowing inputs, allowing infinitary rules, etc.) are new. It is the elementary formal system mechanism, modified one way or another, that provides the underlying uniformity of treatment in this book. (From now on, we drop the qualifier 'generalized'; we use 'elementary formal system' systematically in the generalized sense.)

If one allows elementary formal systems to accept input, one gets natural generalizations of *enumeration operators* (Rogers [1967], pp. 146–147), and these are actually the main objects of study in much of this book. They provide a common generalization of partial recursive functions and of recursively enumerable relations: one can identify functions with operators which, when given a 1-element input, give an output with 0 or 1 elements; one can identify relations with outputs of constant operators. The development is index-free; after the fact we investigate under what conditions an indexing is possible.

We treat *α-recursion theory*, using another generalization of elementary formal systems. Here our treatment is more of a survey nature, to show similarities with the preceeding material. Again, operators are brought in, in this case they are the *α-enumeration operators.*

All the foregoing theories have many properties in common; we abstract these properties into an axiomatic concept which we call a *production system* (no direct relation of Post production systems, though). We derive abstractly, for production systems, under various extra hypotheses, the basic material of elementary recursion theory (up to priority arguments), all of which then applies to many of the theories mentioned above.

We develop a theory of *embedding* between production systems, and of *effective embedding.* This can be thought of as a generalization of Gödel numbering. The approach is algebraic and uses concepts from category theory.

Finally we use elementary formal systems, generalized still further, to extend the notion of enumeration operator to that of *effective operator of higher type*, again for arbitrary structures.

The subject matter of this book is becoming standard; the particular development we have chosen is rather individualistic. The reader will see proofs of equivalence between our presentations and some of those more common in the literature, but the development using generalized elementary formal systems will not be found elsewhere. Its advantage is in the resulting uniformity of treatment. A disadvantage, of course, is that the rest of the literature may look unfamiliar. At the least, we can offer as compensation a very brief survey of how recursion theory came to have the particular generalizations we treat, and some references to original papers.

II. In the 1930's, many people independently tried to formally characterize the intuitive notion of "effectively calculable" for the structure of arithmetic $\langle \mathbb{N}, s \rangle$, where the domain $\mathbb{N}$ is the set of natural numbers, and the successor function s is assumed given. As is well known, these attempts all led to the same theory. This theory is *ordinary recursion theory* and the Church–Turing Thesis asserts that it does capture the intuitive notion it was intended to capture. We are interested in generalizations of ordinary recursion theory; most have arisen by changing the structure involved, by broadening the "allowed machinery of computation" or by some mixture of the two, although it was not always apparent at the start that this is what was happening.

In a sense, recursion theory was generalized from its start as a formal mathematical subject. Some people worked with numbers (Church [1936], the source of Church's Thesis), while others worked with words (Post [1936], and Turing [1936], the source of Turing's Thesis). Now the collection of words over a fixed finite alphabet, and the collection of natural numbers, are isomorphic in ways that most people would consider effective. So it was not thought to be important that two rather different structures were in use. Also the techniques available for developing recursion theory using these two structures did not seem to admit of a further generalization to any of the other structures met in mathematics.

The study of *relative recursion theory* also goes back to the beginnings of the subject. It asks, what is 'effectively calculable' if, in addition to the usual machinery, one allows the use of some property P of the natural numbers (which need not itself be intuitively effective). It can be thought of as a generalization of ordinary recursion theory to the structure $\langle \mathbb{N}, s, P \rangle$. Again, it was unclear how to generalize further, and vary the domain as well as the given relations.

Beginning with the relatively unknown *R-definability* in 1956, several ways were independently invented to develop for arbitrary structures analogs of ordinary recursion theory on the structure of arithmetic, in which the intention was to keep the intuitive notion of 'finitary effective computation' as intact as possible. Since it is arguable just what 'effectively calculable' should mean for an arbitrary structure, which itself may not be capable of being effectively presented (see Grillot [1974] for more on this) it is not surprising that essentially different theories could be produced. What is surprising, perhaps, is that a large number of the approaches turned out to be equivalent: R-definability (Smullyan [1956]); Fraïsse-computability (Fraïsse [1961]); $\forall$-recursiveness, for countable structures (Lacombe [1964, 1964A]); $\aleph_0$-recursiveness (Montague [1968]); search computability (Moschovakis [1969]); and others more recent yet. So whether or not these capture the intuitive notion of 'effective' for arbitrary structures in all cases, at least the convergence of so many separate approaches argues for the naturalness of the idea. Also, for the mild generalizations of ordinary recursion theory mentioned in the previous two paragraphs, these approaches simply give an equivalent of what was already accepted. So, picking out the best-known name from the list as a representative, we think of *search computability* as the primary generalization of ordinary recursion theory to arbitrary structures. We so treat it in this book.

In the 1950's *hyperarithmetic theory* was invented (Davis [1950]; Mostowski [1951]; and Kleene [1955] from where it gets its name) and refined (Spector [1955]). It studies certain subsets of the natural numbers (those on the lowest level of the analytic heirarchy) by studying the way those sets can be characterized. It was not meant to be an 'effective' theory, indeed the sets involved properly include those definable using the full first order language of arithmetic (the arithmetic sets). But it became apparent that hyperarithmetic theory can be thought of as something like ordinary recursion theory with the computation machinery enlarged. In particular, it allows the asking of certain questions that take infinitely long to answer, but otherwise the machinery is essentially that of ordinary recursion theory. And there are parallels in the theorems provable. Consequently various investigators began thinking of hyperarithmetic theory as a sort of generalized recursion theory. Many of the further developments discussed below arose from attempts to make the analogy between hyperarithmetic theory and ordinary recursion theory into something more than an analogy.

The original characterization of the hyperarithmetic sets was a complicated one involving both relative recursion theory and the theory of the recursive ordinals (a recursive ordinal is an ordinal that is the order type of some recursive well-ordering of the natural numbers). As such, it was not clear how to generalize the notion to structures other than that of arithmetic. In the 1970's an equivalent approach using inductive definitions became central, and this generalizes very well. *Hyperelementary theory* is the name that has been given to this generalization (Moschovakis [1974]). And research has shown parallels between the hyperelementary theory of a structure and the search computability theory of that structure; hyperelementary theory behaves rather like search computability with some infinitary computational machinery allowed.

Hyperarithmetic theory is a theory about the natural numbers. As originally developed, it used the recursive ordinals as a tool. In the 1960's, Kreisel and Sacks suggested reversing things; use hyperarithmetic theory developed by other equivalent means, as a tool to create a theory about the recursive ordinals. The similarities between what resulted and ordinary recursion theory were striking. Indeed, in Rogers [1967] this was simply called the Kreisel–Sacks Analog. Soon, however, it became known as *metarecursion theory* (Kreisel–Sacks [1965], Sacks [1967]).

Meanwhile, Takeuti [1960] was working on a recursion theory of the entire class of ordinals, connecting it with the development of L, the universe of constructible sets, first introduced by Gödel as a tool in showing the consistency of the continuum hypothesis in axiomatic set theory.

Independently, and by different means, Kripke [1964] and Platek [1966] developed an analog of ordinary recursion theory for certain ordinals, called *admissible* or *recursively regular*. Working with the first admissible ordinal yields ordinary recursion theory, while the second gives metarecursion theory. And there are many, many other admissible ordinals, including, but hardly limited to, the cardinals. Now known as *α-recursion theory*, this is a rapidly developing area. It can be thought of as being like ordinary recursion theory with the subject matter generalized from the natural numbers to the initial segment α of the ordinals, and with the means of computation also generalized, replacing 'finite' by 'α-finite', defined in an appropriate way. It has been aptly characterized as the recursion theory of a limited god. The work of Takeuti finds its natural generalization here too, and the connections between recursion theory on the ordinals, and the development of the constructible sets plays a big role in the work on admissible ordinals, making for an interesting relationship between recursion-theoretic and model-theoretic techniques.

Search computability generalized ordinary recursion theory to arbitrary structures, while keeping the meaning of 'finitary' unchanged, as far as possible. Hyperelementary theory similarly generalized hyperarithmetic theory to arbitrary structures. In a similar way, α-recursion theory, developed via the constructible universe L, has been generalized to *admissible sets with urelements* (Barwise [1975]), in which arbitrary structures can be dealt with, and the notion of finiteness generalized at the same time. This too is a rapidly developing area. Search computability can be looked at as a special case, in fact. Also, the connection between hyperarithmetic theory and the α-recursion theory of the second admissible ordinal (metarecursion theory) has been vastly generalized to apply to the hyperelementary theory of any structure that, like $\langle \mathbb{N}, s\rangle$, has a pairing function that can be generated using the machinery of hyperelementary theory. (This is most easily found in Moschovakis [1974] and Barwise [1975], in the references under HYP.)

Relatively early, Kleene introduced into ordinary recursion theory the notion of *recursive functional*. A functional maps functions to numbers; a recursive functional, as Kleene defined it, is a functional that most people would consider to be 'effectively calculable' in an appropriate sense. But it wasn't until Kleene [1959] that a way was found to continue this up the type hierarchy. In this influential paper, Kleene introduced the notion of *recursive functional of finite type*, with type 1 being the usual recursive functions, type 2 being the usual recursive functionals, and with type 3 being 'effective' maps from type 2 functions to numbers, and so on up.

Soon after, Platek [1966] generalized higher type recursion theory so that it too could be developed for essentially arbitrary structures.

Once again, hyperarithmetic theory comes in. An early result of Kleene showed that hyperarithmetic theory could be looked at as higher type recursion theory relative to a particular type 2 functional (Kleene [1959].) Attempts to extend this have led to major research developments, see Hinman [1978].

Recently, too, partial recursive operators and functionals, familiar from ordinary recursion theory, have been introduced into α-recursion theory as well. (See DiPaola [1978], [1978A], [1979], [1981], [198+].)

Search computability, hyperelementary theory, admissible set theory with urelements, and higher type recursion theory all can be looked at as natural and successful generalizations of ordinary recursion theory. Since they all agree in some respects, and differ in others, naturally a kind of axiomatic recursion theory has arisen, to treat of these similarities and differences. In fact, there have been several such. The most successful and most widely adopted are those called 'Computation Theories'. These had their ancestry in Strong [1968] and Wagner [1969], but were properly born in Moschovakis [1971], and had their descent through Fenstad [1974]. They are summarized in Fenstad [1975], applied rather thoroughly to higher types in Moldestad [1977], and are given a compact presentation in Fenstad [1980].

III. We assume knowledge of a small amount of the material of an elementary logic course, enough to deal comfortably with models, and truth in them. We also assume knowledge of the content of a basic course in set theory. In particular, we freely use elementary facts about the ordinal numbers, transfinite induction, and the like. Many of our examples make use of properties of the constructible sets, but, outside of Chapter 9, no theory depends on this. The material in Chapter 9 supposes a reader more adept at dealing with the formal nuances of axiomatic set theory. Strictly speaking, we do not suppose any prior knowledge of ordinary recursion theory, though anything the reader may already know in this area will certainly be found helpful. We use a small amount of the terminology of category theory. For the most part, nothing beyond the basic definitions is needed; category theory is, here, merely an organizing device.

IV. We wish to thank some books, some institutions and some people for contributing to the present work. We begin with the books.

(1) "Theory of Formal Systems" by Raymond Smullyan (Smullyan [1961]), in which elementary formal systems first fully appear, and in which an elegant development of ordinary recursion theory, using them, is presented.

(2) "Theory of Recursive Functions and Effective Computability" by Hartley Rogers Jr. (Rogers [1967]), which is an encyclopedic, systematic account of a major portion of recursion theory and its ramifications, as it had developed by the mid 1960's.

(3) "Elementary Induction on Abstract Structures" by Yiannis Moschovakis (Moschovakis [1974]), which presents an elegant generalization of hyperarithmetic theory to arbitrary structures.

(4) "Admissible Sets and Structures" by Jon Barwise (Barwise [1975]), which presents the lovely subject of admissible sets with urelements.

Next, the institutions. The library at Rensselaer Polytechnic Institute, and the Library at the City University of New York Graduate Center, for being good libraries. The mathematics department at Herbert H. Lehman College (CUNY) for being a place where mathematics happens.

Finally, the people: Raymond Smullyan, for inventing elementary formal systems, for telling me about R-definability, and for long discussions about fixed point theorems; Yiannis Moschovakis for helpful suggestions concerning an earlier draft of the manuscript; Einar Fredriksson of North-Holland Publishing Company for making my experience with the company a most pleasant one; and especially Greer Fitting, for expert technical typing, for the secretarial work of keeping track of what was where, and for assistance in matters of style.

CONTENTS

CHAPTER ONE

RECURSION AND ω-RECURSION THEORIES

1. Introduction

It seems best to begin with an informal description of *ordinary* recursion theory, in which natural numbers are the objects dealt with, and all computations are finite. There are many such informal descriptions. (These days they tend to be based on some sort of notion of ideal computer.) The description we use is directly relevant to the formal notions we will be developing, and we will return to it and elaborate it throughout the book.

Suppose you are given a collection of empty boxes, and into these boxes you are to put natural numbers, or ordered pairs of natural numbers, or ordered triples, or the like. Each box is clearly marked: into this box you can only put natural numbers; into this box you can only put ordered pairs of natural numbers; and so on. And suppose you're given a certain list of instructions as to how to fill those boxes. These instructions are to be what is usually called "effective." An example of an instruction that is *not* effective is: put 7 in box A if the Fermat "last theorem" is true, otherwise put 7 in box B. (If the Fermat conjecture is decided by the time you read this, substitute for it any other unsolved mathematical problem.) An example of an instruction that *is* effective is: look through box B and see if you put an ordered pair in there whose first component is n; if you did, then look through box C and see if you put an ordered triple in there whose first component is $n+1$; if you did, then put $n+2$ in box A. Notice the features of this. You can search through a box to see what is there; you can take successors; and you can put things into boxes (you can not take things out). And the conditions are "positive"; you can act on the basis of what *is* in a box but not on the basis of what *is not*.

Now, suppose you are given a list of "effective" instructions, and you proceed to follow them. Pick one of the boxes, say box A. As time goes on you put more and more numbers (or pairs, etc.) into box A. In this way a certain set is being generated. Such a set is called *recursively enumerable*. Much of ordinary recursion theory is the study of the recursively enumerable sets.

Often other characterizations of ordinary recursion theory are given. One common one is that it is the study of what *functions* on the natural numbers are "effectively calculable". It turns out that a function on the natural numbers is effectively calculable just when it is recursively enumerable when thought of as a set of ordered pairs (that is, just when its *graph* is recursively enumerable.) We find it convenient, for now, to take recursive enumerability of sets as the basic notion and treat functions via their graphs. Later on we will modify this position.

Incidentally, one might ask what would be the effect of allowing, not only adding to boxes, but removing from them as well. As it happens, the effect is great. For one thing, one is never certain of what is going to remain in any given box. You may look in box A after you've worked three days and see the number 35 in there, but you have no guarantee that it will still be there in a month; you might find your instructions have forced you to take it out. Even more, the collection of sets and relations that may be generated using this additional freedom properly includes the recursively enumerable ones. It also includes ones which are only "recursively approximable". We do not develop this notion. And we do not allow instructions to remove things from boxes.

Our first goal is to generalize things so that, instead of applying just to the natural numbers, where one is allowed to take successors, one can treat an arbitrary collection of items on which certain specified operations (more generally, relations) are given to work with. The mechanism we introduce for doing this is that of *elementary formal systems.* These originated in Smullyan [1961], for the case of words over a finite alphabet. Here they are generalized to apply to arbitrary structures. The connections between them and our informal description above should be obvious.

2. Structures

Let $\mathscr{A}$ be a set, and let $\mathscr{R}_1, \mathscr{R}_2, \ldots, \mathscr{R}_k$ be relations on $\mathscr{A}$. (We allow one-place relations, and generally call them sets.) The $k+1$ tuple $\langle \mathscr{A}; \mathscr{R}_1, \mathscr{R}_2, \ldots, \mathscr{R}_k \rangle$ is a ***structure with domain*** $\mathscr{A}$. We follow the convention that German letters stand for structures, script letters stand for domains and relations, and the domain of $\mathfrak{A}$ is $\mathscr{A}$, the domain of $\mathfrak{B}$ is $\mathscr{B}$, and so on. We allow trivial structures $\langle \mathscr{A} \rangle$, in which no relations are specified. We say a structure is empty or non-empty according to whether the domain is empty or not. We generally abandon "relation" notation for those relations having a customary symbolism. For example, we may write

$x + y = z$ instead of $+(x, y, z)$ or $\langle x, y, z\rangle \in +$, and we may write $+$ for the addition relation itself.

Certain structures will recur many times in this work, so we introduce special notations for them.

$\mathbb{N} = \{0, 1, 2, \ldots\}$ is the set of natural numbers. Let $y = x^+$ denote the successor relation on $\mathbb{N}$. The structure $\langle \mathbb{N}; y = x^+\rangle$ is the setting for ordinary recursion theory. We denote it by $\mathfrak{S}(\mathbb{N}) = \langle \mathbb{N}; y = x^+\rangle$.

A set is called *hereditarily finite* if it is finite, made up of finite sets, which are made up of finite sets, which etc. The collection of all hereditarily finite sets is denoted L_ω. Let $x \cup \{y\} = z$ denote the relation on L_ω which holds just when z is the result of adding y as an element to the set x. We denote the structure $\langle L_\omega; x \cup \{y\} = z\rangle$ by $\mathfrak{S}(L_\omega)$.

Let $\mathcal{W}(a_1, a_2, \ldots, a_n)$ be the set of words (finite or empty strings or sequences) over the alphabet $a_1, a_2, \ldots, a_n$. Let $x * a = y$ denote the relation on $\mathcal{W}(a_1, a_2, \ldots, a_n)$ which holds just when word y is the result of adding the one letter word a to the end of word x. We denote the structure $\langle \mathcal{W}(a_1, a_2, \ldots, a_n); x * a = y\rangle$ by $\mathfrak{S}(a_1, a_2, \cdots, a_n)$.

3. Elementary formal systems

Suppose a structure $\mathfrak{A} = \langle \mathcal{A}; \mathcal{R}_1, \ldots, \mathcal{R}_k\rangle$ has been specified. One can imagine a set of instructions for filling boxes, as in Section 1, but which uses the members of $\mathcal{A}$ instead of the natural numbers, and which supposes as "known" the relations $\mathcal{R}_1, \ldots, \mathcal{R}_k$ instead of the successor relation. We now define a simple free-variable logical calculus which should be thought of as the formal counterpart of such a set of instructions.

We suppose available an unlimited supply of n-place *predicate-symbols* for each $n > 0$. We use Latin letters to represent them, P, Q, R, etc. The only other symbols of our *alphabet* are an *arrow* and a *comma*. We will be using axiom schemas, so variables are not needed in the formal language itself. They will, however, be used in the metalanguage; we denote them by x, y, v, v_1, v_2 etc.

By an atomic formula over $\mathcal{A}$ we mean an expression of the form $Pa_1, a_2, \ldots, a_n$ where $a_1, a_2, \ldots, a_n \in \mathcal{A}$ and P is an n-place predicate symbol. For convenience we may write $P\boldsymbol{a}$ for $Pa_1, \ldots, a_n$. We also define a *pseudo-atomic formula over* $\mathcal{A}$ to be anything of the form $P\alpha_1, \ldots, \alpha_n$ where each α_i is in $\mathcal{A}$ or is a variable. Pseudo-atomic formulas are expressions of the metalanguage only.

The notion of *formula over* $\mathcal{A}$ is defined by the following rules:

(1) an atomic formula over $\mathscr{A}$ is a formula over $\mathscr{A}$;

(2) if X and Y are formulas over $\mathscr{A}$, so is $X \to Y$.

Formulas are to be thought of as being associated to the right. Thus $A \to B \to C \to D$ should be read as if it were $A \to (B \to (C \to D))$ and thought of as saying A, B and C together imply D.

The metalinguistic notion of *pseudo-formula over* $\mathscr{A}$ is defined analogously, being built up from pseudo-atomic formulas over $\mathscr{A}$. And the notion of a (complete) *instance* of a pseudo-formula over $\mathscr{A}$ has an obvious definition. Any instance of a pseudo-formula over $\mathscr{A}$ is a formula over $\mathscr{A}$.

By the *conclusion* of a (pseudo) formula we mean the final (pseudo) atomic part of it. Thus if A is (pseudo) atomic, A is the conclusion of $X \to A$, and also of A itself.

Recall, we are working with the structure $\mathfrak{A} = \langle \mathscr{A}; \mathscr{R}_1, \ldots, \mathscr{R}_k \rangle$. Now suppose we have selected, once and for all, predicate symbols $R_1, \ldots, R_k$, so that R_j is an n-place predicate symbol if $\mathscr{R}_j$ is an n-ary relation. We call $R_1, \ldots, R_k$ *reserved* predicate symbols. Let $\mathscr{R}_j^*$ consist of all formulas of the form $R_j a$ for which $\mathscr{R}_j a$ holds.

We say a pseudo-formula X is *proper* if none of the reserved predicate symbols $R_1, \ldots, R_k$ occurs in the conclusion of X.

Let $E = \{A_1, \ldots, A_m\}$ be some finite set of pseudo-formulas over $\mathscr{A}$, each proper. By a *derivation from E over* $\mathfrak{A}$ we mean a finite sequence of formulas, $X_1, \ldots, X_j$, such that each term of the sequence either

(1) is a member of $\mathscr{R}_1^* \cup \cdots \cup \mathscr{R}_k^*$, or   ← *i.e. valid set of Predicate (propositon a set of formula (Relation laid down by $\langle R_1, \ldots, R_k \rangle$*

(2) is an instance of some A_i, or

An instance of A_i: Axiom.

(3) comes from two earlier terms by the rule

$$\text{(MP)} \qquad \frac{X \quad X \to Y}{Y} \qquad \text{provided } X \text{ is atomic.}$$

If there is such a derivation ending with X, we say X is *derivable from E over* $\mathfrak{A}$.

E determines, relative to $\mathfrak{A}$, a simple deductive system, called an *elementary formal system over* $\mathfrak{A}$. Each A_i is an *axiom* of that elementary formal system.

Let P be an n-place predicate symbol, and let $\mathscr{P}$ be an n-ary relation on $\mathscr{A}$. We say *P represents* $\mathscr{P}$ *in the elementary formal system determined by E over* $\mathfrak{A}$ if

i.e. $\mathscr{P}$ is a member of $\langle R_1, \ldots, R_k \rangle$

$$a \in \mathscr{P} \quad \text{iff} \quad Pa \text{ is derivable from } E \text{ over } \mathfrak{A}.$$

We say $\mathscr{P}$ *is representable in the elementary formal system determined by E over* $\mathfrak{A}$ if there is some predicate symbol P which represents $\mathscr{P}$.

Finally, we say $\mathscr{P}$ *is recursively enumerable* (*r.e.*) *over* $\mathfrak{A}$ if $\mathscr{P}$ is representable in some elementary formal system over $\mathfrak{A}$. Also $\mathscr{P}$ *is recursive over* $\mathfrak{A}$ if both $\mathscr{P}$ and its complement, $\mathscr{A}^n - \mathscr{P}$, are r.e. over $\mathfrak{A}$.

4. Examples

The relations which are r.e. in our sense over the structure of arithmetic, $\mathfrak{S}(\mathbb{N})$, are the relations on $\mathbb{N}$ usually called r.e. (Rogers [1967]). The relations which are r.e. in our sense over the structure of the hereditarily finite sets, $\mathfrak{S}(L_\omega)$, are simply the relations Σ over L_ω (Barwise [1975]). And the relations which are r.e. in our sense over the word structure, $\mathfrak{S}(a_1, \ldots, a_n)$, are those called formally representable in Smullyan [1961]. This last fact is true essentially by definition (see Smullyan [1961], Ch. 4, §10). Means of verifying our other assertions will be developed in the course of this book. For now, we simply give a few examples of elementary formal systems in use.

First we consider examples involving $\mathfrak{S}(\mathbb{N})$.

I. The addition relation is r.e. over $\mathfrak{S}(\mathbb{N})$, being represented by P in the elementary formal system with axioms:

$$Px, 0, x;$$

$$Px, y, z \to v = y^+ \to w = z^+ \to Px, v, w.$$

Just this once we sketch how one shows that these axioms do what we claim.

CLAIM A. For each $x, y, z \in \mathbb{N}$, if $x + y = z$, then Px, y, z is derivable from the above axioms.

This may be shown by an induction on y.

CLAIM B. For each $x, y, z \in \mathbb{N}$, if Px, y, z is derivable from the above axioms, then $x + y = z$.

This may be shown by an induction on the number of steps in the derivation of Px, y, z. Specifically, one shows that each line of a derivation is "true" in the sense that it comes out true when Px, y, z is interpreted as $x + y = z$, and $v = y^+$ as $v = y + 1$. ($A \to B$ is given the usual implication interpretation: if A, then B.) It follows that if Px, y, z is derivable, then $x + y = z$.

From now on we generally skip such details. Also, from now on we write $x + y = z$ instead of Px, y, z. Generally, for the sake of readability, we will allow such informal notation in our elementary formal system axioms.

II. The "less than" relation is r.e. over $\mathfrak{S}(\mathbb{N})$, having axioms: the above, and

$$x + y = z \rightarrow y = w^{+} \rightarrow x < z.$$

III. "Not equal" is r.e. over $\mathfrak{S}(\mathbb{N})$ having axioms: the above, and

$$x < y \rightarrow x \neq y,$$

$$y < x \rightarrow x \neq y.$$

IV. Addition is *recursive* over $\mathfrak{S}(\mathbb{N})$. To show this, all that remains is to show the complement of the addition relation, is also r.e. over $\mathfrak{S}(\mathbb{N})$. The following axioms suffice: the above, and

$$x + y = z \rightarrow w \neq z \rightarrow \bar{P}x, y, w.$$

Next we move to examples involving $\mathfrak{S}(L_\omega)$. We note that now counterparts of Claim A above are generally established by an induction on the rank or size of something. Recall, the relation we are given to work with is $x \cup \{y\} = z$.

V. $\in$ is r.e. over $\mathfrak{S}(L_\omega)$. Axiom is:

$$x \cup \{y\} = z \rightarrow y \in z.$$

VI. $\subseteq$ is r.e. over $\mathfrak{S}(L_\omega)$. Axioms are:

$$x \subseteq x,$$

$$x \subseteq y \rightarrow y \cup \{z\} = w \rightarrow x \subseteq w.$$

VII. $\notin$ is r.e. over $\mathfrak{S}(L_\omega)$, and hence $\in$ is recursive. Axioms are:

$$x \notin \emptyset,$$

$$x \notin a \rightarrow y \in b \rightarrow y \notin x \rightarrow a \cup \{b\} = c \rightarrow x \notin c,$$

$$x \notin a \rightarrow y \notin b \rightarrow y \in x \rightarrow a \cup \{b\} = c \rightarrow x \notin c.$$

Since this example is a bit more complex we allow ourselves a few comments. The idea behind the axioms is as follows. First, of course, nothing belongs to $\emptyset$. Second, if $x \notin a$, and if we enlarge a to $a \cup \{b\}$ where $x \neq b$, then x is not a member of the larger set either. And, that $x \neq b$ can be said by saying something, y, belongs to one but not to the other.

The counterpart of Claim A is: if $x \notin y$, then $x \notin y$ is derivable from these axioms. This may be shown by a "double" induction along the following lines. First one shows: for each natural number n, if $x \in L_n$ and $y \in L_{n+1}$ and $x \notin y$, then $x \notin y$ is derivable. In doing the induction step of this, it is necessary to do an "inner" induction on the size of y. (Here, L_n has the usual meaning of rank in the hierarchy of constructible sets, as defined in Gödel [1939], where, however, it was denoted M_n. And size has the usual meaning, since for each natural number n, the set L_n is made up of finite sets.)

VIII. "Not equal" is r.e. over $\mathfrak{S}(L_\omega)$. Axioms are: the above, and

$$z \in x \rightarrow z \notin y \rightarrow x \neq y,$$

$$z \notin x \rightarrow z \in y \rightarrow x \neq y.$$

Now some examples involving $\mathfrak{S}(a_1, \ldots, a_n)$. Recall, the "given" relation is $x * a = y$: word x followed by one-letter word a is word y.

IX. Concatenation is r.e. over $\mathfrak{S}(a_1, \ldots, a_n)$. Axioms (we use $\emptyset$ for the empty word) are:

$$Cx, \emptyset, x;$$

$$Cx, y, z \rightarrow y * a = v \rightarrow z * a = w \rightarrow Cx, v, w.$$

X. "Not equal" is r.e. over $\mathfrak{S}(a_1, \ldots, a_n)$ Axioms are: the above, and

$$a_1 \neq a_2,$$

$$a_1 \neq a_3,$$

$$\vdots$$

$$a_{n-1} \neq a_n,$$

$$x * a_1 = z \rightarrow z \neq \emptyset,$$

$$x * a_2 = z \rightarrow z \neq \emptyset,$$

$$\vdots$$

$$x * a_n = z \rightarrow z \neq \emptyset,$$

$\neq a_3,$

$$x * a = u \to Cu, y, t \to x * b = v \to Cv, z, w$$
$$\to a \neq b \to t \neq w,$$
$$x \neq y \to y \neq x.$$

This ends the detailed examples. We now sketch a few others.

By a *finite relation* on $\mathcal{W}(a_1, \ldots, a_n)$ (the collection of words over $a_1, \ldots, a_n$) we mean a binary relation $\sim$ on $\mathcal{W}(a_1, \ldots, a_n)$ which holds in only a finite number of instances. We write $x \sim y$ instead of $\sim x, y$.

Let $\sim$ be some specific finite relation on $\mathcal{W}(a_1, \ldots, a_n)$ and let $*$ be the concatenation relation. Consider the structure $\langle \mathcal{W}(a_1, \ldots, a_n); *, \sim \rangle$. And consider the following elementary formal system axioms.

$$x \sim y \to x \approx y;$$

$$x \approx x,$$
$$x \approx y \to y \approx x,$$
$$x \approx y \to y \approx z \to x \approx z;$$

$$s \approx u \to t \approx v \to x = s * t \to y = u * v \to x \approx y.$$

It is easy to see that the relation represented by $\approx$ is the smallest equivalence relation extending $\sim$ which obeys a substitution principle.

The *word problem for semigroups* is to determine whether, over every such structure, the relation which $\approx$ represents must be recursive. The solution, due to Post [1947] and Markov [1946] is that it need not be.

For our final example, take as domain the collection of points, straight lines, and circles in some fixed Euclidean plane. Take as relations such things as: point x is (is not) on line y; point x is (is not) the center of circle y; the distance between points x and y is (is not) equal to the distance between a and b; etc. Properly set up, the Euclidean constructions will correspond to r.e. relations. For example, the following relation will be r.e.: the triangle with sides x, y and z, and the square with sides a, b, c and d have the same areas. The following relation will not be r.e.: circle x and the square with sides a, b, c and d have the same areas.

5. Remarks

By our definition of elementary formal system, $\boldsymbol{v}$ is in the relation which P represents if there is *some* derivation of $P\boldsymbol{v}$, from E, say. For certain structures it is possible to confine derivations to those which follow some systematic pattern, in which the order of axiom and rule use is specified once and for all, independently of E. This is possible for $\mathfrak{S}(\mathbb{N})$, for instance. It is not possible for a structure whose domain is the set of real numbers, unless one assumes a well ordering of the reals, and is willing to allow infinite derivations. For those structures over which systematic derivations are sufficient, the distinction between systematic and non-systematic becomes, essentially, that between *deterministic* and *non-deterministic* developments. Such a distinction is important in computational complexity discussions, but we do not pursue it here. All our systems are non-deterministic, that is, *any* derivation will do.

A related point concerns what is often called *unordered search*. One possible axiom for an elementary formal system is $Rx, y \to Py$. Now, to derive Pa using this, we must find an x for which Rx, a. If we are in the structure of arithmetic, there is a systematic procedure we can follow: try $R0, a$, then $R1, a$, then $R2, a$, etc. But over other structures such a deterministic procedure may not exist. Nevertheless we assume a search for such an x can always be made, though by what means does not concern us. We are tacitly assuming there is an "unordered search procedure" available.

In Moschovakis [1969], two computation theories on abstract structures are defined: prime computability and search computability. Essentially, the difference between them is that search computability allows a search to be made, while prime computability does not. See Moschovakis [1969], pp. 449–450. In this book we do not treat anything comparable to prime computability. In Chapter 6 it will be seen that there are close relationships between search computability and elementary formal systems.

A final point concerns "constants" in elementary formal systems. We allowed members of the domain to occur in elementary formal system axioms. One might, very properly, object to this on the ground that to use a thing one ought to construct it. Suppose we call an elementary formal system *pure* if there are no members of the domain in the axioms. Then it is an interesting question, for which structures do pure and ordinary elementary formal systems give the same r.e. relations. For example, it is not hard to see that this happens for structure $\langle \mathbb{N}; y = x^+, \{0\}\rangle$. Thus the structure $\mathfrak{S}(\mathbb{N})$ can be modified in a simple way so that pure elementary formal systems are enough. On the other hand, a structure whose domain is the set of real numbers most likely would not allow such a modification.

On this topic Moschovakis presents an interesting argument in Moschovakis [1969A], pp. 627–629. There he argues, in effect, that if constants are accepted as computable, then search (not prime) computability captures the informal notion of computability.

Finally, we should note that in most of our proofs about elementary formal systems, only a finite number of constants are needed. Thus one can generally modify a given structure to "save the phenomena" if one is careful. Still, in this book we do not restrict ourselves to pure elementary formal systems, interesting as such things might be.

6. Informal motivation

Let us imagine, once again, that you are given a collection of boxes, and a list of "effective" instructions for filling them, as in Section 1. Rather than arbitrarily selecting a box and saying you are generating the set in that box, let us say, instead, that one box has officially been designated "output". Now we add a complication: suppose there is also a box labeled "input", and suppose you are not allowed to put items into that box. There may be items there before you start, but you are not allowed to add to them. Otherwise the box functions just as all the others do. You may ask if such and such a thing is there, and if it is, you may act on that information. Of course you are still generating a set; the set being placed in the box labelled output.

Suppose, further, that there is a demon who, before you start, fills the input box. The demon may, quite reasonably, regard you, your list of instructions, and your collection of boxes as a machine. He fills the input box, tells you to start, then observes what turns up in the output box. If he gives this "machine" a different set of inputs, he may get a different set of outputs. What he's got is called an *enumeration operator*. We will give a proper definition below.

Being recursively enumerable fits quite simply into the context of enumeration operators. A relation is recursively enumerable if it is generated when no input is given. Or, equally well, if it is generated regardless of what input is given (you simply have no instructions that tell you to look in the input box). Consequently, we will get many properties of the r.e. relations out of our consideration of enumeration operators.

7. Enumeration operators

Some notation. Suppose $\mathfrak{A}$ is the structure $\langle \mathcal{A}; \mathcal{R}_1, \ldots, \mathcal{R}_k \rangle$, and $\mathcal{R}$ is some relation on $\mathcal{A}$. We write $\langle \mathfrak{A}, \mathcal{R} \rangle$ for the structure $\langle \mathcal{A}; \mathcal{R}_1, \ldots, \mathcal{R}_k, \mathcal{R} \rangle$. Suppose E is a set of axioms for an elementary formal system over $\mathfrak{A}$, and P is a predicate symbol. We write $E \vdash_{\mathfrak{A}} P\boldsymbol{x}$ to mean there is a derivation of

$\boldsymbol{Px}$ from E over $\mathfrak{A}$. Then $\{\boldsymbol{x} \mid E \vdash_{\mathfrak{A}} P\boldsymbol{x}\}$ is the relation which P represents in the elementary formal system with axioms E over $\mathfrak{A}$.

Now, let $\mathfrak{A} = \langle \mathcal{A}; \mathcal{R}_1, \ldots, \mathcal{R}_k \rangle$ be a structure fixed for the rest of this section. Suppose $\mathcal{I}$ is some n-place relation on $\mathcal{A}$. Let E be an elementary formal system over the structure $\langle \mathfrak{A}, \mathcal{I} \rangle$ in which, say, the predicate symbol I has been assigned to $\mathcal{I}$. (Then I can not occur in the conclusion of any axiom in E.) Let O be an m-place predicate symbol. Using the axioms E, O represents a certain m-ary relation on $\mathcal{A}$. Now suppose we keep E fixed, but change the relation $\mathcal{I}$ to $\mathcal{I}'$, still using I to represent it. Then O might represent a different relation on $\mathcal{A}$. In this way a certain operator on $\mathcal{A}$ is created, which we may symbolize by $[E^I_O]$. It uses the axioms E, takes whatever I represents as input, and gives whatever O represents as output. Formally,

$$[E^I_O](\mathcal{I}) = \{\boldsymbol{x} \mid E \vdash_{\langle \mathfrak{A}, \mathcal{I} \rangle} O\boldsymbol{x}\}$$

(where it is understood that the predicate symbol I is assigned to $\mathcal{I}$).

We call the maps $[E^I_O]$ *enumeration operators* over the structure $\mathfrak{A}$.

See Rogers [1967], pp. 146–147 for the usual definition of enumeration operator in ordinary recursion theory. We will show in Chapter 8 that our definition agrees with his in this case.

Let $[E^I_O]$ be an enumeration operator over the structure $\mathfrak{A}$ taking n-ary relations to m-ary relations; that is, I is n-place and O is m-place. We call the pair $\langle n, m \rangle$ the *order* of $[E^I_O]$.

Let $\mathcal{R}$ be an m-ary r.e. relation. Now, given any elementary formal system in which $\mathcal{R}$ is represented we may, by relettering, produce another one, E, in which O represents $\mathcal{R}$ and in which the n-place (n arbitrary) predicate symbol I does not occur. Then the enumeration operator $[E^I_O]$, of order $\langle n, m \rangle$, produces $\mathcal{R}$ given *any* input. We will, from now on, identify the r.e. relations with such constant enumeration operators.

8. Informal motivation

Let us return to our description of recursion theory in terms of boxes, and now imagine that it is not you that is putting items into the boxes, but a being that can live transfinitely long. Then one new kind of instruction might be given to it: have you put every member of the domain into box B; if you have, put n into box A. The infinite-lived being can answer the question: is every number in box B? The formal counterpart of this is our notion of ω-elementary formal systems.

9. ω-elementary formal systems

We add an infinite-premise rule of derivation to the machinery of elementary formal systems over the structure $\mathfrak{A} = \langle \mathcal{A}; \mathcal{R}_1, \ldots, \mathcal{R}_k \rangle$.

First we modify the alphabet by adding the symbol $\forall$. Now an atomic formula is a string $Px_1, \ldots, x_n$ where P is an n-place predicate symbol and each x_i either is in $\mathcal{A}$ or is the symbol $\forall$. We similarly modify the notions of pseudo-atomic formula, pseudo-formula, and formula. Otherwise no syntactical changes are made; *instance* still means instance over $\mathcal{A}$, for example.

Intuitively, $P\boldsymbol{v}, \forall, \boldsymbol{w}$ is to mean $P\boldsymbol{v}, a, \boldsymbol{w}$ holds for each $a \in \mathcal{A}$. Now we give rules governing the formal use of $\forall$. We make the *restriction* that $\forall$ may not occur in the conclusion of any axiom (thus modifying the definition of proper). And we add one more rule of derivation.

ω-RULE. If $P\boldsymbol{v}, \forall, \boldsymbol{w}$ is atomic, then

$$\frac{P\boldsymbol{v}, a, \boldsymbol{w} \quad \text{for each } a \in \mathcal{A}}{P\boldsymbol{v}, \forall, \boldsymbol{w}}.$$

The notion of an *ω-elementary formal system* can be now be formulated. Derivations are now well ordered, possibly infinite, sequences, following the rules of Section 3, but also allowing ω-rule applications. Call a relation $\mathcal{P} \subseteq \mathcal{A}^n$ *ω-r.e. over* $\mathfrak{A}$ if it is representable in some ω-elementary formal system over $\mathfrak{A}$. Also $\mathcal{P}$ is *ω-recursive over* $\mathfrak{A}$ if both $\mathcal{P}$ and $\mathcal{A}^n - \mathcal{P}$ are ω-r.e. over $\mathfrak{A}$.

In Fitting [1978] there is a direct proof that, for the structure $\mathfrak{S}(\mathbb{N})$, the ω-recursive relations are those usually called *hyperarithmetic*. (See Rogers [1967], p. 382 for a definition of this term.) In Section 13 we will show that, for any structure $\mathfrak{A}$, the ω-recursive relations are essentially those which Moschovakis calls *hyperelementary* (Moschovakis [1974], p. 17). Since the hyperelementary relations on $\mathfrak{S}(\mathbb{N})$ are the hyperarithmetic ones, this implies the above result.

The definition of enumeration operator over $\mathfrak{A}$ may also be modified in the obvious way, to define the notion of *ω-enumeration operator.* We skip the details. Also the ω-r.e. relations can be identified with the "constant" ω-enumeration operators.

PROPOSITION 9.1. *Each r.e. relation over $\mathfrak{A}$ is also ω-r.e. over $\mathfrak{A}$. Each enumeration operator over $\mathfrak{A}$ is also an ω-enumeration operator over $\mathfrak{A}$.*

PROOF. If the axioms don't mention $\forall$, the ω-rule is not relevant.

We conclude with some illustrations of the uses of the ω-rule. Consider $\mathfrak{S}(\mathbb{N})$, the structure of arithmetic. Choose some standard indexing of the one place relations (sets) r.e. over this structure. We assume, without going through the details, that the following relations can be shown to be r.e. over $\mathfrak{S}(\mathbb{N})$ (this is done for a particular indexing in Chapter 6):

$$Ux, y \;\Leftrightarrow\; y \text{ belongs to the set with index } x.$$

Now, let F be the set of indexes of *infinite* r.e. sets. F is *not* r.e. over $\mathfrak{S}(\mathbb{N})$ [follows from Ch. 7, §7]. It is, however, ω-r.e. over $\mathfrak{S}(\mathbb{N})$, with the following set of axioms. (In them, Gx, y is intended to mean: in set with index x, there is a member bigger than y.)

axioms for U,

axioms for $<$ (see Section 4),

$y < z \rightarrow Ux, z \rightarrow Gx, y$,

$Gx, \forall \rightarrow Fx$.

For a second, sketchy, example, return to the Euclidean geometry example at the end of Section 4. Even though squaring the circle is not a recursive process, it is ω-recursive. This follows since there are well-known ruler and compass techniques for producing better and better square approximations to a circle, and in the ω-recursion theory we have an ability to take limits. We leave the detailed development to the interested reader.

10. Basic structural properties

Let $\mathfrak{A} = \langle \mathcal{A}; \mathcal{R}_1, \ldots, \mathcal{R}_k \rangle$ be a structure, fixed for this section. Both the enumeration operators and the ω-enumeration operators over $\mathfrak{A}$ have certain simple common structural features, which we establish in this section. Chapter 2 will be devoted to the consequences of these features. We do both enumeration operators and ω-enumeration operators together where possible. The ω-rule plays little role in the results of this section.

If Φ is an operator and $\mathcal{P}$ is a relation, when we write $\Phi(\mathcal{P})$, we mean to imply $\mathcal{P}$ is appropriate as input for Φ, that is, $\mathcal{P}$ is n-ary and Φ is of order $\langle n, m \rangle$.

PROPOSITION 10.1. *Each of the following is r.e., and hence ω-r.e.; over $\mathfrak{A}$.*

(1) $\emptyset$, *the empty set,*

(2) $=_{\mathcal{A}}$, *the equality relation on* $\mathcal{A}$,

(3) *each of* $\mathcal{R}_1, \ldots, \mathcal{R}_k$,

(4) *for each* $a \in \mathcal{A}$, *the set* $\{a\}$.

PROOF. (1) $\emptyset$ is represented by any non-reserved one-place predicate symbol in the elementary formal system with no axioms.

(2) $=_{\mathcal{A}}$ is represented by E in the elementary formal system with axiom Ex, x.

(3) $\mathcal{R}_i$ is represented by R_i, the predicate symbol assigned to it, in the elementary formal system with no axioms.

(4) The set $\{a\}$ is represented by C in the elementary formal system with axiom Ca.

REMARK. If we require only *pure* elementary formal systems, part 4 above is lost; the other parts are not affected, nor are any of the results below.

DEFINITION. (1) $T^n_{i,j}$ is a *transposition operator* of order $\langle n, n\rangle$, which interchanges the ith and the jth place. Specifically,

$$T^n_{i,j}(\mathcal{P}) = \{\langle x_1, \ldots, x_j, \ldots, x_i, \ldots, x_n\rangle \mid \langle x_1, \ldots, x_i, \ldots, x_j, \ldots, x_n\rangle \in \mathcal{P}\}$$

for $\mathcal{P} \subseteq \mathcal{A}^n$. Note: $n = 1, 2, 3, \ldots$ and $1 \leqslant i, j \leqslant n$. We allow $i = j$, in which case $T^n_{i,j}$ is actually an *identity operator*, which we denote I^n.

(2) P^n_i is a *projection operator* of order $\langle n, n-1\rangle$, on the ith coordinate. Specifically,

$$P^n_i(\mathcal{P}) = \{\langle x_1, \ldots, x_{i-1}, x_{i+1}, \ldots, x_n\rangle \mid (\exists x_i)\langle x_1, \ldots, x_{i-1}, x_i, x_{i+1}, \ldots, x_n\rangle \in \mathcal{P}\}$$

for $\mathcal{P} \subseteq \mathcal{A}^n$. Note: $n = 2, 3, 4, \ldots$ and $1 \leqslant i \leqslant n$.

(3) D^n_i is a *dual projection operator* of order $\langle n, n-1\rangle$, on the ith coordinate. Specifically,

$$D^n_i(\mathcal{P}) = \{\langle x_1, \ldots, x_{i-1}, x_{i+1}, \ldots, x_n\rangle \mid (\forall x_i)\langle x_1, \ldots, x_{i-1}, x_i, x_{i+1}, \ldots, x_n\rangle \in \mathcal{P}\}$$

for $\mathcal{P} \subseteq \mathcal{A}^n$. Note: $n = 2, 3, 4, \ldots$ and $1 \leqslant i \leqslant n$.

(4) A^n is a *place-adding operator* of order $\langle n, n+1\rangle$. Specifically

$$A^n(\mathcal{P}) = \mathcal{A} \times \mathcal{P}$$

for $\mathcal{P} \subseteq \mathcal{A}^n$. Note $n = 1, 2, 3, \ldots$.

PROPOSITION 10.2. *Each* $T^n_{i,j}$, P^n_i *and* A^n *is an enumeration operator over* $\mathfrak{A}$, *and hence also an* ω*-enumeration operator over* $\mathfrak{A}$. *Each* D^n_i *is an* ω*-enumeration operator over* $\mathfrak{A}$.

PROOF. (1) $T^n_{i,j} = [E^I_O]$ where I and O are n-place and E consists of

$$Ix_1, \ldots, x_i, \ldots, x_j, \ldots, x_n \to Ox_1, \ldots, x_j, \ldots, x_i, \ldots, x_n.$$

(2) $P^n_i = [E^I_O]$ where I is n-place, O is $n-1$ place, and E consists of

$$Ix_1, \ldots, x_{i-1}, x_i, x_{i+1}, \ldots, x_n \to Ox_1, \ldots, x_{i-1}, x_{i+1}, \ldots, x_n.$$

(3) $A^n = [E^I_O]$ where I is n-place, O is $n+1$ place, and E consists of

$$Ix_1, \ldots, x_n \to Oy, x_1, \ldots, x_n.$$

(4) $D^n_i = [E^I_O]$ where I is n-place, O is $n-1$ place, and E consists of

$$Ix_1, \ldots, x_{i-1}, \forall, x_{i+1}, \ldots, x_n \to Ox_1, \ldots, x_{i-1}, x_{i+1}, \ldots, x_n.$$

DEFINITION. Let Φ and Ψ be operators of the same order, $\langle n, m\rangle$.

$$(\Phi \cap \Psi)(\mathcal{P}) =_{\text{df}} \Phi(\mathcal{P}) \cap \Psi(\mathcal{P}),$$

$$(\Phi \cup \Psi)(\mathcal{P}) =_{\text{df}} \Phi(\mathcal{P}) \cup \Psi(\mathcal{P}).$$

PROPOSITION 10.3. *The collection of* (ω) *enumeration operators over* $\mathfrak{A}$ *is closed under* $\cap$ *and* $\cup$.

PROOF. Say $\Phi = [E^A_B]$ and $\Psi = [F^C_D]$. We may suppose, without loss of generality, that E and F contain no predicate symbols in common, other than the reserved ones. We call E and F *disjoint* if this is the case. Now

(1) $\Phi \cap \Psi = [H^I_O]$ where H consists of

the axioms of E, the axioms of F,

$Ix \to Ax$, $Ix \to Cx$,

$By \to Dy \to Oy$.

(2) $\Phi \cup \Psi = [H^I_O]$ where H consists of

the axioms of E, the axioms of F,

$Ix \to Ax$, $Ix \to Cx$,

$By \to Oy$, $Dy \to Oy$.

DEFINITION. Let Φ and Ψ be (ω) enumeration operators of orders $\langle n, m\rangle$ and $\langle n, m'\rangle$ respectively. By $\Phi \times \Psi$ we mean the map of order $\langle n, m+m'\rangle$ given by

$$(\Phi \times \Psi)(\mathcal{P}) =_{\text{df}} \Phi(\mathcal{P}) \times \Psi(\mathcal{P}).$$

PROPOSITION 10.4. *The collection of* (ω) *enumeration operators over* $\mathfrak{A}$ *is closed under* $\times$.

PROOF. Say $\Phi = [E^A_B]$ and $\Psi = [F^C_D]$ where E and F are disjoint. Then $\Phi \times \Psi = [H^I_O]$ where H consists of

the axioms of E, the axioms of F,

$Ix \to Ax, \quad Ix \to Cx,$

$Bu \to Dv \to Ou, v.$

DEFINITION. Let Φ and Ψ be (ω) enumeration operators of order $\langle m, p\rangle$ and $\langle n, m\rangle$ respectively. By $\Phi\Psi$ we mean the *composition map* of order $\langle n, p\rangle$ given by

$$(\Phi\Psi)(\mathscr{P}) = \Phi(\Psi(\mathscr{P})).$$

PROPOSITION 10.5. *The collection of* (ω) *enumeration operators over* $\mathfrak{A}$ *is closed under composition.*

PROOF. Say $\Phi = [E^A_B]$ and $\Psi = [F^C_D]$ where E and F are disjoint. Then $\Phi\Psi = [H^I_O]$ where H consists of

the axioms of E, the axioms of F,

$Ix \to Cx, \quad Dy \to Ay, \quad Bz \to Oz.$

COROLLARY 10.6. *If* Φ *is an* (ω) *enumeration operator and* $\mathscr{P}$ *is* (ω) *r.e., then* $\Phi(\mathscr{P})$ *is* (ω) *r.e.*

PROOF. Identify (ω) r.e. relations with constant (ω) enumeration operators.

PROPOSITION 10.7. *Each* (ω) *enumeration operator* Φ *over* $\mathfrak{A}$ *is* monotone, *that is,*

$$\mathscr{P} \subseteq \mathscr{R} \;\Rightarrow\; \Phi(\mathscr{P}) \subseteq \Phi(\mathscr{R}).$$

PROOF. If $\mathscr{P} \subseteq \mathscr{R}$, every derivation over $\langle \mathfrak{A}, \mathscr{P}\rangle$ is also a derivation over $\langle \mathfrak{A}, \mathscr{R}\rangle$.

PROPOSITION 10.8. *Each enumeration operator* Φ *over* $\mathfrak{A}$ *is* compact, *that is,*

$$\boldsymbol{v} \in \Phi(\mathscr{P}) \;\Leftrightarrow\; \text{for some finite } \mathfrak{F} \subseteq \mathscr{P},\ \boldsymbol{v} \in \Phi(\mathscr{F}).$$

PROOF. Elementary formal system derivations are finite.

11. Recursion and ω-recursion theories

Let $\mathfrak{A} = \langle \mathcal{A}; \mathcal{R}_1, \ldots, \mathcal{R}_k \rangle$ be a structure. In a natural way both the collection of enumeration operators over $\mathfrak{A}$ and the collection of ω-enumeration operators over $\mathfrak{A}$ can be made into categories. We will study these categories in later chapters. Here we give the definitions.

By $[\mathcal{A}]^n$ $(n > 0)$ we mean the collection of all n-ary relations on $\mathcal{A}$. That is, $[\mathcal{A}]^n$ = power set of $\mathcal{A}^n$. We generally identify $\mathcal{A}^1$ and $\mathcal{A}$.

The *objects* of our category are $[\mathcal{A}]^1, [\mathcal{A}]^2, [\mathcal{A}]^3, \ldots$.

The *morphisms* are the (ω) enumeration operators over $\mathfrak{A}$.

An operator (morphism) of order $\langle n, m \rangle$ has *domain* $[\mathcal{A}]^n$ and *codomain* $[\mathcal{A}]^m$.

Composition of morphisms is ordinary function composition.

The *identity morphisms* are the operators I^n of the previous section.

By rec$(\mathfrak{A})$ we mean the category of enumeration operators over $\mathfrak{A}$ as just described. We call it the *recursion theory on* $\mathfrak{A}$. By *ordinary recursion theory*, we officially mean the recursion theory on $\mathfrak{S}(\mathbb{N})$.

By ω-rec$(\mathfrak{A})$ we mean the category of ω-enumeration operators over $\mathfrak{A}$. We call it the ω-recursion theory on $\mathfrak{A}$. For us, *hyperarithmetic theory* means ω-rec$(\mathfrak{S}(\mathbb{N}))$.

Note that by Proposition 9.1, rec$(\mathfrak{A})$ is a subcategory of ω-rec$(\mathfrak{A})$.

12. The least fixed point theorem

Once again, let $\mathfrak{A} = \langle \mathcal{A}; \mathcal{R}_1, \ldots, \mathcal{R}_k \rangle$ be a structure, fixed for this section. We prove, for both rec$(\mathfrak{A})$ and ω-rec$(\mathfrak{A})$, analogs of an important result of ordinary recursion theory, due originally to Kleene. See Rogers [1967], pp. 193–194 for a proof in the ordinary case. Applications of the generalized version will occur throughout this work; the first is in the next section.

DEFINITION. Let Φ be an (ω) enumeration operator. $\mathcal{P}$ is a *fixed point* of Φ if $\Phi(\mathcal{P}) = \mathcal{P}$. $\mathcal{P}$ is the *least fixed point* if (1) $\mathcal{P}$ is a fixed point, and (2) if $\mathcal{R}$ is any fixed point for Φ then $\mathcal{P} \subseteq \mathcal{R}$.

THEOREM 12.1. *Let Φ be an enumeration operator (an ω-enumeration operator) of order $\langle n, n \rangle$ over $\mathfrak{A}$. Then Φ has a (unique) least fixed point which is r.e. (ω-r.e.) over $\mathfrak{A}$.*

PROOF. The proof is quite lengthy. We divide it into parts for readability.

Define, for each ordinal α, a relation $\mathcal{P}_\alpha$ by

$$\mathcal{P}_\alpha = \Phi\Big(\bigcup_{\gamma<\alpha} \mathcal{P}_\gamma\Big).$$

Part I. If $\alpha \leqslant \beta$ then $\mathcal{P}_\alpha \subseteq \mathcal{P}_\beta$.

Proof of I. Suppose $\alpha \leqslant \beta$. Then

$$\bigcup_{\gamma<\alpha} \mathcal{P}_\gamma \subseteq \bigcup_{\gamma<\beta} \mathcal{P}_\gamma.$$

But Φ is monotone, so

$$\Phi\Big(\bigcup_{\gamma<\alpha} \mathcal{P}_\gamma\Big) \subseteq \Phi\Big(\bigcup_{\gamma<\beta} \mathcal{P}_\gamma\Big)$$

or $\mathcal{P}_\alpha \subseteq \mathcal{P}_\beta$.

Part II. $\mathcal{P}_{\alpha+1} = \Phi(\mathcal{P}_\alpha)$.

Proof of part II. It follows from Part I that

$$\mathcal{P}_\alpha = \bigcup_{\gamma\leqslant\alpha} \mathcal{P}_\gamma$$

or $\mathcal{P}_\alpha = \bigcup_{\gamma<\alpha+1} \mathcal{P}_\gamma$. Hence

$$\Phi(\mathcal{P}_\alpha) = \Phi\Big(\bigcup_{\gamma<\alpha+1} \mathcal{P}_\gamma\Big) = \mathcal{P}_{\alpha+1}.$$

Part III. For some ordinal κ, $\mathcal{P}_\kappa$ is a fixed point for Φ, that is,

$$\Phi(\mathcal{P}_\kappa) = \mathcal{P}_\kappa.$$

Proof of Part III. By Part I, the sequence $\mathcal{P}_\alpha$ is a chain. If it were strictly increasing, that is, if $\alpha < \beta \Rightarrow \mathcal{P}_\alpha \subsetneqq \mathcal{P}_\beta$, then the cardinality of the $\mathcal{P}_\alpha$'s would grow without bound. Since each $\mathcal{P}_\alpha \subseteq \mathcal{A}^n$, this is not possible. Hence for some α, β, we must have $\alpha < \beta$ but $\mathcal{P}_\alpha = \mathcal{P}_\beta$. Now, using Parts I and II,

$$\alpha + 1 \leqslant \beta, \qquad \mathcal{P}_{\alpha+1} \subseteq \mathcal{P}_\beta,$$

$$\Phi(\mathcal{P}_\alpha) \subseteq \mathcal{P}_\beta, \qquad \Phi(\mathcal{P}_\alpha) \subseteq \mathcal{P}_\alpha.$$

Also

$$\alpha < \alpha + 1,$$

$$\mathcal{P}_\alpha \subseteq \mathcal{P}_{\alpha+1}, \qquad \mathcal{P}_\alpha \subseteq \Phi(\mathcal{P}_\alpha).$$

so $\mathcal{P}_\alpha$ is a fixed point.

Part IV. Let $\mathcal{R}$ be any fixed point for Φ. Then for each α, $\mathcal{P}_\alpha \subseteq \mathcal{R}$.

Proof of Part IV. Suppose, for each $\gamma < \alpha$ we have $\mathcal{P}_\gamma \subseteq \mathcal{R}$. Then

$$\bigcup_{\gamma<\alpha} \mathcal{P}_\gamma \subseteq \mathcal{R},$$

$$\Phi\Big(\bigcup_{\gamma<\alpha} \mathcal{P}_\gamma\Big) \subseteq \Phi(\mathcal{R}), \qquad \mathcal{P}_\alpha \subseteq \mathcal{R}.$$

Note. Combining Parts III and IV we have that Φ has a least fixed point of the form $\mathcal{P}_\kappa$. What is left is to show $\mathcal{P}_\kappa$ is r.e. (ω-r.e.).

Say $\Phi = [E^I_O]$. Note that, in E, I can not occur in the conclusion of any axiom. Also, by definition,

$$\boldsymbol{x} \in \Phi(\mathcal{R}) \iff E \vdash_{\langle \mathfrak{A}, \mathcal{R}\rangle} O\boldsymbol{x}.$$

Now, let $E' = E \cup \{O\boldsymbol{x} \to I\boldsymbol{x}\}$. (That is, E' is E + "input = output".) We claim

$$\mathcal{P}_\kappa = \{\boldsymbol{x} \mid E' \vdash_{\mathfrak{A}} O\boldsymbol{x}\},$$

which will finish the proof.

Part V. For each α, (in particular, for κ)

$$\mathcal{P}_\alpha \subseteq \{\boldsymbol{x} \mid E' \vdash_{\mathfrak{A}} O\boldsymbol{x}\}.$$

Proof of Part V. By induction on α. Suppose result known for each $\gamma < \alpha$.

Now, suppose $\boldsymbol{a} \in \mathcal{P}_\alpha$. Then $\boldsymbol{a} \in \Phi(\bigcup_{\gamma<\alpha} \mathcal{P}_\gamma)$, so by definition,

$$E \vdash_{\langle \mathfrak{A}, \bigcup_{\gamma<\alpha}\mathcal{P}_\gamma\rangle} O\boldsymbol{a}. \qquad (*)$$

By induction hypothesis, for all $\boldsymbol{x} \in \bigcup_{\gamma<\alpha} \mathcal{P}_\gamma$, $E' \vdash_{\mathfrak{A}} O\boldsymbol{x}$. But then, for all $\boldsymbol{x} \in \bigcup_{\gamma<\alpha} \mathcal{P}_\gamma$,

$$E' \vdash_{\mathfrak{A}} I\boldsymbol{x} \qquad (**)$$

using our "input-output" axiom.

By (*) and (**), since also $E \subseteq E'$, $E' \vdash_{\mathfrak{A}} O\boldsymbol{a}$. Thus $\mathcal{P}_\alpha \subseteq \{\boldsymbol{x} \mid E' \vdash_{\mathfrak{A}} O\boldsymbol{x}\}$.

Part VI. $\{\boldsymbol{x} \mid E' \vdash_{\mathfrak{A}} O\boldsymbol{x}\} \subseteq \mathcal{P}_\kappa$.

Proof of Part VI. We show this by an induction on the length of derivations from E' over $\mathfrak{A}$.

Suppose: whenever $O\boldsymbol{x}$ is derivable from E' in $< \alpha$ steps, then $\boldsymbol{x} \in \mathcal{P}_\kappa$.

Now suppose $O\boldsymbol{a}$ is derivable from E' in α steps. We show $\boldsymbol{a} \in \mathcal{P}_\kappa$.

Let $\mathcal{R}_I = \{\boldsymbol{x} \mid I\boldsymbol{x}$ occurs before the last line in some (fixed) derivation from E' of $O\boldsymbol{a}\}$.

Let $\mathcal{R}_O = \{\boldsymbol{x} \mid O\boldsymbol{x}$ occurs before the last line in this derivation$\}$.

Now I occurs in the conclusion of only one axiom in E', namely $O\boldsymbol{x} \to I\boldsymbol{x}$. It follows that $\mathcal{R}_I \subseteq \mathcal{R}_O$. Also, by induction hypothesis, $\mathcal{R}_O \subseteq \mathcal{P}_\kappa$. Hence

$$\mathscr{R}_I \subseteq \mathscr{P}_\kappa. \tag{***}$$

Finally, it should be clear that $E \vdash_{\langle \mathfrak{A}, \mathscr{R}_I \rangle} O\boldsymbol{a}$. This says that $\boldsymbol{a} \in \Phi(\mathscr{R}_I)$. Then by (***), since Φ is monotone, $\boldsymbol{a} \in \Phi(\mathscr{P}_\kappa) = \mathscr{P}_\kappa$.

This concludes the proof.

13. Inductive definability

There are many ways open to generalize ordinary recursion theory, and hyperarithmetic theory. One important and extensively developed version is *inductive definability*, and in this section we establish close connections between it and elementary formal systems. The standard reference for this topic is Moschovakis [1974]. We differ in certain details from it, as we will explain below. We note that much use will be made of the connections established here, in providing examples of various axiomatically defined systems later in the book.

Let $\mathscr{A}$ be a set. The notion of *pseudo-atomic formula over* $\mathscr{A}$ is as in Section 3. But here we build things up using more of the conventional machinery of first-order logic.

DEFINITION. By a *positive pseudo-formula over* $\mathscr{A}$ we mean any expression built up in the usual way, starting with the pseudo-atomic formulas over $\mathscr{A}$, but using only $\wedge, \vee, \exists$ and $\forall$ as logical operations.

Let $\mathscr{R}_1, \ldots, \mathscr{R}_k$ be a fixed list of relations on $\mathscr{A}$. Let $R_1, \ldots, R_k$ be distinct predicate symbols permanently assigned to represent $\mathscr{R}_1, \ldots, \mathscr{R}_k$, and let E be yet another, two-place predicate symbol (intended to represent equality). These are fixed for the rest of this section.

DEFINITION. Let P be a predicate symbol distinct from $R_1, \ldots, R_k$ and E. By a *P-formula* we mean a positive pseudo-formula over $\mathscr{A}$ which contains no predicate symbols other than $R_1, \ldots, R_k, E$ and P.

DEFINITION. Let P be an n-place predicate symbol, let $\mathscr{P}$ be an n-ary relation on $\mathscr{A}$, and let $\varphi(P)$ be a P-formula. We say $\varphi(\mathscr{P})$ *is true* if $\varphi(P)$ is true of $\mathscr{A}$ in the customary sense, under the interpretation which assigns $\mathscr{R}_i$ to R_i, $=_{\mathscr{A}}$ to E, and $\mathscr{P}$ to P.

PROPOSITION 13.1 (Monotonicity Property). *Suppose $\varphi(P)$ is a P-formula,*

P is n-place, and $\mathcal{P}$ and $\mathcal{P}'$ are n-ary relations on $\mathcal{A}$. If $\varphi(\mathcal{P})$ is true and $\mathcal{P} \subseteq \mathcal{P}'$, then $\varphi(\mathcal{P}')$ is true.

PROOF. By induction on the degree of φ.

DEFINITION. Let P be n-place and let $\varphi(P, x_1, \ldots, x_n)$ be a P-formula with $x_1, \ldots, x_n$ as free variables. We say φ *defines an operator* Γ_φ of order $\langle n, n\rangle$ on $\mathcal{A}$ as follows:

$$\Gamma_\varphi(\mathcal{P}) = \{\langle a_1, \ldots, a_n\rangle \in \mathcal{A}^n \mid \varphi(\mathcal{P}, a_1, \ldots, a_n) \text{ is true}\}.$$

PROPOSITION 13.2. *Γ_φ is monotone.*

PROOF. By Proposition 13.1.

It follows from Proposition 13.2 that each Γ_φ has a least fixed point. The argument is given as Parts I, II, III and IV of the proof of Theorem 12.1, which used nothing but the monotonicity of the operator in question.

DEFINITION. A relation $\mathcal{P}$ on $\mathcal{A}$ is a *fixed point* if there is a P-formula φ which defines an operator Γ_φ with $\mathcal{P}$ as its least fixed point.

$\mathcal{P}$ is an *existential fixed point* if the formula φ does not contain $\forall$.

DEFINITION. A relation $\mathcal{R}$ is *inductive* (*existential inductive*) *on $\mathcal{A}$ in the relations* $\mathcal{R}_1, \ldots, \mathcal{R}_k$ if there is a fixed point (an existential fixed point) $\mathcal{P}$, and there are $a_1, \ldots, a_j \in \mathcal{A}$ so that

$$\langle x_1, \ldots, x_n\rangle \in \mathcal{R} \Leftrightarrow \langle a_1, \ldots, a_j, x_1, \ldots, x_n\rangle \in \mathcal{P}.$$

Our interest in these notions stems from the following theorem, whose proof will occupy the rest of this section.

THEOREM 13.3. *Let $\mathfrak{A} = \langle \mathcal{A}; \mathcal{R}_1, \ldots, \mathcal{R}_k\rangle$ be a structure with $\mathcal{A}$ infinite.*

(1) *The relations r.e. over $\mathfrak{A}$ are the relations existential inductive on $\mathcal{A}$ in $\mathcal{R}_1, \ldots, \mathcal{R}_k$.*

(2) *The relations ω-r.e. over $\mathfrak{A}$ are the relations inductive on $\mathcal{A}$ in $\mathcal{R}_1, \ldots, \mathcal{R}_k$.*

REMARKS. The above definitions, with some differences in terminology, are to be found in Chapter 1 of Moschovakis [1974], with the following change. In Moschovakis [1974] it is, in effect, assumed that $\neq_{\mathcal{A}}$ is always one of the "given" relations $\mathcal{R}_i$, or equivalently, that $\neq_{\mathcal{A}}$ is inductive (existential inductive) in the given relations. When we need this we will assume it explicitly.

In Moschovakis [1974], a relation $\mathcal{R}$ is called *inductive on the structure* $\mathfrak{A} = \langle \mathcal{A}; \mathcal{R}_1, \ldots, \mathcal{R}_k \rangle$ if $\mathcal{R}$ is inductive on $\mathcal{A}$ in the relations $\mathcal{R}_1, \ldots, \mathcal{R}_k$, $\bar{\mathcal{R}}_1, \ldots, \bar{\mathcal{R}}_k$, $\neq_{\mathcal{A}}$ (where $\bar{\mathcal{R}}_i$ is the complement of $\mathcal{R}_i$, and as remarked above, $\neq_{\mathcal{A}}$ is tacit in Moschovakis [1974]). A relation $\mathcal{R}$ is called *hyperelementary* on $\mathfrak{A}$ if both $\mathcal{R}$ and $\bar{\mathcal{R}}$ are inductive on $\mathfrak{A}$. Note that Moschovakis [1974] thus handles *structures* differently than we do here.

We now turn to the proof of Theorem 13.3. Once again, $\mathfrak{A} = \langle \mathcal{A}; \mathcal{R}_1, \ldots, \mathcal{R}_k \rangle$ is fixed.

LEMMA 13.4. *Let φ be a P-formula, and let Γ_φ be the operator, of order $\langle n, n \rangle$, which φ defines.*

(1) *Γ_φ is an ω-enumeration operator over $\mathfrak{A}$.*

(2) *If φ has no universal quantifiers Γ_φ is an enumeration operator over $\mathfrak{A}$.*

PROOF. By induction on the degree of φ.

First, suppose φ is pseudo-atomic. Say, for definiteness, that φ is $R_i(x, y, a, b)$. Let E be an elementary formal system with the single axiom

$$R_i x, y, a, b \to Ax, y.$$

It is clear that

$$\varphi(\mathcal{P}, x, y) \text{ is true} \iff E \vdash_{\langle \mathfrak{A}, \mathcal{P} \rangle} Ax, y,$$

that is, $\Gamma_\varphi(\mathcal{P}) = [E^B_A](\mathcal{P})$ [where B is any new n-place predicate symbol].

If φ is $E(x, y)$ or $P(x_1, \ldots, x_n)$ the argument is similar.

If φ is not pseudo-atomic, the result follows by the closure properties of Section 10.

It should be clear that if φ has no universal quantifiers, the last part of Proposition 10.2 will never be used; the axioms for the elementary formal system will not involve the symbol $\forall$, and so an enumeration operator will be produced.

LEMMA 13.5. *Every fixed point is ω-r.e. over $\mathfrak{A}$. Every existential fixed point is r.e. over $\mathfrak{A}$.*

PROOF. Suppose $\mathcal{P}$ is a fixed point. Say it is the least fixed point of the operator Γ_φ for some P-formula φ. By Lemma 13.4, Γ_φ is an ω-enumeration operator over $\mathfrak{A}$, so by Theorem 12.1, $\mathcal{P}$ is ω-r.e. over $\mathfrak{A}$. The second part is similar.

PROPOSITION 13.6. *Every relation inductive on $\mathscr{A}$ in $\mathscr{R}_1, \ldots, \mathscr{R}_k$ is ω-r.e. over $\mathfrak{A}$. Every relation existential inductive on $\mathscr{A}$ in $\mathscr{R}_1, \ldots, \mathscr{R}_k$ is r.e. over $\mathfrak{A}$.*

PROOF. Suppose $\mathscr{R}$ is inductive on $\mathscr{A}$ in $\mathscr{R}_1, \ldots, \mathscr{R}_k$. Then there is a fixed point $\mathscr{P}$, and there are $a_1, \ldots, a_j \in \mathscr{A}$, with

$$\boldsymbol{x} \in \mathscr{R} \Leftrightarrow \langle \boldsymbol{a}, \boldsymbol{x} \rangle \in \mathscr{P}.$$

By Lemma 13.5, $\mathscr{P}$ is ω-r.e. over $\mathfrak{A}$. Say it is represented by P in the elementary formal system with axioms E. Add to E the axiom

$$P\boldsymbol{a}, \boldsymbol{x} \rightarrow R\boldsymbol{x}.$$

Then R represents $\mathscr{R}$, so $\mathscr{R}$ is ω-r.e. over $\mathfrak{A}$.
The second part is similar.

DEFINITION. Let E be a set of axioms for an (ω) elementary formal system. We call E *simple* if, besides the predicate symbols $R_1, \ldots, R_k$, there is only one other predicate symbol in E. We call an (ω) r.e. relation simple if it is generated using a simple set of axioms.

LEMMA 13.7. *Now suppose $\mathscr{A}$ is infinite. Let $\mathscr{P}$ be an (ω) r.e. relation over $\mathfrak{A}$. For some simple (ω) r.e. relation $\mathscr{R}$ over $\mathfrak{A}$, and for some $\boldsymbol{a} \in \mathscr{A}^n$,*

$$\boldsymbol{x} \in \mathscr{P} \Leftrightarrow \langle \boldsymbol{a}, \boldsymbol{x} \rangle \in \mathscr{R}.$$

PROOF. $\mathscr{A}$ is infinite. For convenience in notation we suppose it includes the natural numbers. Let E be a set of axioms for the (ω) r.e. relation $\mathscr{P}$, in which it is represented by P. To further keep notation uncluttered, we confine ourselves to a representative special case. We suppose P is two-place, and the only predicate symbols of E other than $R_1, \ldots, R_k$ and P are

B (2-place), C (1-place), D (3-place).

Let R be a new 4-place predicate symbol. And for each axiom A in E, let A' be the result of replacing, in A, every occurrence of

Bu, v	by	$R0, 0, u, v,$
Cu	by	$R1, 0, 0, u,$
Du, v, w	by	$R2, u, v, w,$
Pu, v	by	$R3, 0, u, v.$

Let $E' = \{A' \mid A \in E\}$. Then E' is simple. And it should be clear that whole derivations from E can be translated into derivations from E' and conversely. Hence

$$E' \vdash_{\mathfrak{A}} R3, 0, u, v \iff E \vdash_{\mathfrak{A}} Pu, v.$$

Let $\mathcal{R}$ be the relation R represents. Then $\mathcal{R}$ is (ω) r.e. and

$$\langle u, v \rangle \in \mathcal{P} \iff \langle 3, 0, u, v \rangle \in \mathcal{R}.$$

LEMMA 13.8. *Again assume $\mathcal{A}$ is infinite. Every simple relation ω-r.e. over $\mathfrak{A}$ is a fixed point. Every simple relation r.e. over $\mathfrak{A}$ is an existential fixed point.*

PROOF. Once again, for notational ease, rather than a general argument, we give a representative example. Our example illustrates the first half of the lemma, the second half is similar.

Say $\mathcal{P}$ is a simple ω-r.e. relation over $\mathfrak{A}$, represented by P using the simple set E of axioms:

$$R_1 y, z \to P0, y, z;$$

$$P0, y, w \to R_2 w, z \to P0, y, z;$$

$$P0, \forall, w \to P0, z, w \to P0, 1, z;$$

$$P0, 1, y \to P0, 1, z \to Px, y, z.$$

Notice, we have carefully arranged the axioms so the variables in each conclusion are the same (if present) and in the same order. This can always be managed with appropriate variable changes. Now, based on these axioms, we construct a formula; the pattern of construction should be clear. Let $\varphi(x, y, z)$ be

$$[R_1(y, z) \wedge x = 0]$$

$$\vee (\exists w)[P(0, y, w) \wedge R_2(w, z) \wedge x = 0]$$

$$\vee (\exists w)[(\forall t)P(0, t, w) \wedge P(0, z, w) \wedge x = 0 \wedge y = 1]$$

$$\vee [P(0, 1, y) \wedge P(0, 1, z)].$$

Using φ, we get an operator Γ_φ, and we claim the least fixed point of Γ_φ is precisely $\mathcal{P}$. We outline a proof of this.

(1) Suppose $\langle a, b, c \rangle \in \Gamma_\varphi(\mathcal{P})$. Then $\varphi(\mathcal{P}, a, b, c)$ is true. Then one of the four "clauses" of φ must apply, say the fourth. Then we have $\langle 0, 1, b \rangle \in \mathcal{P}$ and $\langle 0, 1, c \rangle \in \mathcal{P}$. But $\mathcal{P}$ is represented by P, so $P0, 1, b$ and $P0, 1, c$ must be derivable from E. Now, an instance of the fourth axiom of E is

$$P0, 1, b \to P0, 1, c \to Pa, b, c.$$

Hence, trivially, Pa, b, c is derivable from E, so $\langle a, b, c\rangle \in \mathscr{P}$. We have sketched a proof that $\Gamma_\varphi(\mathscr{P}) \subseteq \mathscr{P}$.

(2) Let $\mathscr{D}$ be some three-place relation on $\mathscr{A}$. Suppose that, whenever Px, y, z has a derivation from E of length $< \alpha$, then $\langle x, y, z\rangle \in \mathscr{D}$. It is easy to see that, as a consequence, if Pa, b, c has a derivation from E of length α, then $\varphi(\mathscr{D}, a, b, c)$ is true, and so $\langle a, b, c\rangle \in \Gamma_\varphi(\mathscr{D})$.

(3) Take $\mathscr{D}$ above to be $\mathscr{P}$. Then item (2) gives us $\mathscr{P} \subseteq \Gamma_\varphi(\mathscr{P})$. Hence $\mathscr{P}$ is a fixed point of Γ_φ.

(4) Take $\mathscr{D}$ to be any fixed point of Γ_φ. Then item (2) and a straightforward transfinite induction give us $\mathscr{P} \subseteq \mathscr{D}$.

Hence $\mathscr{P}$ is the *least* fixed point of Γ_φ, and we are done.

PROPOSITION 13.9. *Let $\mathscr{A}$ be infinite. Every relation ω-r.e. over $\mathfrak{A}$ is inductive on $\mathscr{A}$ in $\mathscr{R}_1, \ldots, \mathscr{R}_k$. Every relation r.e. over $\mathfrak{A}$ is existential inductive on $\mathscr{A}$ in $\mathscr{R}_1, \ldots, \mathscr{R}_k$.*

PROOF. Immediate from Lemmas 13.7 and 13.8.

14. *R*-definability

There have been many attempts to define a notion of recursion theory suitable for arbitrary structures, and they have led in a variety of directions. But a surprising number of them have turned out to be in the same family, in that they give essentially equivalent results. This may be counted an argument for naturalness, if not for philosophical correctness. The best known member of this family is *search computability*, from Moschovakis in [1969]. We will establish its relationship with elementary formal systems in Chapter 6. The earliest member of this family known to the author is *R-definability* due to R. Smullyan, in 1956. It has been quite neglected in the literature. This is not surprising. Professor Smullyan informs us that, in 1957 he wrote a paper developing elementary formal systems for arbitrary structures, much as we have done (but not considering inputs). But he subsequently lost the paper. Professor Myhill also informs us that he continued the investigation, establishing, for instance, that for the structure of the reals, the only r.e. sets are those that are r.e. unions of intervals with algebraic end points. This too was never published. All that appeared in print was a brief abstract (Smullyan [1956]) describing an equivalent semantic approach. This semantic version is still of considerable interest, and we now present it in some detail, under the name of R-definability. We also introduce what we call ω-R-definability, which relates to inductive definability (Section 13) as R-definability relates to search computability.

As usual, let $\mathfrak{A} = \langle \mathcal{A}; \mathcal{R}_1, \ldots, \mathcal{R}_k \rangle$ be a structure, fixed for this section, and let $R_1, \ldots, R_k$ be the predicate symbols assigned to $\mathcal{R}_1, \ldots, \mathcal{R}_k$.

By an *interpretation* I over $\mathfrak{A}$ we mean an assignment, to each n-place relation symbol P, of some n-ary relation $I(P)$ on $\mathcal{A}$, in such a way that for each R_i $(i = 1, 2, \ldots, k)$ we have $I(R_i) = \mathcal{R}_i$.

Let $Pa_1, \ldots, a_n$ be an atomic formula over $\mathcal{A}$ with no occurrences of $\forall$. We call it *true under the interpretation* I if $\langle a_1, \ldots, a_n \rangle$ is in the relation $I(P)$.

We say $P\boldsymbol{a}, \forall, \boldsymbol{b}$ is *true under* I if, for each $c \in \mathcal{A}$, $P\boldsymbol{a}, c, \boldsymbol{b}$ is true under I. Similarly for more than one occurrence of $\forall$.

Let $X_1 \to X_2 \to \cdots \to X_{n-1} \to X_n$ be an (ω) elementary formal system formula over $\mathcal{A}$, with each X_i atomic. We call it *true under* I if one of $X_1, \ldots, X_{n-1}$ is not true under I, or if X_n is true under I.

Let F be a pseudo-formula over $\mathcal{A}$. We call it *true under* I if each substitution instance of it over $\mathcal{A}$ is true under I.

Let E be a set of axioms for an (ω) elementary formal system over $\mathfrak{A}$. We call I a *model for* E if each member of E is true under I.

Let P be a predicate symbol. By P^E we mean the intersection of all relations $I(P)$ such that I is a model for E. Also, we define an interpretation I^E by

$$I^E(P) = P^E.$$

LEMMA 14.1. *Let A be an atomic formula over $\mathcal{A}$, with or without occurrences of $\forall$. A is true under I^E if and only if A is true in every model for E.*

THEOREM 14.2. *I^E is, itself, a model for E.*

PROOF. Suppose $X_1 \to X_2 \to \cdots \to X_{n-1} \to X_n$ is an instance of a member of E, where each X_i is atomic. Suppose each of $X_1, \ldots, X_{n-1}$ is true under I^E. We show X_n is also true under I^E.

Let I be an arbitrary model for E. By the lemma, each of $X_1, \ldots, X_{n-1}$ must be true under I. Since I is a model for E, it follows that X_n is true under I. Since I is arbitrary, X_n is true in every model for E, hence X_n is true under I^E.

DEFINITION. The model I^E is called the *minimal model* for E over $\mathfrak{A}$.

DEFINITION. A relation $\mathcal{P}$ on $\mathcal{A}$ is R-definable (ω-R-*definable*) over $\mathfrak{A}$ if, for some predicate symbol P and some elementary formal system (ω-elementary formal system) E, we have $I^E(P) = \mathcal{P}$.

THEOREM 14.3. *The relations R-definable over $\mathfrak{A}$ are the relations r.e. over $\mathfrak{A}$. The relations ω-R-definable over $\mathfrak{A}$ are the relations ω-r.e. over $\mathfrak{A}$.*

PROOF. Let us write

$$E \models_{\mathfrak{A}} P\boldsymbol{a} \quad \text{if } P\boldsymbol{a} \text{ is true in the minimal model for } E \text{ over } \mathfrak{A}.$$

The theorem will be established if we show

$$E \models_{\mathfrak{A}} P\boldsymbol{a} \iff E \vdash_{\mathfrak{A}} P\boldsymbol{a}.$$

This is obviously a "completeness" result of a simple kind. We only sketch the proof.

Suppose $E \vdash_{\mathfrak{A}} P\boldsymbol{a}$. One may show, by an induction on derivation lengths (which are ordinals for ω-elementary formal systems) that each line of a derivation from E is true in any model for E, in particular, in the minimal model. Hence $P\boldsymbol{a}$ is true in the minimal model; $E \models_{\mathfrak{A}} P\boldsymbol{a}$.

Suppose not-$E \vdash_{\mathfrak{A}} P\boldsymbol{a}$. Define an interpretation I as follows. For each relation symbol R, set

$$I(R) = \{\boldsymbol{c} \mid E \vdash_{\mathfrak{A}} R\boldsymbol{c}\}.$$

I turns out to be a model for E, so there is a model in which $P\boldsymbol{a}$ is not true. Then $P\boldsymbol{a}$ must be false in the minimal model; not-$E \models_{\mathfrak{A}} P\boldsymbol{a}$.

REMARKS. We allowed members of $\mathcal{A}$ to occur in elementary formal system axioms. Actually, in Smullyan [1956] this was not done. Axioms were pure, in our sense.

One might ask what would happen if, in the development of R-definability, one allowed all formulas, not just those of forms arising out of ω-elementary formal systems. Actually, the development still can be carried out, though a *minimal* model may no longer exist. See Fitting [1978] for an investigation of this point. Also, it might be asked, for what syntactic forms must a minimal model exist. Are those arising from elementary formal systems as general a class as possible? This is still open.

CHAPTER TWO

PRODUCTION SYSTEMS

1. Introduction

Let us return still another time to our description of ordinary recursion theory in terms of boxes. It should not be surprising that sometimes what we can do depends strongly on the behavior of the natural numbers, while at other times our actions depend only on the behavior of the boxes. Another way of saying this is that some results of ordinary recursion theory make use of the fact that the natural numbers are the subject, while others are independent of the choice of subject matter. For example, in ordinary recursion theory, every infinite recursively enumerable set includes an infinite recursive subset; proofs of this involve the fact that there is a (standard) well-ordering of the natural numbers which is recursive. This is something not shared by all structures. Another result of ordinary recursion theory is that the intersection of two recursively enumerable sets is again recursively enumerable. A proof of this can be given entirely in terms of boxes; what is being put into them is of no particular significance. In brief, a proof runs as follows. Suppose you have a list of instructions on how to fill box A; suppose you have another list of instructions on how to fill box B. Combine these two lists of instructions, and then add the following: if you find you've put something into box A, and you've also put it into box B, then put it into box C. The total list of instructions tells you how to fill box C, and it should be clear that what goes into box C is the intersection of the contents of A and B. This is, essentially, the proof of part of Proposition 1.10.3.

In Ch. 1, §10, we derived a number of items that only depend on the "box mechanism". It makes sense to study these items in the abstract. So we have formulated a list of axioms which characterize what we call a *production system*, and which includes most of the properties of recursion and ω-recursion theories demonstrated in Ch. 1, §10.

Not surprisingly, a production system is a collection of operators meeting certain closure conditions. For each structure $\mathfrak{A}$, both rec($\mathfrak{A}$)

and ω-rec($\mathfrak{A}$) will be examples of production systems. But in these examples, operators are *total*, in the sense that an operator Φ of order $\langle n, m\rangle$ can be given any n-ary relation as input. One might want to restrict inputs in various ways. For example, one might want to consider the behavior of enumeration operators only on recursively enumerable inputs. Consequently, in specifying a production system, one must specify what sets and relations are *allowable* as input for operators. If all sets and relations are allowable, we have a *total* production system, otherwise a *partial* one. Partial production systems will play an important role in Chapters 8 and 9.

Our production system axioms should be compared with the definition of admissible subcategory of $R(X^*)$ in Eilenberg–Elgot [1970]. Every total production system satisfies their axioms, but not conversely.

2. Production systems

Let $\mathcal{A}$ be a set, fixed for this section.

We begin with some terminology and notational conventions.

Let $\mathscr{C}$ be some designated collection of sets and relations on $\mathcal{A}$. We call the members of $\mathscr{C}$ *allowable* (inputs). For each $n = 1, 2, 3, \ldots$ we write $[\mathcal{A}]^n_{\mathscr{C}}$ for the collection of n-ary relations in $\mathscr{C}$.

If $\Phi : [\mathcal{A}]^n_{\mathscr{C}} \to [\mathcal{A}]^m_{\mathscr{C}}$, we call Φ a *$\mathscr{C}$-operator of order* $\langle n, m\rangle$. Note that $\mathscr{C}$-operators map allowable relations to allowable relations.

Let Φ be a $\mathscr{C}$-operator of order $\langle n, m\rangle$. We call Φ *constant with value* $\mathscr{R}$ if, for each $\mathscr{P} \in [\mathcal{A}]^n_{\mathscr{C}}$ we have $\Phi(\mathscr{P}) = \mathscr{R}$. If $\mathcal{O}$ is a collection of $\mathscr{C}$ operators, we say a relation $\mathscr{R}$ is *generated in* $\mathcal{O}$ if $\mathscr{R}$ is the value of some constant operator in $\mathcal{O}$.

Let $T^n_{i,j}$ be a transposition operator, as defined in Ch. 1, §10. By $T^n_{i,j} \upharpoonright \mathscr{C}$ we mean $T^n_{i,j}$ restricted to $[\mathcal{A}]^n_{\mathscr{C}}$. Similarly for $P^n_i \upharpoonright \mathscr{C}$, $D^n_i \upharpoonright \mathscr{C}$ and $A^n \upharpoonright \mathscr{C}$, for projection, dual projection and place-adding operators respectively. We also use the definitions of $\cap$, $\cup$, $\times$ and composition from Chapter 1, with the understanding that the operators involved will only be defined on allowable inputs.

DEFINITION. Let $\mathbf{A} = \langle \mathscr{C}, \mathcal{O}\rangle$, where $\mathscr{C}$ is a collection of sets and relations on $\mathcal{A}$, and $\mathcal{O}$ is a collection of $\mathscr{C}$-operators. We call $\mathbf{A}$ a *production system on* $\mathcal{A}$ if

(1) $\emptyset$, the empty subset of $\mathcal{A}$, is generated in $\mathcal{O}$.

(2) $=_{\mathcal{A}}$, the equality relation on $\mathcal{A}$, is generated in $\mathcal{O}$.

(3) for each $a \in \mathscr{A}$, the set $\{a\}$ is generated in $\mathcal{O}$.
(4) $T^n_{i,j} \restriction \mathscr{C}$ is in $\mathcal{O}$ for each transposition operator.
(5) $P^n_i \restriction \mathscr{C}$ is in $\mathcal{O}$ for each projection operator.
(6) $A^n \restriction \mathscr{C}$ is in $\mathcal{O}$ for each place-adding operator.
(7) $\mathcal{O}$ is closed under $\cap$ and $\cup$.
(8) $\mathcal{O}$ is closed under $\times$.
(9) $\mathcal{O}$ is closed under composition.

If $D^n_i \restriction \mathscr{C}$ is in $\mathcal{O}$ for each dual projection operator, we say $\boldsymbol{A}$ is an *ω-production system* on $\mathscr{A}$.

Let $\mathfrak{A} = \langle \mathscr{A}; \mathscr{R}_1, \dots, \mathscr{R}_k \rangle$ be a structure and let $\boldsymbol{A} = \langle \mathscr{C}, \mathcal{O} \rangle$ be a production (ω-production) system on $\mathscr{A}$. We call it a production (ω-production) system *on the structure* $\mathfrak{A}$ if each of $\mathscr{R}_1, \dots, \mathscr{R}_k$ is generated in $\mathcal{O}$.

NOTATIONAL CONVENTIONS. We will systematically follow the convention that $\boldsymbol{A}$ is a production system on $\mathscr{A}$, or $\mathfrak{A}$; $\boldsymbol{B}$ is a production system on $\mathscr{B}$, or $\mathfrak{B}$, and so on.

If $\boldsymbol{A} = \langle \mathscr{C}, \mathcal{O} \rangle$ is a production system, we will call the members of $\mathscr{C}$ *allowable in* **A**, and the members of $\mathcal{O}$ *operators in* **A**. If Φ is an operator in $\boldsymbol{A}$ we will write $\Phi \in \boldsymbol{A}$. We will say $\mathscr{R}$ is *generated in* **A** if it is generated in $\mathcal{O}$. We will generally write $[\mathscr{A}]^n$ instead of $[\mathscr{A}]^n_{\mathscr{C}}$. We will generally write $T^n_{i,j}$ instead of $T^n_{i,j} \restriction \mathscr{C}$, and similarly for the projection, dual projection and place-adding operators.

DEFINITION. Let $\boldsymbol{A} = \langle \mathscr{C}, \mathcal{O} \rangle$ be a production system on $\mathscr{A}$. If $\mathscr{C}$ is the collection of *all* sets and relations on $\mathscr{A}$, we say $\boldsymbol{A}$ is a *total production system.* Otherwise $\boldsymbol{A}$ is a *partial production system.* An unqualified *production system* means either partial or total.

EXAMPLES. Of course both rec$(\mathfrak{A})$ and ω-rec$(\mathfrak{A})$ are total production systems on the structure $\mathfrak{A}$, and ω-rec$(\mathfrak{A})$ is also an ω-production system.

If we consider rec$(\mathfrak{A})$ but restrict all enumeration operators to r.e. inputs (that is, allowable = r.e.) one still has a production system, and cardinality considerations show that if $\mathscr{A}$ is infinite, it will be a *partial* production system. Similarly for ω-rec$(\mathfrak{A})$.

There is one other family of total production systems which should be kept in mind, the *first order theories,* defined as follows. Let $\mathfrak{A}$ be a structure. Let P be an n-place predicate symbol, not one of $R_1, \dots, R_k$

(assigned to the given relations of $\mathfrak{A}$) or E (representing $=_{\mathfrak{A}}$). Let $\varphi(x_1, \ldots, x_m)$ be any first order formula, with $x_1, \ldots, x_m$ free, and with only $P, R_1, \ldots, R_k$ and E as predicate symbols. Define an operator Φ_φ of order $\langle n, m\rangle$ by

$$\Phi_\varphi(\mathscr{P}) = \{\langle x_1, \ldots, x_m\rangle \mid \varphi(x_1, \ldots, x_m) \text{ is true in the structure } \langle \mathfrak{A}, \mathscr{P}\rangle\}.$$

The collection of all such operators gives a total ω-production system on $\mathfrak{A}$. We call it fo$(\mathfrak{A})$.

3. Elementary results

For this section, $\mathbf{A}$ is some fixed production system, on $\mathscr{A}$. All the operators introduced below should be understood as being defined only on the allowable relations of $\mathbf{A}$.

PROPOSITION 3.1. *Let $\mathscr{R} \in [\mathscr{A}]^n$. The following are equivalent*:

(1) *For some m, there is a constant operator in $\mathbf{A}$ of order $\langle m, n\rangle$ and value $\mathscr{R}$ (i.e. $\mathscr{R}$ is generated in $\mathbf{A}$).*

(2) *For each m, there is a constant operator in $\mathbf{A}$ of order $\langle m, n\rangle$ and value $\mathscr{R}$.*

(3) *If $\mathscr{P} \in [\mathscr{A}]^k$ is generated in $\mathbf{A}$, there is an operator $\Phi \in \mathbf{A}$ of order $\langle k, n\rangle$ such that $\Phi(\mathscr{P}) = \mathscr{R}$.*

(4) *There is an operator $\Phi \in \mathbf{A}$ of order $\langle 1, n\rangle$ such that $\Phi(\emptyset) = \mathscr{R}$.*

PROOF. First we show (1) $\Rightarrow$ (2). Suppose there is a constant operator $\Phi \in \mathbf{A}$ of order $\langle m, n\rangle$ and value $\mathscr{R}$. We show that for each k there is a constant operator of order $\langle k, n\rangle$ and value $\mathscr{R}$. Well,

If $k = m$, the operator is Φ.

If $k > m$, the operator is $\Phi P_1^{m+1} \cdots P_1^{k-2} P_1^{k-1} P_1^k$.

If $k < m$, the operator is $\Phi A^{m-1} \cdots A^{k+2} A^{k+1} A^k$.

Next, that (2) $\Rightarrow$ (3) and that (3) $\Rightarrow$ (4) are trivial. Finally we show (4) $\Rightarrow$ (1). Suppose $\Phi \in \mathbf{A}$ is of order $\langle 1, n\rangle$ and $\Phi(\emptyset) = \mathscr{R}$. By Axiom (1), $\emptyset$ is generated in $\mathbf{A}$, so there is a constant operator Ψ in $\mathbf{A}$, of order $\langle m, 1\rangle$ say, and value $\emptyset$. Then $\Phi\Psi$ is a constant operator, in $\mathbf{A}$, of order $\langle m, n\rangle$ and value $\mathscr{R}$.

REMARK. $=_{\mathscr{A}}$ is generated in $\mathbf{A}$ and $P_1^2 \in \mathbf{A}$, so it follows by the above that $\mathscr{A} \triangleq P_1^2\ (=_{\mathscr{A}})$ is generated in $\mathbf{A}$.

DEFINITION. I^n is the *identity operator* of order $\langle n, n\rangle$. Specifically, for $\mathcal{P} \in [\mathcal{A}]^n$, $I^n(\mathcal{P}) = \mathcal{P}$.

PROPOSITION 3.2. *Each I^n is in* $\mathbf{A}$.

PROOF. $I^n = T^n_{i,i}$.

DEFINITION. $E^n_{i,j}$ is the *equality operator* of order $\langle n, n\rangle$ on the ith and jth coordinates. Specifically, for $\mathcal{P} \in [\mathcal{A}]^n$,

$$E^n_{i,j}(\mathcal{P}) = \{\langle x_1, \ldots, x_n\rangle \in \mathcal{P} \mid x_i = x_j\}.$$

PROPOSITION 3.3. *Each $E^n_{i,j}$ is in* $\mathbf{A}$.

PROOF. If $i = j$, $E^n_{i,j}$ is I^n which is in $\mathbf{A}$.

Next we show $E^n_{n-1,n} \in \mathbf{A}$. Now, $=_{\mathcal{A}}$ is generated in $\mathbf{A}$, so by Proposition 3.1 there is a constant operator $\Phi \in \mathbf{A}$ of order $\langle n, 2\rangle$ and value $=_{\mathcal{A}}$. Then the operator $A^{n-1} \cdots A^3A^2\Phi$ is constant, of order $\langle n, n\rangle$, and value

$$\underbrace{\mathcal{A} \times \mathcal{A} \times \cdots \times \mathcal{A}}_{n-2} \times =_{\mathcal{A}},$$

and so

$$E^n_{n-1,n} = (A^{n-1} \cdots A^3A^2\Phi) \cap I^n.$$

Finally, for any $i \neq j$,

$$E^n_{i,j} = T^n_{j,n}T^n_{i,n-1}E^n_{n-1,n}T^n_{j,n}T^n_{i,n-1}.$$

LEMMA 3.4. *There are "place-duplicating" operators in* $\mathbf{A}$. *That is, for each n, i, j, there is an operator $\Phi \in \mathbf{A}$ of order $\langle n, n+1\rangle$ that inserts a duplicate of the ith place at position j. Specifically*

$$\Phi(\mathcal{P}) = \{\langle x_1, \ldots, x_{n+1}\rangle \mid x_j = x_i \text{ and } \langle x_1, \ldots, x_{j-1}, x_{j+1}, \ldots, x_{n+1}\rangle \in \mathcal{P}\}.$$

PROOF. We show the special case where $i = 2$, $j = 1$; the general case then follows using transpositions. Thus, we must show there is an operator $\Phi \in \mathbf{A}$ such that

$$\Phi(\mathcal{P}) = \{\langle y, x_1, \ldots, x_n\rangle \mid y = x_1 \text{ and } \langle x_1, \ldots, x_n\rangle \in \mathcal{P}\}.$$

Well, $\Phi = E^n_{1,2}A^n$.

The following definition and proposition are essentially from Eilenberg, Elgot [1970].

DEFINITION. Let $[n]=\{1,2,\dots,n\}$. Given a function $f:[n]\to[k]$ we define an operator L_f of order $\langle k,n\rangle$, called a *logical operator*, by

$$L_f(\mathcal{P})=\{\langle x_{f(1)},\dots,x_{f(n)}\rangle \mid \langle x_1,\dots,x_k\rangle\in\mathcal{P}\}.$$

EXAMPLES. (1) Let $f:[n]\to[n]$ be defined by

$$f(x)=x \quad \text{for } x\neq i,j,$$

$$f(i)=j, \qquad f(j)=i.$$

Then, for $\mathcal{P}\in[\mathcal{A}]^n$,

$$\begin{aligned} L_f(\mathcal{P}) &= \{\langle x_{f(1)},\dots,x_{f(n)}\rangle \mid \langle x_1,\dots,x_n\rangle\in\mathcal{P}\}\\ &= \{\langle x_1,\dots,x_j,\dots,x_i,\dots,x_n\rangle \mid \langle x_1,\dots,x_i,\dots,x_j,\dots,x_n\rangle\in\mathcal{P}\}\\ &= T^n_{i,j}(\mathcal{P}). \end{aligned}$$

(2) Let $f:[n]\to[n+1]$ be defined by $f(x)=x$. Then, for $\mathcal{P}\in[\mathcal{A}]^{n+1}$, $L_f(\mathcal{P})=P^{n+1}_{n+1}(\mathcal{P})$.

Other projection operators may be shown to be logical operators in a similar way.

(3) Let $f:[n+1]\to[n]$ be defined by

$$f(x)=x \quad \text{for } x\neq n+1, \qquad f(n+1)=n.$$

Then for $\mathcal{P}\in[\mathcal{A}]^n$, $L_f(\mathcal{P})$ is the result of duplicating place n at position $n+1$. The other place duplicating operators are also logical operators.

The proof of the following theorem shows that all the logical operators are built up by composition from the above examples.

PROPOSITION 3.5. *Each logical operator is in* **A**.

PROOF. Suppose $f:[n]\to[p]$ and $g:[p]\to[k]$. It is straightforward that $L_{gf}=L_fL_g$.

Suppose $f:[n]\to[n]$ is a permutation. Then it is a composition of transpositions, $f=t_1t_2\cdots t_j$. But then $L_f=L_{t_j}\cdots L_{t_2}L_{t_1}$, and each L_{t_i} is in **A** by Example (1), hence $L_f\in\mathbf{A}$.

Suppose $f:[n]\to[k]$ is onto. Then $n\geq k$. If $n=k$, f is a permutation, and we are done. If $n>k$ there must exist maps

$$[n]\xrightarrow{f_1}[n-1]\xrightarrow{f_2}[n-2]\xrightarrow{f_3}\cdots\xrightarrow{f_{n-k}}[k]$$

each onto, with f being the composition. Consider one of these maps,

$$g : [r+1] \to [r].$$

This can be put in the form

$$g = P_2 h P_1$$

where P_1 is a permutation of $[r+1]$, P_2 is a permutation of $[r]$, and $h : [r+1] \to [r]$ is given by

$$h(x) = x \quad \text{for } x \neq r+1, \qquad h(r+1) = r.$$

Permutations were treated above, and $L_h \in \boldsymbol{A}$ by Example (3). Thus each $L_{f_i} \in \boldsymbol{A}$, and so $L_f \in \boldsymbol{A}$.

Suppose $f : [n] \to [k]$ is 1–1. Then $n \leq k$. Again, if $n = k$ we have a permutation, and are done. If $n < k$, there must exist maps

$$[n] \xrightarrow{f_1} [n+1] \xrightarrow{f_2} [n+2] \xrightarrow{f_3} \cdots \xrightarrow{f_{k-n}} [k]$$

each 1–1, with f being the composition. Consider one of these maps

$$g : [r] \to [r+1].$$

This can be put in the form

$$g = P_2 h P_1$$

where P_1 is a permuation of $[r]$, P_2 is a permutation of $[r+1]$, and $h : [r] \to [r+1]$ is given by $h(x) = x$. Then, using Example (2) above, we are again done.

Finally, suppose $f : [n] \to [k]$ is neither 1–1 nor onto. Call two members $x, y \in [n]$ *equivalent* if $f(x) = f(y)$. This is an equivalence relation which partitions $[n]$ into a finite number of equivalence classes; say they are $C_1, C_2, \ldots, C_p$ (the numbering is arbitrary). Let $g : [n] \to [p]$ be defined by

$$g(x) = i \quad \text{if } x \in C_i.$$

Then g is onto. Let $h : [p] \to [k]$ be defined by

$$h(i) = y \quad \text{if, for any } x \in C_i,\ f(x) = y.$$

Then h is 1–1. And $f = hg$, so by the above, we are done.

Finally, nothing so far has used Axiom (3), that constants are generated. The following embodies that assumption in a convenient form.

DEFINITION. Let $a \in \mathscr{A}$, and $n = 2, 3, 4, \ldots$. S_a^n is the *a-section operator* of order $\langle n, n-1 \rangle$ defined as follows. For $\mathscr{P} \in [\mathscr{A}]^n$,

$$S^n_a(\mathscr{P}) = \{\langle x_2, \ldots, x_n\rangle \mid \langle a, x_2, \ldots, x_n\rangle \in \mathscr{P}\}.$$

PROPOSITION 3.6. *Each S^n_a is in $\mathbf{A}$.*

PROOF. $\{a\}$ is generated in $\mathbf{A}$. Using Proposition 3.1, let Φ be a constant operator in $\mathbf{A}$ of order $\langle n, 1\rangle$ and value $\{a\}$. Then,

$$S^n_a = P^n_1 P^{n+1}_1 E^{n+1}_{1,2}(\Phi \times I^n).$$

4. Reducibility

In ordinary recursion theory there are several notions of *reducibility* that are studied. The idea is, $\mathscr{P}$ is reducible to $\mathscr{R}$, in some sense, if means of getting certain information about $\mathscr{R}$ can effectively be converted into means of getting comparable information about $\mathscr{P}$. We formulate, for production systems, three reducibility notions, relate them to some familiar concepts in ordinary recursion and hyperarithmetic theory, then derive their elementary properties.

For this section, $\mathbf{A}$ is a fixed production system, on the domain $\mathscr{A}$, $\mathscr{P}$, $\mathscr{Q}$, $\mathscr{R}$, etc. are relations on $\mathscr{A}$. If $\mathscr{P} \subseteq \mathscr{A}^n$, by $\bar{\mathscr{P}}$ we mean $\mathscr{A}^n - \mathscr{P}$.

DEFINITION. (1) $\mathscr{P} \leqslant_{\mathbf{A}} \mathscr{R}$ if $\mathscr{P} = \Phi(\mathscr{R})$ for some $\Phi \in \mathbf{A}$.

(2) $\mathscr{P}$ is *generated in* $\mathscr{R}$ if $\mathscr{P} \leqslant_{\mathbf{A}} \mathscr{R} \times \bar{\mathscr{R}}$.

(3) $\mathscr{P}$ is *co-generated in* $\mathscr{R}$ if $\bar{\mathscr{P}}$ is generated in $\mathscr{R}$.

(4) $\mathscr{P}$ is *bi-generated in* $\mathscr{R}$ if $\mathscr{P}$ is both generated and co-generated in $\mathscr{R}$.

Background for these notions is as follows.

I. In Chapter 8, it will be shown that, for ordinary recursion theory, rec($\mathfrak{S}(\mathsf{N})$), our notion of enumeration operator is equivalent to the usual one. Given this fact, in ordinary recursion theory, $\mathscr{P} \leqslant_{\mathbf{A}} \mathscr{R}$ means $\mathscr{P}$ is *enumeration reducible to* $\mathscr{R}$, usually symbolized by $\mathscr{P} \leqslant_e \mathscr{R}$. See Rogers [1967], pp. 145–147. It then follows from Theorem XXIV in Rogers [1967], p. 151, and Corollary 4.10 below, that in rec($\mathfrak{S}(\mathsf{N})$), *bi-generated in* is the same as *recursive in*. Similarly, *generated in* and *recursively enumerable in* are the same in ordinary recursion theory. We note that B. Horowitz has developed relative (ordinary) recursion theory using elementary formal systems directly. See Horowitz [1975], [1975A], [198+].

II. In hyperarithmetic theory, ω-rec($\mathfrak{S}(\mathsf{N})$), *bi-generated in* coincides with the standard notion of *hyperarithmetic in* (Rogers [1967], p. 410), and *generated in* coincides with *being* Π^1_1 *in*, (Rogers [1967], p. 409).

We outline a proof of this, based heavily on work in Moschovakis [1974]. This result will not be needed here, except for motivation.

It is not hard to see that saying $\mathscr{P}$ is generated in $\mathscr{R}$ in ω-rec($\mathfrak{S}(\mathsf{N})$) is the same as saying $\mathscr{P}$ is ω-r.e. in the structure $\langle \mathsf{N}; y = x^+, \mathscr{R}, \bar{\mathscr{R}}\rangle$. By Theorem 1.13.3 this is the same as saying $\mathscr{P}$ is inductive over N in the relations $y = x^+$, $\mathscr{R}$ and $\bar{\mathscr{R}}$, that is, that $\mathscr{P}$ is inductive on the structure $\langle \mathsf{N}; y = x^+, \mathscr{R}\rangle$ in the terminology of Moschovakis [1974], p. 17. ($y \neq x^+$ is easily shown to be r.e. in $\langle \mathsf{N}; y = x^+\rangle$.) Now, for each $\mathscr{R}$, this structure is acceptable, as defined in Moschovakis [1974], p. 22, hence being inductive on it is equivalent to being Π^1_1 on it by Moschovakis [1974], pp. 20 and 132. Finally, $\mathscr{P}$ being Π^1_1 on the structure $\langle \mathsf{N}; y = x^+, \mathscr{R}\rangle$ is equivalent to the usual definition of $\mathscr{P}$ being Π^1_1 in $\mathscr{R}$. The claim about being hyperarithmetic in follows similarly.

REMARKS. Generally, to keep notation uncluttered, we will write $\leqslant$ for $\leqslant_{\boldsymbol{A}}$.

Note that if $\mathscr{P} \leqslant \mathscr{R}$, then $\mathscr{P}$ and $\mathscr{R}$ are automatically *allowable* relations. It should also be noted that:

$$\mathscr{P} \text{ is generated in } \emptyset \iff \mathscr{P} \leqslant \emptyset \times \bar{\emptyset} \iff \mathscr{P} \leqslant \emptyset$$

$$\iff \mathscr{P} \text{ is generated in } \boldsymbol{A} \quad \text{by Proposition 3.1.}$$

PROPOSITION 4.1. $\leqslant$ *is transitive and reflexive on the allowable relations.*

PROOF. Transitive since the operators of $\boldsymbol{A}$ are closed under composition. Reflexive since there are identity operators.

PROPOSITION 4.2. $\mathscr{P}$ *is generated in* $\boldsymbol{A}$ *if and only if* $\mathscr{P} \leqslant \mathscr{R}$ *for all allowable* $\mathscr{R}$.

PROOF. If $\mathscr{P}$ is generated in $\boldsymbol{A}$ then $\mathscr{P} \leqslant \mathscr{R}$ for all $\mathscr{R}$ by Proposition 3.1.

If $\mathscr{P} \leqslant \mathscr{R}$ for all $\mathscr{R}$, then $\mathscr{P} \leqslant \emptyset$, so $\mathscr{P}$ is generated in $\boldsymbol{A}$ by Proposition 3.1 again.

PROPOSITION 4.3. (1) *If* $\mathscr{P} \times \mathscr{R}$ *is allowable then* $\mathscr{P}$ *and* $\mathscr{R}$ *are allowable and*

$$\mathscr{P} \leqslant \mathscr{P} \times \mathscr{R}, \qquad \mathscr{R} \leqslant \mathscr{P} \times \mathscr{R}.$$

(2) *If* $\mathscr{P} \leqslant \mathscr{S}$ *and* $\mathscr{R} \leqslant \mathscr{S}$, *then* $\mathscr{P} \times \mathscr{R}$ *is allowable and*

$$\mathcal{P} \times \mathcal{R} \leqslant \mathcal{S}.$$

PROOF. (1) Say $\mathcal{P}$ is n-ary and $\mathcal{R}$ is k-ary. Define

$$f : [n] \to [n+k]$$

by $f(x) = x$ for all $x \in [n]$, and consider the corresponding logical operator L_f.

$$\begin{aligned} L_f(\mathcal{P} \times \mathcal{R}) &= \{\langle x_{f(1)}, \ldots, x_{f(n)}\rangle \mid \langle x_1, \ldots, x_{n+k}\rangle \in \mathcal{P} \times \mathcal{R}\} \\ &= \{\langle x_1, \ldots, x_n\rangle \mid \langle x_1, \ldots, x_{n+k}\rangle \in \mathcal{P} \times \mathcal{R}\} = \mathcal{P}. \end{aligned}$$

Thus if $\mathcal{P} \times \mathcal{R}$ is allowable, so is $\mathcal{P}$, and $\mathcal{P} \leqslant \mathcal{P} \times \mathcal{R}$. The result for $\mathcal{R}$ is similar.

(2) Say $\mathcal{P} = \Phi(\mathcal{S})$ and $\mathcal{R} = \Psi(\mathcal{S})$ for $\Phi, \Psi \in \mathbf{A}$. Then

$$\mathcal{P} \times \mathcal{R} = (\Phi \times \Psi)(\mathcal{S}),$$

DEFINITION. $\mathcal{P} \sim \mathcal{R}$ if $\mathcal{P} \leqslant \mathcal{R}$ and $\mathcal{R} \leqslant \mathcal{P}$.

REMARK. In ordinary recursion theory the counterpart of this notion is often symbolized by $\equiv_e$. See Rogers [1967], p. 147.

PROPOSITION 4.4. (1) *If $\mathcal{P} \times \mathcal{R}$ is allowable, so is $\mathcal{R} \times \mathcal{P}$, and*

$$\mathcal{P} \times \mathcal{R} \sim \mathcal{R} \times \mathcal{P}.$$

(2) *If $\mathcal{P}$ is allowable, so is $\mathcal{P} \times \mathcal{P}$, and*

$$\mathcal{P} \times \mathcal{P} \sim \mathcal{P}.$$

(3) *If $\mathcal{R} \times \mathcal{S}$ is allowable and $\mathcal{P} \leqslant \mathcal{R}$, then $\mathcal{P} \times \mathcal{S}$ is allowable, and*

$$\mathcal{P} \times \mathcal{S} \leqslant \mathcal{R} \times \mathcal{S}.$$

(4) *If $\mathcal{R} \times \mathcal{S}$ is allowable and $\mathcal{P} \leqslant \mathcal{R}$, $\varphi \leqslant \mathcal{S}$, then $\mathcal{P} \times \varphi$ is allowable, and*

$$\mathcal{P} \times \varphi \leqslant \mathcal{R} \times \mathcal{S}.$$

(5) *If $\mathcal{R} \times \mathcal{S}$ is allowable, and $\mathcal{P} \sim \mathcal{R}$ and $\varphi \sim \mathcal{S}$, then $\mathcal{P} \times \varphi$ is allowable, and*

$$\mathcal{P} \times \varphi \sim \mathcal{R} \times \mathcal{S}.$$

PROOF. (1) Suppose $\mathcal{P} \times \mathcal{R}$ is allowable. By Proposition 4.3 (1),

$$\mathcal{P} \leqslant \mathcal{P} \times \mathcal{R}, \qquad \mathcal{R} \leqslant \mathcal{P} \times \mathcal{R}.$$

Then by 4.3 (2), $\mathscr{R} \times \mathscr{P}$ is allowable and

$$\mathscr{R} \times \mathscr{P} \leqslant \mathscr{P} \times \mathscr{R}.$$

(2) By reflexivity, $\mathscr{P} \leqslant \mathscr{P}$. Then by 4.3 (2), $\mathscr{P} \times \mathscr{P}$ is allowable and $\mathscr{P} \times \mathscr{P} \leqslant \mathscr{P}$. Then by 4.3 (1), $\mathscr{P} \leqslant \mathscr{P} \times \mathscr{P}$.

(3) Suppose $\mathscr{R} \times \mathscr{S}$ is allowable. By 4.3 (1),

$$\mathscr{R} \leqslant \mathscr{R} \times \mathscr{S}, \qquad \mathscr{S} \leqslant \mathscr{R} \times \mathscr{S}.$$

But $\mathscr{P} \leqslant \mathscr{R}$, so by transitivity, $\mathscr{P} \leqslant \mathscr{R} \times \mathscr{S}$. The result follows by 4.3 (2).

(4) and (5) are similar.

PROPOSITION 4.5. *$\mathscr{P}$ is bi-generated in $\mathscr{R}$ if and only if $\mathscr{P} \times \bar{\mathscr{P}} \leqslant \mathscr{R} \times \bar{\mathscr{R}}$.*

PROOF. (1) Suppose $\mathscr{P}$ is bi-generated in $\mathscr{R}$. Then $\mathscr{P} \leqslant \mathscr{R} \times \bar{\mathscr{R}}$, and $\bar{\mathscr{P}} \leqslant \mathscr{R} \times \bar{\mathscr{R}}$. Now use Proposition 4.3 (2).

(2) Suppose $\mathscr{P} \times \bar{\mathscr{P}} \leqslant \mathscr{R} \times \bar{\mathscr{R}}$. By 4.3 (1), $\mathscr{P} \leqslant \mathscr{P} \times \bar{\mathscr{P}}$, so $\mathscr{P} \leqslant \mathscr{R} \times \bar{\mathscr{R}}$, and thus $\mathscr{P}$ is generated in $\mathscr{R}$. Similarly for $\bar{\mathscr{P}}$.

PROPOSITION 4.6. (1) *$\mathscr{P}$ generated in $\mathscr{Q}$, $\mathscr{Q}$ bi-generated in $\mathscr{R}$ imply $\mathscr{P}$ generated in $\mathscr{R}$.*

(2) *$\mathscr{P} \leqslant \mathscr{Q}$, $\mathscr{Q}$ generated in $\mathscr{R}$ imply $\mathscr{P}$ generated in $\mathscr{R}$.*

PROOF. (1) $\mathscr{P} \leqslant \mathscr{Q} \times \bar{\mathscr{Q}}$ and $\mathscr{Q} \times \bar{\mathscr{Q}} \leqslant \mathscr{R} \times \bar{\mathscr{R}}$ imply $\mathscr{P} \leqslant \mathscr{R} \times \bar{\mathscr{R}}$.

(2) $\mathscr{P} \leqslant \mathscr{Q}$ and $\mathscr{Q} \leqslant \mathscr{R} \times \bar{\mathscr{R}}$ imply $\mathscr{P} \leqslant \mathscr{R} \times \bar{\mathscr{R}}$.

DEFINITION. Let Φ be an operator in **A**. By the *dual* of Φ we mean the map Ψ given by $\Psi(\mathscr{P}) = \overline{\Phi(\bar{\mathscr{P}})}$.

PROPOSITION 4.7. (1) *If $\mathscr{P}$ is allowable, the collection of relations $\leqslant \mathscr{P}$ is closed under $\cap$, $\cup$, $\times$, and all operators of* **A**.

(2) *If $\mathscr{P} \times \bar{\mathscr{P}}$ is allowable, the collection of relations generated in $\mathscr{P}$ is closed under $\cap$, $\cup$, $\times$ and all operators of* **A**.

(3) *If $\mathscr{P} \times \bar{\mathscr{P}}$ is allowable, the collection of relations bi-generated in $\mathscr{P}$ is closed under complementation, $\cap$, $\cup$, $\times$, and under all operators of* **A** *whose duals are also in* **A**. *(This includes transposition and place-adding operators, and in ω-production systems, projection and dual projection as well.)*

PROOF. (1) Suppose $\mathscr{Q}, \mathscr{R} \leqslant \mathscr{P}$. *Say* $\mathscr{Q} = \Phi(\mathscr{P})$ and $\mathscr{R} = \Psi(\mathscr{P})$. Then $\mathscr{Q} \cap \mathscr{R} = (\Phi \cap \Psi)(\mathscr{P})$ so $\mathscr{Q} \cap \mathscr{R} \leqslant \mathscr{P}$. Closure under $\cup$ and $\times$ is similar.

Suppose $\mathscr{R} \leqslant \mathscr{P}$. Say $\mathscr{R} = \Phi(\mathscr{P})$. Then $\Psi(\mathscr{R}) = \Psi(\Phi(\mathscr{P})) = (\Psi\Phi)(\mathscr{P})$, so $\Psi(\mathscr{R}) \leqslant \mathscr{P}$.

(2) is like (1) but with $\mathscr{P} \times \bar{\mathscr{P}}$ in place of $\mathscr{P}$.

(3) Closure under complementation is immediate from the definition. Then closure under $\cap$ and $\cup$ follows from (2), using DeMorgan's laws.

Some remarks before treating $\times$. $\mathscr{A}$ is generated in $\mathbf{A}$. It follows that $\mathscr{A}$ is generated in $\mathscr{P}$, so by (2), for each n, $\mathscr{A}^n$ is generated in $\mathscr{P}$.

Now, suppose $\mathscr{Q}, \mathscr{R}$ are bi-generated in $\mathscr{P}$. Then $\mathscr{Q}, \mathscr{R}$ are generated in $\mathscr{P}$, hence so is $\mathscr{Q} \times \mathscr{R}$ by (2). Also $\bar{\mathscr{Q}}, \bar{\mathscr{R}}$ are generated in $\mathscr{P}$. Say $\mathscr{Q}$ is n-ary and $\mathscr{R}$ is m-ary. Now,

$$\overline{\mathscr{Q} \times \mathscr{R}} = (\bar{\mathscr{Q}} \times \mathscr{A}^m) \cup (\mathscr{A}^n \times \bar{\mathscr{R}})$$

and by the above remarks, and (2), this must be generated in $\mathscr{P}$. Thus $\mathscr{Q} \times \mathscr{R}$ is bi-generated in $\mathscr{P}$.

Closure under those operators of $\mathbf{A}$ whose duals are also in $\mathbf{A}$ follows easily from (2) and closure under complementation. Finally, transposition and place-adding operators are self-dual, while projection and dual projection operators are duals of each other.

DEFINITION. For this, we suppose $\mathscr{A}$ has at least two members, two of which are denoted 0 and 1.

Let $\mathscr{P} \subseteq \mathscr{A}^n$. The *characteristic function* of $\mathscr{P}$ is

$$c_{\mathscr{P}}(\boldsymbol{v}) = \begin{cases} 1 & \text{if } \boldsymbol{v} \in \mathscr{P}, \\ 0 & \text{if } \boldsymbol{v} \notin \mathscr{P}. \end{cases}$$

We identify functions with their graphs. That is, a function is a single valued relation.

LEMMA 4.7. *Suppose $c_{\mathscr{P}}$ is allowable. Then so is $\mathscr{P} \times \bar{\mathscr{P}}$ and*

$$\mathscr{P} \times \bar{\mathscr{P}} \leqslant c_{\mathscr{P}}.$$

PROOF. Say $\mathscr{P}$ is n-ary. Then

$$\mathscr{P} = (T^n_{1,n} S^{n+1}_1 T^{n+1}_{1,n+1})(c_{\mathscr{P}}) \quad \text{and} \quad \bar{\mathscr{P}} = (T^n_{1,n} S^{n+1}_0 T^{n+1}_{1,n+1})(c_{\mathscr{P}}).$$

The result follows by Proposition 4.3 (2).

LEMMA 4.8. *Suppose $\mathscr{P} \times \bar{\mathscr{P}}$ is allowable. Then so is $c_{\mathscr{P}}$ and*

$$c_{\mathscr{P}} \leqslant \mathscr{P} \times \bar{\mathscr{P}}.$$

PROOF. Say $\mathscr{P}$ is n-ary. Let Φ_0 and Φ_1 be constant operators in $\boldsymbol{A}$ of order $\langle n, 1\rangle$, with Φ_0 of value $\{0\}$ and Φ_1 of value $\{1\}$. Set

$$\Psi_0 = I^n \times \Phi_0, \qquad \Psi_1 = I^n \times \Phi_1.$$

Then $\Psi_0, \Psi_1 \in \boldsymbol{A}$ are of order $\langle n, n+1\rangle$ and

$$\Psi_0(\mathscr{R}) = \mathscr{R} \times \{0\}, \qquad \Psi_1(\mathscr{R}) = \mathscr{R} \times \{1\}.$$

Now, $\mathscr{P}, \bar{\mathscr{P}} \leqslant \mathscr{P} \times \bar{\mathscr{P}}$. Say $J, K \in \boldsymbol{A}$ are such that

$$J(\mathscr{P} \times \bar{\mathscr{P}}) = \mathscr{P}, \qquad K(\mathscr{P} \times \bar{\mathscr{P}}) = \bar{\mathscr{P}}.$$

Then

$$c_{\mathscr{P}} = \{\Psi_1 J \cup \Psi_0 K)(\mathscr{P} \times \bar{\mathscr{P}})$$

which completes the proof.

PROPOSITION 4.9. *$c_{\mathscr{P}}$ is allowable if and only if $\mathscr{P} \times \bar{\mathscr{P}}$ is allowable and if so,*

$$c_{\mathscr{P}} \sim \mathscr{P} \times \bar{\mathscr{P}}.$$

COROLLARY 4.10. *$c_{\mathscr{P}} \leqslant c_{\mathscr{R}}$ if and only if $\mathscr{P}$ is bi-generated in $\mathscr{R}$.*

PROOF. By the above and Proposition 4.5.

5. Pointwise generated functions

Let $\boldsymbol{A}$ be a production system, and let $f : \mathscr{A}^n \to \mathscr{A}^m$ be a function. There are two reasonable meanings that might be assigned to the loose notion of f being "computable" in $\boldsymbol{A}$. One is that the graph of f should be generated as a whole; this is the notion we used in the last section for characteristic functions. The other is that there should be an operator that evaluates f point by point. In this section we properly define these notions, and prove a result connecting them which will be used a great deal in later chapters.

THEOREM 5.1. (1) *If $\mathscr{R}$ is generated in $\boldsymbol{A}$ then there is an operator $\Phi \in \boldsymbol{A}$ such that, for allowable $\mathscr{P}$,*

$$\boldsymbol{v} \in \Phi(\mathscr{P}) \Leftrightarrow \text{ for some } \boldsymbol{w} \in \mathscr{P}, \langle \boldsymbol{w}, \boldsymbol{v}\rangle \in \mathscr{R}.$$

(2) *Let $\mathscr{Q} \neq \emptyset$ be allowable in $\boldsymbol{A}$. If $\mathscr{R} \leqslant \mathscr{Q}$, then there is an operator $\Phi \in \boldsymbol{A}$ such that, for allowable $\mathscr{Q} \times \mathscr{P}$,*

$$\boldsymbol{v} \in \Phi(\mathscr{Q} \times \mathscr{P}) \Leftrightarrow \text{ for some } \boldsymbol{w} \in \mathscr{P}, \langle \boldsymbol{w}, \boldsymbol{v}\rangle \in \mathscr{R}.$$

PROOF. We do (2) first. Say $\mathscr{Q}$ is m-ary, $\mathscr{P}$ is n-ary, and $\mathscr{R}$ is $n+k$-ary. This means Φ is to be of order $\langle m+n, k\rangle$. The idea is: the set of $\boldsymbol{v}$ such that, for some $\boldsymbol{w} \in \mathscr{P}$, $\langle \boldsymbol{w}, \boldsymbol{v}\rangle \in \mathscr{R}$, can be constructed by forming

$$\underset{n+k \text{ place}}{\mathscr{R}} \cap (\underset{n \text{ place}}{\mathscr{P}} \times \mathscr{A}^k);$$

then "knocking off" the first n components.

$\mathscr{R} \leqslant \mathscr{Q}$. Say $\mathscr{R} = J(\mathscr{Q})$. Also, let L_f be the logical operator such that

$$L_f : \langle x_1, \ldots, x_m, x_{m+1}, \ldots, x_{m+n}\rangle \rightarrow \langle x_1, \ldots, x_m\rangle.$$

Then $\mathscr{R} = (JL_f)(\mathscr{Q} \times \mathscr{P})$.

Further, let L_g be the logical operator such that

$$L_g : \langle x_1, \ldots, x_m, x_{m+1}, \ldots, x_{m+n}\rangle \rightarrow \langle x_{m+1}, \ldots, x_{m+n}\rangle.$$

Then $\mathscr{P} = L_g(\mathscr{Q} \times \mathscr{P})$.

Now $\mathscr{A}$ is generated, hence so is $\mathscr{A}^k$. Let K be a constant operator in $\mathbf{A}$ of order $\langle m+n, k\rangle$ and value $\mathscr{A}^k$. Then, in particular,

$$\mathscr{A}^k = K(\mathscr{Q} \times \mathscr{P}).$$

Combining these,

$$\mathscr{R} \cap (\mathscr{P} \times \mathscr{A}^k) = [(JL_f) \cap (L_g \times K)](\mathscr{Q} \times \mathscr{P}).$$

Finally, let L_h be the logical operator such that

$$L_h : \langle x_1, \ldots, x_n, x_{n+1}, \ldots, x_{n+k}\rangle \rightarrow \langle x_{n+1}, \ldots, x_{n+k}\rangle.$$

Then if we set

$$\Phi = L_h[(JL_f) \cap (L_g \times K)]$$

it should be clear that Φ is the desired operator.

Now (1). Say $\mathscr{P}$ is n-ary and $\mathscr{R}$ is $n \neq k$-ary. Let K now be a constant operator in $\mathbf{A}$ of order $\langle n, k\rangle$ and value $\mathscr{A}^k$. Also $\mathscr{R}$ is now generated; let J be a constant operator in $\mathbf{A}$ of order $\langle n, n+k\rangle$ and value $\mathscr{R}$. Finally, let L_h be as in the proof of (1). Then the desired operator is

$$\Phi = L_h[J \cap (I^n \times K)].$$

DEFINITION. Let $\mathscr{R}$ be an $n+k$-ary relation, and $\mathscr{P}$ be n-ary.

$$\mathscr{R}''(\mathscr{P}) = \{\boldsymbol{v} \mid \text{for some } \boldsymbol{w} \in \mathscr{P}, \langle \boldsymbol{w}, \boldsymbol{v}\rangle \in \mathscr{R}\}.$$

COROLLARY 5.2. (1) *If $\mathscr{R}$ is generated in $\mathbf{A}$ then there is an operator $\Phi \in \mathbf{A}$ such that $\Phi(\mathscr{P}) = \mathscr{R}''(\mathscr{P})$.*

(2) *If* $\mathcal{Q} \neq \emptyset$ *is allowable in* $\mathbf{A}$ *and* $\mathcal{R} \leqslant \mathcal{Q}$, *then there is an operator* $\Phi \in \mathbf{A}$ *such that, for allowable* $\mathcal{Q} \times \mathcal{P}$, $\Phi(\mathcal{Q} \times \mathcal{P}) = \mathcal{R}''(\mathcal{P})$.

COROLLARY 5.3. (1) *If* $\mathcal{R}$ *is generated in* $\mathbf{A}$ *then there is an operator* $\Phi \in \mathbf{A}$ *such that*

$$x \in \mathcal{R} \;\Rightarrow\; \Phi(\{x\}) = \{0\}, \qquad x \notin \mathcal{R} \;\Rightarrow\; \Phi(\{x\}) = \emptyset.$$

(2) *If* $\mathcal{Q} \neq \emptyset$ *is allowable in* $\mathbf{A}$ *and* $\mathcal{R} \leqslant \mathcal{Q}$, *then there is an operator* $\Phi \in \mathbf{A}$ *such that*

$$x \in \mathcal{R} \;\Rightarrow\; \Phi(\mathcal{Q} \times \{x\}) = \{0\}, \qquad x \notin \mathcal{R} \;\Rightarrow\; \Phi(\mathcal{Q} \times \{x\}) = \emptyset.$$

PROOF. (1) $\mathcal{R}$ is generated, and $\{0\}$ is generated, hence so is $\mathcal{R} \times \{0\} = \mathcal{S}$. Take Φ to be $\mathcal{S}''$ and use Corollary 5.2 (1).

(2) is similar. We also need the observation that $\mathcal{Q} \times \{x\}$ is allowable for all $x \in \mathcal{A}^n$. This may be seen as follows. For $x \in \mathcal{A}^n$, $\{x\}$ is a Cartesian product of one-element sets of the form $\{a\}$ where $a \in \mathcal{A}$, hence it is generated in $\mathbf{A}$, and hence $\{x\} \leqslant \mathcal{Q}$. $\mathcal{Q} \leqslant \mathcal{Q}$, so $\mathcal{Q} \times \{x\}$ is allowable by Proposition 4.3.

COROLLARY 5.4. *Let f be a partial function from* $\mathcal{A}^n$ *to* $\mathcal{A}^k$ *(the domain need not be all of* $\mathcal{A}^n$*).*

(1) *If the graph of f is generated in* $\mathbf{A}$, *then for some* $\Phi \in \mathbf{A}$,

$$\Phi(\{v\}) = \begin{cases} \{f(v)\} & v \in \operatorname{dom} f, \\ \emptyset & v \notin \operatorname{dom} f. \end{cases}$$

(2) *Let* $\mathcal{Q}$ *be allowable in* $\mathbf{A}$. *If the graph of f is* $\leqslant \mathcal{Q}$ *(written* $f \leqslant \mathcal{Q}$*), then for some* $\Phi \in \mathbf{A}$,

$$\Phi(\mathcal{Q} \times \{v\}) = \begin{cases} \{f(v)\} & v \in \operatorname{dom} f, \\ \emptyset & v \notin \operatorname{dom} f. \end{cases}$$

PROOF. Indentifying f and its graph, we have

$$f''(\{v\}) = \begin{cases} \{f(v)\} & v \in \operatorname{dom} f, \\ \emptyset & v \notin \operatorname{dom} f, \end{cases}$$

and the theorem follows easily.

DEFINITION. Let f be a partial function from $\mathcal{A}^n$ to $\mathcal{A}^k$.

(1) We say f is *generated in* $\mathbf{A}$ if the graph of f is generated in $\mathbf{A}$.

(2) We say f is *generated pointwise* in $\mathbf{A}$ if there is an operator $\Phi \in \mathbf{A}$ of order $\langle n, k \rangle$ such that

$$\Phi(\{\boldsymbol{v}\}) = \begin{cases} \{f(\boldsymbol{v})\} & \boldsymbol{v} \in \operatorname{dom} f, \\ \emptyset & \boldsymbol{v} \notin \operatorname{dom} f. \end{cases}$$

COROLLARY 5.5. *If a partial function f is generated in $\mathbf{A}$, it is generated pointwise in $\mathbf{A}$.*

REMARKS. In Chapter 8, we will show there is a large class of production systems in which these two notions coincide. In general, they do not, and for the time being, we give first place to the notion of being generated, identifying f with its graph. In Chapter 10, when we come to discuss higher types, we will take pointwise generation as basic.

6. Production systems with equality

One of the production system axioms says that equality is generated. In this section we consider some effects of requiring that equality be *bi*-generated. We look specifically at the consequences for finite sets and for functions.

DEFINITION. Let $\mathbf{A}$ be a production system. We say it is a system *with equality* if $\neq_{\mathscr{A}}$ is generated in $\mathbf{A}$.

DEFINITION. For $1 \leq i, j \leq n$ we define a map of order $\langle n, n \rangle$ by

$$\bar{E}^n_{i,j}(\mathscr{R}) = \{\langle x_1, \ldots, x_n \rangle \in \mathscr{R} \mid x_i \neq x_j\}.$$

PROPOSITION 6.1. *If $\mathbf{A}$ is a production system with equality, each $\bar{E}^n_{i,j} \in \mathbf{A}$.*

PROOF. Similar to that of Proposition 3.3.

PROPOSITION 6.2. *Let $\mathbf{A}$ be any production system. Let $\mathscr{F} \subseteq \mathscr{A}^n$ be finite. Then $\mathscr{F}$ is generated in $\mathbf{A}$.*

PROOF. If $a_1, \ldots, a_n \in \mathscr{A}$, each of $\{a_1\}, \ldots, \{a_n\}$ is generated, by Axiom (3). Then $\{\langle a_1, \ldots, a_n \rangle\}$ is generated, since there is closure under $\times$. Finally, any finite subset of $\mathscr{A}^n$ can be built up since there is closure under $\cup$.

COROLLARY 6.3. *Let* $\mathbf{A}$ *be any production system, and let* $\mathcal{F}$ *be a finite relation on* $\mathcal{A}$.

(1) $\mathcal{F} \leqslant \mathcal{P}$ *for any allowable* $\mathcal{P}$.

(2) $\Phi(\mathcal{F})$ *is generated for* $\Phi \in \mathbf{A}$.

(3) $\mathcal{P} \leqslant \mathcal{F}$ *if and only if* $\mathcal{P}$ *is generated in* $\mathbf{A}$.

LEMMA 6.4. *Let* $\mathbf{A}$ *be a production system with equality and let* $\mathcal{P} \in [\mathcal{A}]^n$. *If* $\mathcal{R}$ *has one fewer member than* $\mathcal{P}$, *then* $\mathcal{R}$ *is allowable in* $\mathbf{A}$ *and* $\mathcal{R} \leqslant \mathcal{P}$.

PROOF. For simplicity of notation, say $n = 3$. Suppose $\mathcal{P} = \mathcal{R} \cup \{\boldsymbol{a}\}$. By Proposition 6.2, $\{\boldsymbol{a}\}$ is generated in $\boldsymbol{A}$; let $\Phi \in \boldsymbol{A}$ be constant, of order $\langle n, n\rangle$, and value $\{\boldsymbol{a}\}$. Then

$$\mathcal{R} = P_4^4 P_5^5 P_6^6 (\bar{E}_{1,4}^6 \cup \bar{E}_{2,5}^6 \cup \bar{E}_{3,6}^6)(I^3 \times \Phi)(\mathcal{P}).$$

PROPOSITION 6.5. *Let* $\mathbf{A}$ *be a production system with equality, and let* $\mathcal{P}, \mathcal{R} \subseteq \mathcal{A}^n$ *differ by a finite number of elements. If* $\mathcal{P}$ *is allowable in* $\mathbf{A}$ *then*

(1) $\mathcal{R}$ *is allowable in* $\mathbf{A}$ *and* $\mathcal{R} \leqslant \mathcal{P}$.

(2) $\mathcal{P} \sim \mathcal{R}$.

(3) *If* $\mathcal{P} \times \bar{\mathcal{P}}$ *is allowable in* $\mathbf{A}$ *then* $\mathcal{P}$ *and* $\mathcal{R}$ *are each bi-generated in the other.*

PROOF. Let $\mathcal{P}$ be allowable in $\mathbf{A}$.

If $\mathcal{R}$ has fewer members than $\mathcal{P}$, $\mathcal{R}$ is allowable in $\mathbf{A}$ and $\mathcal{R} \leqslant \mathcal{P}$ by repeated application of Lemma 6.4.

If $\mathcal{R}$ has more members than $\mathcal{P}$, $\mathcal{R} = \mathcal{P} \cup \mathcal{F}$ where $\mathcal{F}$ is finite. Then $\mathcal{F} \leqslant \mathcal{P}$ by Corollary 6.3 and $\mathcal{P} \leqslant \mathcal{P}$, so $\mathcal{R} = \mathcal{P} \cup \mathcal{F} \leqslant \mathcal{P}$ by Proposition 4.7.

Thus we have (1). (2) follows by switching the roles of $\mathcal{P}$ and $\mathcal{R}$.

Finally we have (3) upon noting that, if $\mathcal{P}$ and $\mathcal{R}$ differ by a finite number of elements, the same is true of $\bar{\mathcal{P}}$ and $\bar{\mathcal{R}}$. Proposition 4.4 (5) is also needed.

COROLLARY 6.6. *Let* $\mathbf{A}$ *be a production system with equality, and let* $\mathcal{F}$ *be a finite relation on* $\mathcal{A}$.

(1) $\mathcal{P} \leqslant \mathcal{R} \cup \mathcal{F}$ *if and only if* $\mathcal{P} \leqslant \mathcal{R}$.

(2) $\mathcal{P}$ *is generated in* $\mathcal{R} \cup \mathcal{F}$ *if and only if* $\mathcal{P}$ *is generated in* $\mathcal{R}$.

PROOF. (2) Uses Proposition 4.6.

Next we look at the behavior of functions. We assume $\mathscr{A}$ has more than one member, so characteristic functions are meaningful.

LEMMA 6.7. *Let* **A** *be a production system with equality. If f is a* total *function from* $\mathscr{A}^n$ *to* $\mathscr{A}^m$, *then* $\bar{f} \leq f$.

PROOF. We present the simplest case only: say f maps $\mathscr{A}$ to $\mathscr{A}$. Now, $\mathscr{A}$ is generated; let Φ be constant, of order $\langle 2, 1\rangle$ and value $\mathscr{A}$. Then

$$\bar{f} = P_2^3 \bar{E}_{2,3}^3 (I^2 \times \Phi)(f).$$

THEOREM 6.8. *Let* **A** *be a production system with equality, and let f and g be total functions. Then* $f \leq g$ *if and only if* $f \times \bar{f} \leq g \times \bar{g}$.

PROOF. (1) Suppose $f \leq g$. Then g is allowable. Now $g \leq g$ and by Lemma 6.7, $\bar{g} \leq g$, so $g \times \bar{g}$ is allowable by Proposition 4.3. Further, by that, $g \leq g \times \bar{g}$ and $f \leq g$, so $f \leq g \times \bar{g}$. Also $\bar{f} \leq f$, so $\bar{f} \leq g \times \bar{g}$. Then $f \times \bar{f} \leq g \times \bar{g}$.

(2) Suppose $f \times \bar{f} \leq g \times \bar{g}$. Well, $f \leq f \times \bar{f}$ so $f \leq g \times \bar{g}$. Also $\bar{g} \leq g$ and $g \leq g$, so $g \times \bar{g} \leq g$. Then $f \leq g$.

COROLLARY 6.9. *Let* **A** *be a production system with equality. If f and g are total functions,* $f \leq g$ *if and only if f is bi-generated in g.*

PROOF. By Proposition 4.5.

COROLLARY 6.10. *Let* **A** *be a production system with equality.*

(1) $c_{\mathscr{P}}$ *is bi-generated in* $c_{\mathscr{R}}$ *if and only if* $\mathscr{P}$ *is bi-generated in* $\mathscr{R}$.

(2) *If* $c_{\mathscr{P}}$ *is allowable then* $c_{\mathscr{P}}$ *and* $\mathscr{P}$ *are each bi-generated in the other.*

PROOF. (1) Follows by the above and Corollary 4.10.

For (2), first note that if $c_{\mathscr{P}}$ is allowable, so is $\mathscr{P} \times \bar{\mathscr{P}}$ by Lemma 4.7. Since $c_{\mathscr{P}}$ is total, $\overline{c_{\mathscr{P}}} \leq c_{\mathscr{P}}$. But $c_{\mathscr{P}} \leq \mathscr{P} \times \bar{\mathscr{P}}$ by Lemma 4.8. It follows that $c_{\mathscr{P}} \times \overline{c_{\mathscr{P}}} \leq \mathscr{P} \times \bar{\mathscr{P}}$. Also $\mathscr{P} \times \bar{\mathscr{P}} \leq c_{\mathscr{P}}$ by Lemma 4.7, so $\mathscr{P} \times \bar{\mathscr{P}} \leq c_{\mathscr{P}} \times \overline{c_{\mathscr{P}}}$.

7. Closure under least fixed point

We characterize the r.e. and ω-r.e. relations directly in terms of production systems, by-passing elementary formal systems. The terminology of inductive definability is from Ch. 1, §13.

PROPOSITION 7.1. *Let* $\mathbf{A}$ *be a production system on the structure* $\mathfrak{A} = \langle \mathcal{A}; \mathcal{R}_1, \ldots, \mathcal{R}_k \rangle$. *Let* φ *be a* P*-formula (for this structure) where* P *is* n*-place and* φ *has* $x_1, \ldots, x_n$ *as free variables. Let* Γ_φ *be the monotone operator that* φ *defines. Then*

(1) *If* φ *has no universal quantifiers,* $\Gamma_\varphi \in \mathbf{A}$.

(2) *If* $\mathbf{A}$ *is an* ω*-production system,* $\Gamma_\varphi \in \mathbf{A}$.

PROOF. By an induction on the degree of φ, which we omit.

REMARK. A simple checking of the axioms shows that the *least* total production system on $\mathfrak{A}$ is precisely the collection of Γ_φ defined by P-formulas without universal quantifiers (where, however, to get an operator of order $\langle n, m \rangle$ we must allow P to be n-place but φ to have m free variables). Similarly the least total ω-production system on $\mathfrak{A}$ is the collection of all Γ_φ defined by arbitrary P-formulas φ. Both these are subsystems of $\text{fo}(\mathfrak{A})$.

DEFINITION. We say a production system $\mathbf{A}$ is *closed under the taking of least fixed points* if, for every monotone $\Phi \in \mathbf{A}$ of order $\langle n, n \rangle$, the least fixed point of Φ is generated in $\mathbf{A}$.

PROPOSITION 7.2. *Let* $\mathbf{A}$ *be a production system over the structure* $\mathfrak{A}$ *with infinite domain, and suppose* $\mathbf{A}$ *is closed under the taking of least fixed points.*

(1) *Every relation r.e. over* $\mathfrak{A}$ *is generated in* $\mathbf{A}$.

(2) *If* $\mathbf{A}$ *is an* ω*-production system, every relation* ω*-r.e. over* $\mathfrak{A}$ *is generated in* $\mathbf{A}$.

PROOF. We show (2); (1) is similar.

Suppose $\mathcal{R}$ is ω-r.e. over $\mathfrak{A}$. Then by Theorem 1.13 (3), $\mathcal{R}$ is inductive on $\mathcal{A}$ in $\mathcal{R}_1, \ldots, \mathcal{R}_k$. Thus there is a relation $\mathcal{P}$ which is a fixed point, and there are $a_1, \ldots, a_j \in \mathcal{A}$ so that

$$\boldsymbol{x} \in \mathcal{R} \iff \langle \boldsymbol{a}, \boldsymbol{x} \rangle \in \mathcal{P}.$$

$\mathcal{P}$ is the least fixed point of an operator Γ_φ, but by Proposition 7.1, $\Gamma_\varphi \in \mathbf{A}$, hence $\mathcal{P}$ is generated in $\mathbf{A}$ since $\mathbf{A}$ is closed under the taking of least fixed points. It follows that $\mathcal{R}$ is generated in $\mathbf{A}$ since it may be obtained from $\mathcal{P}$ by the use of section operators.

COROLLARY 7.3. *Let* $\mathfrak{A}$ *be a structure with infinite domain.* $\mathcal{R}$ *is r.e.* (ω*-r.e.*) *over* $\mathfrak{A}$ *if and only if* $\mathcal{R}$ *is generated in every production* (ω*-production*) *system over* $\mathfrak{A}$ *which is closed under the taking of least fixed points.*

PROOF. If $\mathscr{R}$ is generated in every production system over $\mathfrak{A}$ which is closed under the taking of least fixed points, then $\mathscr{R}$ is generated in rec($\mathfrak{A}$) so $\mathscr{R}$ is r.e. over $\mathfrak{A}$. The converse is the above proposition.

8. A Kleene–Post result

In Kleene–Post [1954] a theorem was shown which implied that in ordinary recursion theory there were two sets, neither recursive in the other. In this section we show an analog of this for a class of total production systems which includes all recursion theories with countable domains. Our proof uses a small amount of the methods of forcing. Even so, it is essentially the original Kleene–Post proof.

In Friedberg [1957] and Muchnik [1956] the Kleene–Post result was improved to show that in ordinary recursion theory there were two *r.e.* sets neither recursive in the other. The proof of this introduced what are called priority arguments into the subject, and much research has gone into generalizing the class of structures in which priority arguments can be used. We do not pursue this, quite extensive, topic here. See Stoltenberg-Hansen [1979] for a treatment of priority methods in an axiomatic setting closely related to that of Chapter 8.

Since domains are to be countable in this section, we may as well use the natural numbers. We will sometimes make usc of special features of $\mathbb{N}$, but from the outside, so to speak, not within our production systems.

Let Φ be an operator on $\mathbb{N}$ of order $\langle n, k\rangle$, say. Recall, Φ is *compact* if $\boldsymbol{x} \in \Phi(\mathscr{P}) \Rightarrow \boldsymbol{x} \in \Phi(\mathscr{F})$ for some finite $\mathscr{F} \subseteq \mathscr{P}$. And Φ is *monotone* if $\mathscr{P} \subseteq \mathscr{R} \Rightarrow \Phi(\mathscr{P}) \subseteq \Phi(\mathscr{R})$.

LEMMA 8.1. *Suppose Φ is an operator on $\mathbb{N}$ which is compact and monotone. Let $\mathscr{P}_1 \subseteq \mathscr{P}_2 \subseteq \mathscr{P}_3 \subseteq \cdots$ be a chain of n-ary relations, and let $\mathscr{P} = \bigcup_i \mathscr{P}_i$. Then*

$$\boldsymbol{x} \in \Phi(\mathscr{P}) \Leftrightarrow \text{for some } j, \boldsymbol{x} \in \Phi(\mathscr{P}_j).$$

Let $\mathscr{I}$ be a collection of operators on $\mathbb{N}$. We do not require that $\mathscr{I}$ be a production system, but we assume operators in $\mathscr{I}$ are total. We write $\mathscr{P} \leqslant_{\mathscr{I}} \mathscr{R}$ if $\mathscr{P} = \Phi(\mathscr{R})$ for some $\Phi \in \mathscr{I}$. We say $\mathscr{P}$ and $\mathscr{R}$ are $\mathscr{I}$-*incomparable* if neither $\mathscr{P} \leqslant_{\mathscr{I}} \mathscr{R}$ nor $\mathscr{R} \leqslant_{\mathscr{I}} \mathscr{P}$.

By a 0–1 *function* we mean a function whose domain is an initial segment of $\mathbb{N}$ (not necessarily proper) and whose range is $\{0, 1\}$. A 0–1 function f

whose domain is a proper initial segment of $\mathbb{N}$ will be called a *finite* 0–1 *function.* If the domain of f is all of $\mathbb{N}$, f is simply a *characteristic function.*

THEOREM 8.2. *Let $\mathcal{I}$ be a countable set of compact, monotone operators on $\mathbb{N}$, each of order $\langle 2,2\rangle$. There exist $\mathcal{I}$-incomparable* 0–1 *functions.*

The proof of this, the main result of the section, occupies the next several pages. After it is established, we derive some corollaries.

By a *forcing condition* we mean an ordered pair of finite 0–1 functions, $\Gamma = \langle f_\Gamma, g_\Gamma\rangle$, so that f_Γ and g_Γ have the same domains.

Let $\mathcal{G}$ be the collection of all forcing conditions. If $\Gamma, \Delta \in \mathcal{G}$, we write $\Gamma \subseteq \Delta$ if $f_\Gamma \subseteq f_\Delta$ and $g_\Gamma \subseteq g_\Delta$, and we call Δ an *extension* of Γ. Note that $\subseteq$ is reflexive and transitive on forcing conditions.

$\mathbb{N}^2$ is the set $\mathbb{N}\times\mathbb{N}$.

$\mathcal{I}$ is countable; let $\Phi_0, \Phi_1, \Phi_2, \ldots$ be an enumeration of it.

Next we set up a certain formal first order language.

Let f and g be two *formal symbols.* By an *atomic formula* we mean any expression of one of the forms ($n, k \in \mathbb{N}$ are variables).

$$\langle n,k\rangle \in f, \qquad \langle n,k\rangle \in \Phi_j(f);$$
$$\langle n,k\rangle \in g, \qquad \langle n,k\rangle \in \Phi_j(g).$$

(Formally, $\langle n,k\rangle \in f$ is simply some two-place atomic formula, and so is $\langle n,k\rangle \in \Phi_j(f)$ for each $j = 0,1,2,\ldots$.)

Formulas are built up from these atomic formulas using $\wedge$, $\vee$, $\sim$, $\supset$ and $\exists$ in the usual way. Note that we omit $\forall$.

We define a relation $\models$ between forcing conditions and formulas as follows.

On closed atomic formulas: for each $\Gamma \in \mathcal{G}$

$$\Gamma \models \langle n,k\rangle \in f \qquad \text{if } \langle n,k\rangle \in f_\Gamma,$$
$$\Gamma \models \langle n,k\rangle \in g \qquad \text{if } \langle n,k\rangle \in g_\Gamma,$$
$$\Gamma \models \langle n,k\rangle \in \Phi_j(f) \qquad \text{if } \langle n,k\rangle \in \Phi_j(f_\Gamma),$$
$$\Gamma \models \langle n,k\rangle \in \Phi_j(g) \qquad \text{if } \langle n,k\rangle \in \Phi_j(g_\Gamma).$$

Next, $\models$ is extended to all closed formulas as follows.

$$\Gamma \models (X \wedge Y) \quad \Leftrightarrow \quad \Gamma \models X \text{ and } \Gamma \models Y.$$
$$\Gamma \models (X \vee Y) \quad \Leftrightarrow \quad \Gamma \models X \text{ or } \Gamma \models Y.$$
$$\Gamma \models \sim X \quad \Leftrightarrow \quad \text{for every } \Delta \in \mathcal{G} \text{ such that } \Gamma \subseteq \Delta, \text{ not-}\Delta \models X.$$

$$\Gamma \models (X \supset Y) \quad \Leftrightarrow \quad \text{for every } \Delta \in \mathcal{G} \text{ such that } \Gamma \subseteq \Delta, \text{ if } \Gamma \models X \text{ then } \Gamma \models Y.$$

$$\Gamma \models (\exists x) X(x) \Leftrightarrow \Gamma \models X(c) \text{ for some } c \in \mathbb{N}^2.$$

This makes $\langle \mathcal{G}, \subseteq, \models, \mathbb{N}^2 \rangle$ into a constant domain Kripke Intuitionistic logic model, in the sense of Fitting [1969] (except for the omission of $\forall$).

LEMMA 8.3. *Let $\Gamma \in \mathcal{G}$ and X be a formula with no free variables. There is an extension Δ of Γ, in $\mathcal{G}$, such that $\Delta \models (X \vee \sim X)$.*

PROOF. If some extension Ω of Γ is such that $\Omega \models X$, set $\Delta = \Omega$. If for every extension Ω of Γ, not-$\Omega \models X$, then $\Gamma \models \sim X$, so set $\Delta = \Gamma$.

LEMMA 8.4. *Let $\Gamma \in \mathcal{G}$ and X be a formula with no free variables. If $\Gamma \models X$ and $\Delta \in \mathcal{G}$ is an extension of Γ, then $\Delta \models X$.*

PROOF. By induction on the degree of X.

$\mathcal{C} \subseteq \mathcal{G}$ is called a *complete sequence* if $\mathcal{C}$ is a chain of forcing conditions and, for every closed formula X, there is some $\Gamma \in \mathcal{C}$ such that $\Gamma \models (X \vee \sim X)$.

LEMMA 8.5. *There is a complete sequence.*

PROOF. Enumerate the closed formulas: $X_1, X_2, X_3, \ldots$. (It is here that we use the countability of $\mathbb{N}$ and $\mathcal{I}$.) Suppose $\Gamma_1, \Gamma_2, \ldots, \Gamma_k$ have been defined so that $\Gamma_1 \subseteq \Gamma_2 \subseteq \cdots \subseteq \Gamma_k$ and for each $1 \leqslant i \leqslant k$, $\Gamma_i \models (X_i \vee \sim X_i)$. By Lemma 8.3, there is an extension Δ of Γ_k such that $\Delta \models (X_{k+1} \vee \sim X_{k+1})$; set $\Gamma_{k+1} = \Delta$. In this way $\mathcal{C}$ is generated. (Note that this is non-constructive.)

Let $\mathcal{C}$ be a fixed complete sequence. Set

$$f_{\mathcal{C}} = \bigcup \{f_\Gamma \mid \Gamma \in \mathcal{C}\}, \qquad g_{\mathcal{C}} = \bigcup \{g_\Gamma \mid \Gamma \in \mathcal{C}\}.$$

Then both $f_{\mathcal{C}}$ and $g_{\mathcal{C}}$ are 0–1 functions.

LEMMA 8.6 (Truth Lemma). *Let $\varphi(f, g)$ be any closed formula. $\varphi(f_{\mathcal{C}}, g_{\mathcal{C}})$ is true (over $\mathbb{N}^2$) if and only if for some $\Gamma \in \mathcal{C}$, $\Gamma \models \varphi(f, g)$.*

PROOF. By an induction on the degree of φ. Begin with φ atomic.

If φ is $\langle n, k \rangle \in f$ or $\langle n, k \rangle \in g$ the result is simple. Now suppose φ is $\langle n, k \rangle \in \Phi_j(f)$.

Case 1. $\langle n, k\rangle \in \Phi_j(f_{\mathscr{C}})$ is true. Then by Lemma 8.1, for some $\Gamma \in \mathscr{C}$, $\langle n, k\rangle \in \Phi_j(f_\Gamma)$. But then $\Gamma \vDash \langle n, k\rangle \in \Phi_j(f)$.

Case 2. For some $\Gamma \in \mathscr{C}$, $\Gamma \vDash \langle n, k\rangle \in \Phi_j(f)$. Then $\langle n, k\rangle \in \Phi_j(f_\Gamma)$ but Φ_j is monotone and $f_\Gamma \subseteq f_{\mathscr{C}}$, so $\langle n, k\rangle \in \Phi_j(f_{\mathscr{C}})$.

Similarly if φ is $\langle n, k\rangle \in \Phi_j(g)$. Higher degree formulas are left to the reader. (It is here that our omission of $\forall$ is used. The usual definition of $\Gamma \vDash (\forall x)X(x)$ in Kripke models does not allow this theorem to extend to formulas containing $\forall$.)

LEMMA 8.7 (Combinatorial Lemma). *Let $\Gamma \in \mathscr{G}$ and $j \in \mathbb{N}$.*

(1) *For some extension Δ of Γ, and for some n,*

$$\Delta \vDash \{[\langle n,0\rangle \in \Phi_j(g) \wedge \sim \langle n,0\rangle \in f] \vee [\sim \langle n,0\rangle \in \Phi_j(g) \wedge \langle n,0\rangle \in f]\}.$$

(2) *For some extension Δ of Γ, and for some n,*

$$\Delta \vDash \{[\langle n,0\rangle \in \Phi_j(f) \wedge \sim \langle n,0\rangle \in g] \vee [\sim \langle n,0\rangle \in \Phi_j(f) \wedge \langle n,0\rangle \in g]\}.$$

PROOF. We show only (1). Let n be an integer new to Γ (this exists since Γ is a pair of *finite* 0–1 functions).

Case 1: There is an extension Γ^* of Γ such that $\Gamma^* \vDash \langle n,0\rangle \in \Phi_j(g)$. Now, whether this happens at Γ^* or not depends only on the second component of Γ^*. Let Δ be the same as Γ^* in the second component but, if $\langle n,0\rangle$ is the first component of Γ^*, take it out and replace it by $\langle n,1\rangle$. (If $\langle n,0\rangle$ isn't in the first component of Γ^*, $\langle n,1\rangle$ is, so just set $\Delta = \Gamma^*$.) Then also $\Delta \vDash \langle n,0\rangle \in \Phi_j(g)$. But $\Delta \vDash \langle n,1\rangle \in f$, and it follows that no extension of Δ can force $\langle n,0\rangle \in f$; hence $\Delta \vDash \sim \langle n,0\rangle \in f$.

Case 2: For every extension Γ^* of Γ we have not-$\Gamma^* \vDash \langle n,0\rangle \in \Phi_j(g)$. Then $\Gamma \vDash \sim \langle n,0\rangle \in \Phi_j(g)$. Let Δ be some extension of Γ with $\langle n,0\rangle$ in the first component. (This exists since n was new to Γ.) Then $\Delta \vDash \langle n,0\rangle \in f$. But since Δ extends Γ, $\Delta \vDash \sim \langle n,0\rangle \in \Phi_j(g)$.

COROLLARY 8.8. *For each $\Gamma \in \mathscr{G}$ and $j \in \mathbb{N}$:*

(1) *For some extension Δ of Γ and some n,*

$$\Delta \vDash \sim [\langle n,0\rangle \in f \equiv \langle n,0\rangle \in \Phi_j(g)].$$

(2) *For some extension Δ of Γ and some n,*

$$\Delta \vDash \sim [\langle n,0\rangle \in g \equiv \langle n,0\rangle \in \Phi_j(f)].$$

PROOF. (Here $X \equiv Y$ abbreviates $(X \supset Y) \wedge (Y \supset X)$.) This may be derived directly from Lemma 8.7. Or one may proceed as follows. If X is a theorem of Intuitionistic Logic, $\Gamma \models X$ for every $\Gamma \in \mathcal{G}$ (see Fitting [1969]). Also if $\Gamma \models X$ and $\Gamma \models X \supset Y$ then $\Gamma \models Y$. The result then follows using Lemma 8.7 and the Intuitionistic Logic theorem.

$$[(A \wedge \sim B) \vee (\sim A \wedge B)] \supset \sim[(A \supset B) \wedge (B \supset A)].$$

COROLLARY 8.9 (Key Corollary). *For each* $\Gamma \in \mathcal{G}$ *and each* $j \in \mathbb{N}$,
(1) *not*-$\Gamma \models \sim(\exists x) \sim [x \in f \equiv x \in \Phi_j(g)]$,
(2) *not*-$\Gamma \models \sim(\exists x) \sim [x \in g \equiv x \in \Phi_j(f)]$.

PROOF. We show (1). Let $\Gamma \in \mathcal{G}$. By Corollary 8.8, for some extension Δ of Γ and some n,

$$\Delta \models \sim [\langle n, 0\rangle \in f \equiv \langle n, 0\rangle \in \Phi_j(g)].$$

Hence

$$\Delta \models (\exists x) \sim [x \in f \equiv x \in \Phi_j(g)].$$

If we had that

$$\Gamma \models \sim(\exists x) \sim [x \in f \equiv x \in \Phi_j(g)],$$

then for every extension Γ^* of Γ,

$$\text{not-}\Gamma^* \models (\exists x) \sim [x \in f \equiv x \in \Phi_j(g)],$$

which is a contradiction since Δ is an extension of Γ.

THEOREM 8.10. $f_{\mathcal{C}}$ *and* $g_{\mathcal{C}}$ *are* $\mathcal{J}$-*incomparable.*

PROOF. Suppose $f_{\mathcal{C}} \leq_{\mathcal{J}} g_{\mathcal{C}}$ (the other way round is similar). Say $f_{\mathcal{C}} = \Phi_j(g_{\mathcal{C}})$. Then the following is true:

$$\sim(\exists x) \sim [x \in f_{\mathcal{C}} \equiv x \in \Phi_j(g_{\mathcal{C}})].$$

By the Truth Lemma (8.6), for some $\Gamma \in \mathcal{C}$,

$$\Gamma \models \sim(\exists x) \sim [x \in f \equiv x \in \Phi_j(g)]$$

which contradicts the Key Corollary (8.9).

We have completed the proof of Theorem 8.2. We derive some corollaries.

COROLLARY 8.11. *Let* **A** *be a total production system with a countable collection of operators, in which every operator is compact and monotone. Then, in* **A**, *there are two sets, neither bi-generated in the other.*

PROOF. In a production system, singletons are generated. Since there are countably many operators in $\boldsymbol{A}$, the domain of $\boldsymbol{A}$ must also be countable. Then by Theorem 8.2, there are two 0–1 functions $f_{\mathscr{C}}$ and $g_{\mathscr{C}}$ which are $\boldsymbol{A}$-incomparable. If $f_{\mathscr{C}}$ were finite, by Corollary 6.3, we would have $f_{\mathscr{C}} \leq g_{\mathscr{C}}$, which is impossible. So $f_{\mathscr{C}}$, and similarly $g_{\mathscr{C}}$, are not finite. Hence they are characteristic functions, say $f_{\mathscr{C}}$ for $\mathscr{P}$ and $g_{\mathscr{C}}$ for $\mathscr{R}$. Then by Corollary 4.10, neither $\mathscr{P}$ nor $\mathscr{R}$ is bi-generated in the other.

COROLLARY 8.12. *In any recursion theory with a countable domain, there are two sets, neither recursive in the other.*

CHAPTER THREE

EMBEDDINGS

1. Introduction

In group theory one is interested in not only the groups themselves, but also in their relationships with each other: in the homomorphisms. Well, we're not only interested in production systems, but in their interrelationships. What takes the place of group homomorphism is what we call an *embedding*. Roughly speading, an embedding is a coding of the domain of one production system into that of another which "carries along" the production system machinery.

The collection of all production systems, with embeddings as morphisms, constitutes a category, with the collections of all recursion theories, and of all ω-recursion theories as full subcategories. This chapter is the first of a sequence of three which study such categories, both for their general algebraic features, and for features of particular embeddings of interest to us.

It is standard practice, in ordinary recursion theory, to discuss effectiveness on structures other than that of arithmetic, by using the device of Gödel numbering. Natural numbers are assigned, in some systematic way, to the objects of interest, then one works with the objects indirectly by manipulating these "code" numbers within ordinary recursion theory. Since, in our approach, every structure has its own recursion theory, we do not need such a device here. Instead, standard Gödel numbering techniques generally provide us with examples of embeddings between recursion theories. Thus they are still present, but in a different role.

2. Codings

DEFINITIONS. Let θ be a mapping from $\mathscr{A}$ into non-empty subsets of $\mathscr{B}$ such that, if $v \neq v'$ then $\theta(v) \cap \theta(v') = \emptyset$. We call θ a *coding* of $\mathscr{A}$ into $\mathscr{B}$. The members of $\theta(v)$ are *codes* for v. If, for each $v \in \mathscr{A}$, $\theta(v)$ has exactly one member, we say θ is a 1–1 *coding*. If each member of $\mathscr{B}$ is a code, we say θ is *onto*. We generally write $\theta : \mathscr{A} \to \mathscr{B}$ to indicate θ is a coding of $\mathscr{A}$ into $\mathscr{B}$ (sometimes with explanatory words, if confusion with an ordinary function seems likely).

Let $\alpha : \mathscr{A} \to \mathscr{B}$ and $\beta : \mathscr{B} \to \mathscr{C}$ be codings. By the *composition*, $\beta\alpha$ of α and β we mean the coding satisfying

$$x \in (\beta\alpha)(v) \Leftrightarrow x \in \beta(w) \quad \text{for some } w \in \alpha(v).$$

Thus $x \in (\beta\alpha)(v)$ if x is a "β-code" for some "α-code" for v. The composition of 1–1 codings is 1–1. The composition of onto codings is onto.

For each set $\mathscr{A}$ we define a coding $\text{inj}_{\mathscr{A}} : \mathscr{A} \to \mathscr{A}$ by

$$\text{inj}_{\mathscr{A}}(v) = \{v\} \quad \text{for } v \in \mathscr{A}.$$

$\text{inj}_{\mathscr{A}}$ is 1–1, onto, and clearly serves as an identity under composition.

A (large) category is defined by taking, as objects, all sets and as morphisms, all codings, with composition defined as above, and the inj as identity morphisms. We call this category **Dom**. In it, $\emptyset$ is an initial object. But in it the connection between monic, epi, 1–1 and onto is not quite as expected. (See MacLane [1971] for terminology.)

PROPOSITION 2.1. *In* **Dom**, *every morphism is monic.*

PROOF. Suppose $\alpha : \mathscr{A} \to \mathscr{B}$ and $\beta, \gamma : \mathscr{C} \to \mathscr{A}$ are codings. Suppose also that $\alpha\beta = \alpha\gamma$. Let $c \in \mathscr{C}$; we show $\beta(c) = \gamma(c)$, and hence $\beta = \gamma$.

Let $x \in \beta(c)$. Then $x \in \mathscr{A}$ so $\alpha(x)$ is defined. Choose some $y \in \alpha(x)$. Then $y \in \alpha(x)$ for $x \in \beta(c)$ so $y \in (\alpha\beta)(c)$. Then $y \in (\alpha\gamma)(c)$. This means $y \in \alpha(z)$ for some $z \in \gamma(c)$. Now $y \in \alpha(z)$ and $y \in \alpha(x)$ and sets of codes for distinct objects must be disjoint, so $x = z$. Since $z \in \gamma(c)$, $x \in \gamma(c)$.

Thus $\beta(c) \subseteq \gamma(c)$. The reverse inclusion is similar, so $\beta = \gamma$.

PROPOSITION 2.2. *In* **Dom** *a morphism* α *is epi if and only if* α *is* 1–1 *and onto.*

PROOF. First suppose α is epi, say $\alpha : \mathscr{A} \to \mathscr{B}$. Let $x \in \mathscr{A}$ and suppose y, $z \in \alpha(x)$. Define codings $\beta, \gamma : \mathscr{B} \to \mathscr{B}$ by $\beta = \text{inj}_{\mathscr{B}}$, and γ is $\text{inj}_{\mathscr{B}}$ except on y and z, where $\gamma(y) = \{z\}$ and $\gamma(z) = \{y\}$. Then $\beta\alpha = \gamma\alpha$. Since α is epi, $\beta = \gamma$ and this implies $y = z$. Thus α is 1–1.

Again, suppose $\alpha : \mathscr{A} \to \mathscr{B}$ is epi, but that α is not onto. Let $y \in \mathscr{B}$ be not an α-code. Let $z \notin \mathscr{B}$, and let $\mathscr{C} = \mathscr{B} \cup \{z\}$. Define codings

$\beta, \gamma : \mathscr{B} \to \mathscr{C}$ by $\beta(x) = \{x\}$ for all $x \in \mathscr{B}$, and $\gamma(x) = \{x\}$ for all x except y while $\gamma(y) = \{y, z\}$. Again $\beta\alpha = \gamma\alpha$, so $\beta = \gamma$, and this implies $y = z$, a contradiction. Thus α must be onto.

If $\mathscr{A}$ is 1–1 and onto it is easy to check that it is epi.

Let $\theta : \mathscr{A} \to \mathscr{B}$ be a coding. We extend θ to n-tuples as follows. If $\boldsymbol{a} = \langle a_1, \ldots, a_n \rangle$ and $\boldsymbol{b} = \langle b_1, \ldots, b_n \rangle$, we write $\boldsymbol{b} \in \theta(\boldsymbol{a})$ if $b_1 \in \theta(a_1) \wedge \cdots \wedge b_n \in \theta(a_n)$. Also we define maps associated with θ, but on relations, as follows.

Let $\mathscr{R} \subseteq \mathscr{A}^n$. By $\theta_n(\mathscr{R})$ we mean $\{\boldsymbol{b} \in \mathscr{B}^n \mid \text{for some } \boldsymbol{a} \in \mathscr{R}, \boldsymbol{b} \in \theta(\boldsymbol{a})\}$.

Let $\mathscr{S} \subseteq \mathscr{B}^n$. By $\theta_n^{-1}(\mathscr{S})$ we mean $\{\boldsymbol{a} \in \mathscr{A}^n \mid \text{for some } \boldsymbol{b} \in \mathscr{S}, \boldsymbol{b} \in \theta(\boldsymbol{a})\}$.

Then both θ_n and θ_n^{-1} are mappings:

$$\theta_n : \text{power set } (\mathscr{A}^n) \to \text{power set } (\mathscr{B}^n),$$

$$\theta_n^{-1} : \text{power set } (\mathscr{B}^n) \to \text{power set } (\mathscr{A}^n).$$

The following facts may now be established. Most of them depend on the "disjointness" feature of codings: $a \neq b$ implies $\theta(a)$ and $\theta(b)$ are disjoint.

PROPOSITION 2.3. *Let $\alpha : \mathscr{A} \to \mathscr{B}$ and $\beta : \mathscr{B} \to \mathscr{C}$ be codings. Let $\mathscr{P}, \mathscr{R} \subseteq \mathscr{A}^n$ and $\mathscr{S} \subseteq \mathscr{B}^n$.* Then

(1) $(\alpha_n^{-1}\alpha_n)(\mathscr{R}) = \mathscr{R}$.

(2) *If α is* 1–1, $(\alpha_n\alpha_n^{-1})(\mathscr{S}) \subseteq \mathscr{S}$.

(3) *If α is onto*, $\mathscr{S} \subseteq (\alpha_n\alpha_n^{-1})(\mathscr{S})$.

(4) $(\beta\alpha)_n = \beta_n\alpha_n$.

(5) $(\beta\alpha)_n^{-1} = \alpha_n^{-1}\beta_n^{-1}$.

(6) $\alpha_n(\mathscr{P} \cap \mathscr{R}) = \alpha_n(\mathscr{P}) \cap \alpha_n(\mathscr{R})$

(7) $\alpha_n(\mathscr{P} \cup \mathscr{R}) = \alpha_n(\mathscr{P}) \cup \alpha_n(\mathscr{R})$

(8) *if* $\varphi \subseteq \mathscr{A}^m$, $\alpha_{n+m}(\mathscr{P} \times \varphi) = \alpha_n(\mathscr{P}) \times \alpha_m(\varphi)$.

3. Pre-embeddings

DEFINITION. Let $\boldsymbol{A}$ and $\boldsymbol{B}$ be production systems, and let $\theta : \mathscr{A} \to \mathscr{B}$ be a coding. We call θ a *pre-embedding* if

(1) for each $\mathscr{P} \in [\mathscr{A}]^n$ allowable in $\boldsymbol{A}$, $\theta_n(\mathscr{P})$ is allowable in $\boldsymbol{B}$;

(2) for each $\mathscr{R} \in [\mathscr{B}]^n$ allowable in $\boldsymbol{B}$, $\theta_n^{-1}(\mathscr{R})$ is allowable in $\boldsymbol{A}$.

REMARKS. If $\boldsymbol{A}$ and $\boldsymbol{B}$ are total production systems, recursion or ω-recursion theories for example, every coding from $\mathscr{A}$ to $\mathscr{B}$ is automatically a pre-embedding. Pre-embeddings will play a non-trivial role in Chapter 9, when many partial production systems will be studied.

PROPOSITION 3.1. *The composition of pre-embeddings is a pre-embedding.*

PROOF. Let **A**, **B** and **C** be production systems and $\alpha : \mathscr{A} \to \mathscr{B}$ and $\beta : \mathscr{B} \to \mathscr{C}$ be pre-embeddings. For $\mathscr{R} \subseteq \mathscr{A}^n$, by Proposition 2.3 (4), $(\beta\alpha)_n(\mathscr{R}) = \beta_n(\alpha_n(\mathscr{R}))$. If $\mathscr{R}$ is allowable in **A**, $\alpha_n(\mathscr{R})$ is allowable in **B** since α is a pre-embedding; then $\beta_n(\alpha_n(\mathscr{R}))$ is allowable in **C**. $(\beta\alpha)_n^{-1}$ is treated similarly using 2.3 (5).

DEFINITION. Let θ be a pre-embedding from **A** to **B**. We use $[\mathscr{A}]^n$ for the collection of n-ary relations allowable in **A**; and similarly for $[\mathscr{B}]^n$.

If we have a map $\Phi : [\mathscr{A}]^n \to [\mathscr{A}]^m$, by Φ^θ we mean the map from $[\mathscr{B}]^n$ to $[\mathscr{B}]^m$ given by $\Phi^\theta = \theta_m \Phi \theta_n^{-1}$.

Similarly, if we have a map $\Psi : [\mathscr{B}]^n \to [\mathscr{B}]^m$, by $\Psi^{\theta^{-1}}$ we mean the map from $[\mathscr{A}]^n$ to $[\mathscr{A}]^m$ given by $\Psi^{\theta^{-1}} = \theta_m^{-1} \Psi \theta_n$.

PROPOSITION 3.2. *Let θ be a pre-embedding from **A** to **B**. Then for $\Phi \in$ **A**,*

$$(\Phi^\theta)^{\theta^{-1}} = \Phi.$$

PROOF. Say Φ is of order $\langle n, m\rangle$. Then, using Proposition 2.3,

$$(\Phi^\theta)^{\theta^{-1}} = \theta_m^{-1} \Phi^\theta \theta_n = \theta_m^{-1} \theta_m \Phi \theta_n^{-1} \theta_n = \Phi.$$

PROPOSITION 3.3. *Let θ be a pre-embedding from **A** to **B**.*

(1) *θ preserves composition of operators.*

(2) *θ^{-1} preserves identity operators.*

PROOF. We only prove (1).

Say $\Phi \in$ **A** is of order $\langle n, m\rangle$ and $\Psi \in$ **A** is of order $\langle m, p\rangle$. Then, using Proposition 2.3 again,

$$\Psi^\theta \Phi^\theta = \theta_p \Psi \theta_m^{-1} \theta_m \Phi \theta_n^{-1} = \theta_p \Psi \Phi \theta_n^{-1} = (\Psi\Phi)^\theta.$$

PROPOSITION 3.4. *Let θ be a pre-embedding from **A** to **B** which is* 1–1 *and onto.*

(1) *θ preserves identity operators.*

(2) *θ^{-1} preserves composition of operators.*

PROOF. Similar to the above.

PROPOSITION 3.5. *Let θ be a pre-embedding from **A** to **B**, and let $\Phi, \Psi \in$ **A**. Then (when defined),*

(1) $(\Phi \cap \Psi)^\theta = \Phi^\theta \cap \Psi^\theta$,
(2) $(\Phi \cup \Psi)^\theta = \Phi^\theta \cup \Psi^\theta$,
(3) $(\Phi \times \Psi)^\theta = \Phi^\theta \times \Psi^\theta$.

PROOF. We show 1 only. Say Φ and Ψ are of order $\langle n, m\rangle$. Then using Proposition 2.3, for $\mathscr{P} \in [\mathscr{B}]^n$,

$$\begin{aligned}(\Phi \cap \Psi)^\theta(\mathscr{P}) &= [\theta_m(\Phi \cap \Psi)\theta_n^{-1}](\mathscr{P}) \\ &= \theta_m(\Phi \cap \Psi)(\theta_n^{-1}(\mathscr{P})) \\ &= \theta_m[\Phi(\theta_n^{-1}(\mathscr{P})) \cap \Psi(\theta^{-1}(\mathscr{P}))] \\ &= \theta_m(\Phi(\theta_n^{-1}(\mathscr{P}))) \cap \theta_m(\Psi(\theta_n^{-1}(\mathscr{P}))) \\ &= \Phi^\theta(\mathscr{P}) \cap \Psi^\theta(\mathscr{P}) \\ &= (\Phi^\theta \cap \Psi^\theta)(\mathscr{P}).\end{aligned}$$

4. Embeddings and co-embeddings

DEFINITION. Let $\boldsymbol{A}$ and $\boldsymbol{B}$ be production systems, and let θ be a pre-embedding from $\boldsymbol{A}$ to $\boldsymbol{B}$.

We call θ an *embedding* of $\boldsymbol{A}$ in $\boldsymbol{B}$, and write $\theta : \boldsymbol{A} \to \boldsymbol{B}$, if for each operator $\Phi \in \boldsymbol{A}$, Φ^θ is an operator in $\boldsymbol{B}$.

We call θ^{-1} a *co-embedding* of $\boldsymbol{B}$ in $\boldsymbol{A}$, and write $\theta^{-1} : \boldsymbol{B} \to \boldsymbol{A}$, if for each operator $\Psi \in \boldsymbol{B}$, $\Psi^{\theta^{-1}}$ is an operator in $\boldsymbol{A}$.

Embeddings and co-embeddings are 1–1 or onto if the underlying codings are 1–1 or onto.

PROPOSITION 4.1. *Let $\alpha : \boldsymbol{A} \to \boldsymbol{B}$ and $\beta : \boldsymbol{B} \to \boldsymbol{C}$ be embeddings. Then the composition, $\beta\alpha$, is an embedding. In fact, $\Phi^{\beta\alpha} = (\Phi^\alpha)^\beta$. Similarly the composition of co-embeddings is a co-embedding.*

PROOF. For embeddings only. If we show $\Phi^{\beta\alpha} = (\Phi^\alpha)^\beta$, we are done, since both α and β map operators to operators. Now, say $\Phi \in \boldsymbol{A}$ is of order $\langle n, m\rangle$. We have

$$\begin{aligned}\Phi^{\beta\alpha} &= (\beta\alpha)_m \Phi(\beta\alpha)_n^{-1} \\ &= \beta_m \alpha_m \Phi \alpha_n^{-1} \beta_n^{-1} \\ &= \beta_m \Phi^\alpha \beta_n^{-1} = (\Phi^\alpha)^\beta\end{aligned}$$

PROPOSITION 4.2. *Let θ be a pre-embedding from **A** to **B**.*

(1) *If θ is an embedding, θ takes relations generated in **A** to relations generated in **B**.*

(2) *If θ^{-1} is a co-embedding, θ^{-1} takes relations generated in **B** to relations generated in **A**.*

PROOF. We only prove (1). Let $\mathcal{R} \in [\mathcal{A}]^n$ be generated in **A**. Then for some $\Phi \in \mathbf{A}$, Φ is constant, of value $\mathcal{R}$. Then $\Phi^\theta \in \mathbf{B}$, and trivially, Φ^θ is constant, of value $\theta_n(\mathcal{R})$.

REMARKS. Just because θ is an embedding, it doesn't follow that θ^{-1} is a co-embedding. The following example illustrates this.

Let $\mathbb{N}$ be the natural numbers, and let $\mathfrak{A} = \langle \mathbb{N} \rangle$, a trivial structure. Define a coding $\theta : \mathbb{N} \to \mathbb{N}$ by $\theta(n) = \{n\}$, that is, $\theta = \text{inj}_{\mathbb{N}}$. It is easy to see directly, or by use of Theorem 5.1 below, that $\theta : \text{rec}(\mathfrak{A}) \to \text{rec}(\mathfrak{S}(\mathbb{N}))$ is an embedding. Let $\mathscr{E}$ be the set of even numbers. $\mathscr{E}$ is r.e. in $\text{rec}(\mathfrak{S}(\mathbb{N}))$. But $\theta_1^{-1}(\mathscr{E}) = \mathscr{E}$, and this is certainly not r.e. in $\text{rec}(\mathfrak{A})$ since there is not enough machinery there to generate it. Then by the above proposition, θ^{-1} is not a co-embedding.

DEFINITION. Let $\mathscr{C}$ be a collection of structures. We say T is a *theory assignment for* $\mathscr{C}$ if T assigns to each structure $\mathfrak{A} \in \mathscr{C}$ some production system $T(\mathfrak{A})$ on the structure $\mathfrak{A}$.

5. Theory assignments

EXAMPLES. Let $\mathscr{C}$ be the entire class of structures. Then rec(-), ω-rec(-) and f.o.(-) are theory assignments for $\mathscr{C}$.

In Chapter 9 we will consider theory assignments that are meaningful for some, but not all structures, at which time $\mathscr{C}$ will play a bigger role.

Let T be a theory assignment for $\mathscr{C}$ and let $\mathfrak{A}$, $\mathfrak{B} \in \mathscr{C}$. If we have a coding $\theta : \mathcal{A} \to \mathcal{B}$, it may be quite difficult to tell whether θ is an *embedding* of $T(\mathfrak{A})$ in $T(\mathfrak{B})$. But for certain theory assignments we have a simple condition.

To keep notation simple, if $\theta : \mathfrak{A} \to \mathfrak{B}$, and $\mathcal{R}$ is an n-ary relation on $\mathcal{A}$, we will often write $\mathcal{R}^\theta$ for $\theta_n(\mathcal{R})$.

DEFINITION. Let $\mathscr{C}$ be a collection of structures, let T be a theory assignment for $\mathscr{C}$, and let $\mathscr{E}$ be a collection of codings. We call T *elementary with respect to* $\mathscr{E}$ if the following is the case: For any two structures $\mathfrak{A}, \mathfrak{B} \in \mathscr{C}$, where $\mathfrak{A} = \langle \mathscr{A}; \mathscr{R}_1, \ldots, \mathscr{R}_k \rangle$, and for any coding $\theta : \mathscr{A} \to \mathscr{B}$ which is in $\mathscr{E}$, θ is an *embedding of* $T(\mathfrak{A})$ in $T(\mathfrak{B})$ whenever

(1) θ is a pre-embedding of $T(\mathfrak{A})$ in $T(\mathfrak{B})$,

(2) Each of $(=_{\mathscr{A}})^\theta, \mathscr{R}_1^\theta, \ldots, \mathscr{R}_k^\theta$ is generated in $T(\mathfrak{B})$.

REMARKS. rec(-), ω-rec(-) and f.o.(-) assign *total* production systems to structures. For such theory assignments clause (1) is automatically satisfied. We will consider partial production systems in some detail in Chapters 8 and 9.

THEOREM 5.1. *The following are elementary theory assignments for the class of all structures with respect to the class of all codings*:

(1) rec(-), (2) ω-rec(-), f.o.(-).

COROLLARY 5.2. *If* $\theta : \text{rec}(\mathfrak{A}) \to \text{rec}(\mathfrak{B})$ *is an embedding,* θ *is also an embedding of* ω-rec($\mathfrak{A}$) *in* ω-rec($\mathfrak{B}$).

PROOF. Every r.e. relation on $\mathfrak{B}$ is also ω-r.e.

PROOF OF THEOREM 5.1. We leave the proof of (3) to the reader.

Suppose now that T(-) is either rec(-) or ω-rec(-).

Suppose $\mathfrak{A} = \langle \mathscr{A}; \mathscr{R}_1, \ldots, \mathscr{R}_k \rangle$, $\theta : \mathscr{A} \to \mathscr{B}$ is a coding (hence automatically a pre-embedding since recursion and ω-recursion theories are total), and each of $(=_{\mathscr{A}})^\theta, \mathscr{R}_1^\theta, \ldots, \mathscr{R}_k^\theta$ is generated in $T(\mathfrak{B})$. We show $\theta : T(\mathfrak{A}) \to T(\mathfrak{B})$ is an embedding. We define a certain translation procedure taking elementary formal systems over $\mathfrak{A}$ into elementary formal systems over $\mathfrak{B}$. The idea is to replace members of $\mathscr{A}$ by all codes for them, and the relations $\mathscr{R}_1, \ldots, \mathscr{R}_k$ by $\mathscr{R}_1^\theta, \ldots, \mathscr{R}_k^\theta$, of course adjoining extra axioms to handle these over $\mathfrak{B}$. In order to define our translation procedures more easily, we first introduce some notation.

Let A be a pseudo-formula (recall; contains variables) and let C be a one-place predicate symbol. By A^C (axiom A restricted to C) we mean

$$Cx_1 \to Cx_2 \to \cdots \to Cx_n \to A$$

where $x_1, \ldots, x_n$ are all the variables of A (in some fixed order). Below C will represent the set of codes.

Let $\approx$ be a two place relation symbol (intended below to represent $(=_{\mathscr{A}})^\theta$). Let P be some n-place predicate symbol. By an *equality axiom* for P we mean

$$\approx x_1, y_1 \rightarrow \approx x_2, y_2 \rightarrow \cdots \rightarrow \approx x_n, y_n \rightarrow Px_1, \ldots, x_n \rightarrow Py_1, \ldots, y_n$$

where all variables are distinct.

We note that the set of codes, in $\mathscr{B}$, is the set of first (or second) components of $(=_{\mathscr{A}})^\theta$.

Now, let E be some elementary formal system over $\mathfrak{A}$ (accepting inputs). Let us suppose that the predicate symbols $R_1, \ldots, R_k$, have been assigned to $\mathscr{R}_1, \ldots, \mathscr{R}_k$, that the predicate symbols assigned to the given (displayed) relations of $\mathfrak{B}$ are different, and do not occur in E. Let $\approx$ (two place) and C (one place) be new predicate symbols. We define an elementary formal system E^θ over $\mathfrak{B}$ as follows.

First, say A is an axiom of E. For each member of $\mathscr{A}$ which occurs in A, choose one θ-code for it from $\mathscr{B}$. Let A' be the result of replacing, in A, the members of $\mathscr{A}$ by these codes.

Now, the axioms of E^θ are:

(1) For each of $i = 1, 2, \ldots, k$, axioms for $(\mathscr{R}_i)^\theta$ in which we use R_i again as the representing predicate symbol.

(2) Axioms for $(=_{\mathscr{A}})^\theta$ in which $\approx$ is the representing predicate symbol.

(3) Equality axioms for each predicate symbol of E.

(4) $\approx x, y \rightarrow Cx$.

(5) $(A')^C$ for each A in E.

Now, if $[E^I_O]$ is an operator in $T(\mathfrak{A})$ of order $\langle n, m\rangle$, $[(E^\theta)^I_O]$ is an operator in $T(\mathfrak{B})$, also of order $\langle n, m\rangle$. And it should be clear that whatever $[E^I_O]$ does, $[(E^\theta)^I_O]$ does the corresponding thing to codes. That is, for $\mathscr{P} \in [\mathscr{A}]^n$,

$$\theta_m [E^I_O](\mathscr{P}) = [(E^\theta)^I_O]\theta_n (\mathscr{P}).$$

Further, it should be clear that $[(E^\theta)^I_O]$ "ignores" non-codes, and behaves the same whether it has all, or only some codes for a thing. Now, if $\mathscr{R} \in [\mathscr{B}]^n$, $\theta_n\theta_n^{-1}(\mathscr{R})$ is the result of discarding from $\mathscr{R}$ all non-codes, then adjoining all other codes for things with any codes there. Hence we may formally state our observation about $[(E^\theta)^I_O]$ as follows: for $\mathscr{R} \in [\mathscr{B}]^n$,

$$[(E^\theta)^I_O](\mathscr{R}) = [(E^\theta)^I_O]\theta_n\theta_n^{-1}(\mathscr{R}).$$

Combining our two observations, we have

$$\begin{aligned}[(E^\theta)^I_O](\mathscr{R}) &= [(E^\theta)^I_O]\theta_n\theta_n^{-1}(\mathscr{R}) \\ &= \theta_m [E^I_O]\theta_n^{-1}(\mathscr{R}) = [E^I_O]^\theta (\mathscr{R}).\end{aligned}$$

Since, by construction, $[(E^\theta)^I_O]$ is an operator in $T(\mathfrak{B})$, so is $[E^I_O]^\theta$. Since operators map to operators, θ is an embedding.

6. Examples

We give examples of embeddings between the recursion theories over the structures named in Ch. 1, §2. Since we are working with recursion theories, which are total, every coding is automatically a pre-embedding. Note that by Corollary 5.2, there are similar embeddings between the corresponding ω-recursion theories.

I. *Numbers to sets.* Define a coding $\theta : \mathbb{N} \to L_\omega$ by: $\theta(n) = \{n^\#\}$ where $n^\#$ is the Von Neumann ordinal for n. Just this once we go through the details, verifying that θ is an embedding of rec($\mathfrak{S}(\mathbb{N})$) into rec($\mathfrak{S}(L_\omega)$).

The image, under θ, of the successor relation on the natural numbers, is represented in rec($\mathfrak{S}(L_\omega)$) by S using the axioms

$$S\emptyset, \{\emptyset\},$$

$$Sx, y \to x \cup \{x\} = x' \to y \cup \{y\} = y' \to Sx', y'.$$

Also, the image of $=_{\mathbb{N}}$ is represented in rec($\mathfrak{S}(L_\omega)$) by $\approx$ using the above and

$$\approx \emptyset, \emptyset$$

$$\approx x, y \to Sx, x' \to Sy, y' \to \approx x', y'.$$

Then by Theorem 5.1, θ is an embedding of rec($\mathfrak{S}(\mathbb{N})$) into rec($\mathfrak{S}(L_\omega)$). It is 1–1 but not onto.

II. *Sets to words*, with an alphabet $a_1, a_2, \ldots, a_n$ of more than one letter.

Let us say two of the letters are } and {. Using these letters, each hereditarily finite set has "names" in a natural way. { } "names" $\emptyset$. If $n_1, n_2, \ldots, n_k$ are "names" for the sets $s_1, s_2, \ldots, s_k$, then $\{n_1 n_2 \cdots n_k\}$ is a "name" for the set $\{s_1, s_2, \ldots, s_k\}$. Now, define a coding $\theta : L_\omega \to \mathcal{W}(\{, \}, a_3, \ldots, a_n)$ by

$\theta(s)$ is the collection of all "names" for s.

θ is an embedding of rec($\mathfrak{S}(L_\omega)$) into rec($\mathfrak{S}(a_1, \ldots, a_n)$), neither 1–1 nor onto.

III. *Words to numbers.* Say, for convenience, that our letters are $1, 2, \ldots, n$. Define a coding $\theta : \mathcal{W}(1, 2, \ldots, n) \to \mathbb{N}$ by

$$\theta(a_1 a_2 \cdots a_j) = \{2^{a_1} \cdot 3^{a_2} \cdot \cdots \cdot p_j^{a_j}\}$$

where p_j is the jth prime, and the a_i are letters. Then

$\theta:\text{rec}(\mathfrak{S}(1,2,\ldots,n))\to\text{rec}(\mathfrak{S}(\mathbb{N}))$ is an embedding, 1–1 but not onto. (It is usually called a Gödel numbering of $\mathcal{W}(a_1,\ldots,a_n)$.)

IV. *Numbers to words,* with an alphabet of one letter. Define a coding $\theta:\mathbb{N}\to\mathcal{W}(a)$ by

$$\theta(n)=\{\underbrace{aa\cdots a}_{n}\}$$

Then $\theta:\text{rec}(\mathfrak{S}(\mathbb{N}))\to\text{rec}(\mathfrak{S}(a))$ is a 1–1, onto embedding.

V. *Words to sets.* (a) By composing III with I, there is an embedding of $\text{rec}(\mathfrak{S}(a_1,\ldots,a_n))$ into $\text{rec}(\mathfrak{S}(L_\omega))$.

(b) A more direct example of such an embedding is the following. Associate with each letter a_i some arbitrary "code" set $a_i^\#$. Associate with the word $w=w_1w_2\cdots w_n$ (where each w_i is a letter) the finite function f_w given by $f_w(i)=w_i^\#$. Set $\theta(w)=\{f_w\}$. This too gives an embedding, 1–1, but not onto.

VI. *Words to words.* Clearly, the choice of letters is not important, only the number of them.

Define $\theta:\mathcal{W}(a_1,\ldots,a_n)\to\mathcal{W}(a_1,\ldots,a_n,a_{n+1},\ldots,a_k)$ by $\theta(w)=\{w\}$. This is a coding, 1–1 but not onto, and gives an embedding on the recursion theories.

Similarly, by composing III, IV and the above, there are also embeddings that shrink alphabets. There are more directly defined examples of such embeddings in Smullyan [1961].

VII. *Sets to numbers.* (a) Compose II and III.

(b) Another example is the following. Define a coding $\theta:L_\omega\to\mathbb{N}$ by $\theta(F_n)=\{n\}$ where F_n is the nth constructible set, using the definition in Gödel [1939]. This is 1–1 and onto, and $\theta:\text{rec}(\mathfrak{S}(L_\omega))\to\text{rec}(\mathfrak{S}(\mathbb{N}))$ is an embedding.

7. The category of production systems

We define a (large) category, of production systems, as follows. The objects are all production systems. The morphisms are all embeddings (domain and co-domain specified, since the same coding may work for many production systems). By Proposition 4.1, the composition of embeddings is an embedding. It is easy to see that the injection codings are embeddings, and serve as identity morphisms. Thus we have a category, which we call **Prod**.

In **Prod**, $\emptyset$ serves as initial object. Certain subcategories of **Prod** are of obvious interest to us. The full sub-category having all recursion theories as objects is one; we call it **Rec**. Likewise there is the full subcategory having all ω-recursion theories as objects; we call it $\boldsymbol{\omega}$**-Rec**. And there is the full subcategory of first order theories. We call it **F. O.** It is the study of these and related categories that largely occupies the next few chapters.

We note that, by taking production systems as objects, but using co-embeddings as morphisms, we also obtain a category. We may call it **Co-prod**, but we will have little to say about it.

CHAPTER FOUR

COMBINING PRODUCTION SYSTEMS

1. Introduction

It is common mathematical practice to combine "like" mathematical objects into another of the same sort. Thus, one can form the direct sum of two rings, and obtain a ring, or the product of two topological spaces, and obtain a topological space. In this chapter we show how two recursion theories may be combined into a recursion theory, using *either* disjoint union, $\dot{\cup}$, or Cartesian product, $\times$. Similarly for ω-recursion theories, and for first order theories. This will play an important role in the next chapter when it will be necessary to discuss two recursion theories "jointly."

We begin by defining $\dot{\cup}$ and $\times$ for domains, and for codings. Both $\dot{\cup}$ and $\times$ have certain common features, and we spend some time discussing them simultaneously. The language of category theory is convenient here. We use no deep results, only the terminology, which is helpful as an organizing device.

After treating domains, we define $\dot{\cup}$ and $\times$ between "like" production systems. It is here that *theory assignments* take on a major role. If T is a theory assignment which meets certain conditions, we can define $\dot{\cup}$ and $\times$ for $T(\mathfrak{A})$ and $T(\mathfrak{B})$. Again, nice algebraic structures result, which are best described using category theoretic language.

Let us write $\square$ for either $\dot{\cup}$ or $\times$, where it does not matter which. Having defined $\square$ for the production systems given by the theory assignment T, if $\boldsymbol{A} = T(\mathfrak{A})$, then $\boldsymbol{A} \square \boldsymbol{A}$ is meaningful. This production system may or may not have an embedding back into $\boldsymbol{A}$. If it does, and if such an embedding meets certain conditions, we call $\boldsymbol{A}$ *separable* (with respect to $\square$). Ordinary recursion theory is separable, as are many other recursion and ω-recursion theories. Being separable with respect to $\times$ is a consequence of having an effective pairing function. Being separable with respect to $\dot{\cup}$ is a generalization of the feature of ordinary recursion theory that the domain can effectively be split into the even numbers and the odd numbers. Separable production systems are rather well behaved, and the chapter concludes with a study of them.

2. Combining domains and codings

We want to combine production systems; here we begin by combining domains. The two techniques we discuss are disjoint union, $\dot{\cup}$, and Cartesian product, $\times$.

I. *Disjoint union.* The disjoint union of two sets is the union of disjoint copies of them. Any systematic way of choosing copies generally will serve. We officially adopt the following:

DEFINITION. $\mathscr{A} \dot{\cup} \mathscr{B} = (\mathscr{A} \times \{0\}) \cup (\mathscr{B} \times \{1\})$.

Note that, under this definition, $\mathscr{A} \dot{\cup} \mathscr{B}$ and $\mathscr{B} \dot{\cup} \mathscr{A}$ are, in general, quite distinct. But they are obviously "isomorphic." Consider the 1–1, onto coding $\gamma : \mathscr{A} \dot{\cup} \mathscr{B} \to \mathscr{B} \dot{\cup} \mathscr{A}$ given by

$$\gamma(\langle x, 0\rangle) = \{\langle x, 1\rangle\}, \qquad \gamma(\langle x, 1\rangle) = \{\langle x, 0\rangle\}.$$

Similarly $\mathscr{A} \dot{\cup} (\mathscr{B} \dot{\cup} \mathscr{C})$ and $(\mathscr{A} \dot{\cup} \mathscr{B}) \dot{\cup} \mathscr{C}$ are different, but are isomorphic under a 1–1, onto coding, $\mathscr{A}$.

The empty domain, $\emptyset$, acts like a unit for this operation. That is, $\emptyset \dot{\cup} \mathscr{A}$ and $\mathscr{A}$, though different, are isomorphic under an obvious 1–1, onto coding, λ. Similarly, there is a 1–1, onto coding ρ from $\mathscr{A} \dot{\cup} \emptyset$ to $\mathscr{A}$.

There is also a simple way of defining $\dot{\cup}$ on *codings* so that it "keeps the parts separate."

DEFINITION. Suppose we have codings $f : \mathscr{A} \to \mathscr{A}'$ and $g : \mathscr{B} \to \mathscr{B}'$. By $f \dot{\cup} g : \mathscr{A} \dot{\cup} \mathscr{B} \to \mathscr{A}' \dot{\cup} \mathscr{B}'$ we mean the coding given by

$$(f \dot{\cup} g)(\langle x, 0\rangle) = \{\langle y, 0\rangle \mid y \in f(x)\},$$

$$(f \dot{\cup} g)(\langle x, 1\rangle) = \{\langle y, 1\rangle \mid y \in g(x)\}.$$

It is easy to check that $\dot{\cup}$ has the following pleasant properties.

$$(f' \dot{\cup} g')(f \dot{\cup} g) = (f'f) \dot{\cup} (g'g)$$

and

$$\mathrm{inj}_{\mathscr{A}} \dot{\cup} \mathrm{inj}_{\mathscr{B}} = \mathrm{inj}_{\mathscr{A} \dot{\cup} \mathscr{B}}$$

II. *Cartesian product.*

DEFINITION. $\mathscr{A} \times \mathscr{B} = \{\langle x, y\rangle \mid x \in \mathscr{A} \text{ and } y \in \mathscr{B}\}$.

Paralleling the discussion of $\dot{\cup}$ we first note that, though $\mathscr{A} \times \mathscr{B}$ and $\mathscr{B} \times \mathscr{A}$ are generally different, there is a 1–1, onto coding connecting them:

$$\gamma : \mathscr{A} \times \mathscr{B} \to \mathscr{B} \times \mathscr{A} \quad \text{where } \gamma(\langle x, y \rangle) = \{\langle y, x \rangle\}.$$

Similarly $\mathscr{A} \times (\mathscr{B} \times \mathscr{C})$ and $(\mathscr{A} \times \mathscr{B}) \times \mathscr{C}$ are isomorphic, under α, say.

One element sets act as units for $\times$. To be specific, we use $\{0\}$. Then $\mathscr{A} \times \{0\}$, $\{0\} \times \mathscr{A}$ and $\mathscr{A}$ are all isomorphic in an obvious way.

We also define an operation of $\times$ on codings so that it works with each component separately.

DEFINITION. Suppose we have codings $f : \mathscr{A} \to \mathscr{A}'$ and $g : \mathscr{B} \to \mathscr{B}'$. By $f \times g : \mathscr{A} \times \mathscr{B} \to \mathscr{A}' \times \mathscr{B}'$ we mean the coding given by

$$(f \times g)(\langle x, y \rangle) = \{\langle a, b \rangle \mid a \in f(x) \text{ and } b \in g(y)\}.$$

Then one may check that:

$$(f' \times g')(f \times g) = (f'f) \times (g'g)$$

and

$$\mathrm{inj}_{\mathscr{A}} \times \mathrm{inj}_{\mathscr{B}} = \mathrm{inj}_{\mathscr{A} \times \mathscr{B}}.$$

3. Symmetric monoidal categories

We have defined $\dot{\cup}$ and $\times$ on **Dom**, both for objects and for morphisms. As it happens, $\dot{\cup}$ is *not* co-product, nor is $\times$ product in this category. Still it is clear that there is much common to what may be said about $\dot{\cup}$ and $\times$. In fact, both operations provide us with examples of *symmetric monoidal categories*, a notion which we define below. We will see several other examples of symmetric monoidal categories in this chapter as well, so we spend some time developing their properties in the abstract. See MacLane [1971], Ch. 7 for a fuller development of these ideas.

Let $\mathscr{C}$ be some category and let $\Box$ be some bifunctor (two-place functor) on $\mathscr{C}$. Then, in particular, for arbitrary morphisms,

$$(f' \Box g')(f \Box g) = (f'f) \Box (g'g),$$

and for identity morphisms,

$$1_{\mathscr{A}} \Box 1_{\mathscr{B}} = 1_{\mathscr{A} \Box \mathscr{B}}.$$

DEFINITION. A *monoidal category* $\mathscr{C} = \langle \mathscr{C}; \Box, e, \alpha, \lambda, \rho \rangle$ is: a category $\mathscr{C}$, a bifunctor $\Box : \mathscr{C} \times \mathscr{C} \to \mathscr{C}$, a (unit) object e, and three isomorphisms.

$$\alpha = \alpha_{\mathscr{A},\mathscr{B},\mathscr{C}} : \mathscr{A} \Box (\mathscr{B} \Box \mathscr{C}) \cong (\mathscr{A} \Box \mathscr{B}) \Box \mathscr{C}$$

natural in $\mathscr{A}$, $\mathscr{B}$, and $\mathscr{C}$, and

$$\lambda = \lambda_{\mathscr{A}} : e \,\Box\, \mathscr{A} \cong \mathscr{A} \quad \text{and} \quad \rho = \rho_{\mathscr{A}} : \mathscr{A} \,\Box\, e \cong \mathscr{A}$$

natural in $\mathscr{A}$. Further, the commutativity of enough diagrams is postulated so that the commutativity of "all" diagrams involving these notions follows. See MacLane [1971], Ch. 7, §2 for a proper statement of just what this means.

A monoidal category $\mathscr{C}$ is *symmetric* if it also has isomorphisms

$$\gamma = \gamma_{\mathscr{A},\mathscr{B}} : \mathscr{A} \,\Box\, \mathscr{B} \cong \mathscr{B} \,\Box\, \mathscr{A}$$

natural in $\mathscr{A}$ and $\mathscr{B}$, such that

$$\gamma_{\mathscr{A},\mathscr{B}} \circ \gamma_{\mathscr{B},\mathscr{A}} = 1,$$

$$\rho_{\mathscr{B}} = \lambda_{\mathscr{B}} \circ \gamma_{\mathscr{B},e} : \mathscr{B} \,\Box\, e \cong \mathscr{B}.$$

And again, "all" diagrams involving these notions commute.

I. *Disjoint union.* Consider the operation $\dot{\cup}$ as defined in the last section, along with the isomorphisms (1–1, onto codings) mentioned there. We get a symmetric monoidal category $\langle \mathbf{Dom}; \dot{\cup}, \emptyset, \alpha, \lambda, \rho, \gamma \rangle$ which we denote $\langle \mathbf{Dom}, \dot{\cup} \rangle$ for short.

II. *Cartesian product.* In working with $\times$ we find it desirable to rule out the empty domain. The reason is, there are simple, useful codings of $\mathscr{A}$ into $\mathscr{A} \times \mathscr{B}$, except when $\mathscr{B}$ is empty. So, let **Sdom** be the full subcategory of **Dom** in which the objects are the non-empty objects of **Dom**. Let $\times$ be as defined in the last section. Then, using the isomorphisms mentioned there, $\langle \mathbf{Sdom}; \times, \{0\}, \alpha, \lambda, \rho, \gamma \rangle$ is also a symmetric monoidal category which we denote $\langle \mathbf{Sdom}, \times \rangle$ for short.

4. A few technical results

Let $\mathscr{C}$ be a symmetric monoidal category, with $\Box$ as the monoidal operation. We develop a few simple techniques for working with $\Box$ on the morphisms of $\mathscr{C}$. The first allows the breaking up of $f \,\Box\, g$ into simpler combinations. We note that we use $1_{\mathscr{A}}$ for the identity morphism on $\mathscr{A}$, which happens to be $\mathrm{inj}_{\mathscr{A}}$ in our examples.

PROPOSITION 4.1. *Let $f : \mathscr{A} \to \mathscr{A}'$ and $g : \mathscr{B} \to \mathscr{B}'$ be morphisms in $\mathscr{C}$. The following diagram commutes*:

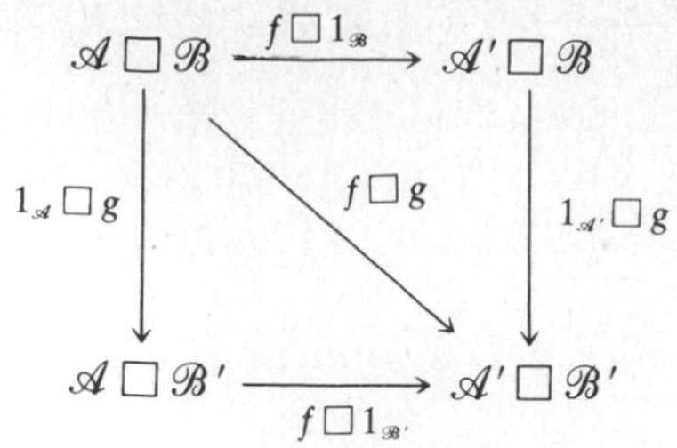

PROOF. $\square$ is a bifunctor, so we have the following calculations:

$$(1_{\mathscr{A}'} \square g)(f \square 1_{\mathscr{B}}) = (1_{\mathscr{A}'} f) \square (g 1_{\mathscr{B}}) = f \square g$$

and

$$(f \square 1_{\mathscr{B}'})(1_{\mathscr{A}} \square g) = (f 1_{\mathscr{A}}) \square (1_{\mathscr{B}'} g) = f \square g.$$

Next, for the two examples of Section 3 we define some special codings.

DEFINITION. Let $\mathscr{A}$ and $\mathscr{B}$ be objects in $\mathscr{C}$, which is either $\langle \mathbf{Dom}, \dot{\cup} \rangle$ or $\langle \mathbf{Sdom}, \times \rangle$. We define morphisms

$$i = i_{\mathscr{A},\mathscr{B}} : \mathscr{A} \to \mathscr{A} \square \mathscr{B}, \qquad j = j_{\mathscr{A},\mathscr{B}} : \mathscr{B} \to \mathscr{A} \square \mathscr{B}$$

as follows.

I. If $\mathscr{C} = \langle \mathbf{Dom}, \dot{\cup} \rangle$ then

$$i(v) = \{\langle v, 0 \rangle\} \quad v \in \mathscr{A},$$

$$j(v) = \{\langle v, 1 \rangle\} \quad v \in \mathscr{B}.$$

II. If $\mathscr{C} = \langle \mathbf{Sdom}, \times \rangle$ then

$$i(v) = \{v\} \times \mathscr{B} \quad v \in \mathscr{A},$$

$$j(v) = \mathscr{A} \times \{v\} \quad v \in \mathscr{B}.$$

These codings have the following naturality features.

PROPOSITION 4.2. *Let* $\mathscr{C}$ *be one of* $\langle \mathbf{Dom}, \dot{\cup} \rangle$ *or* $\langle \mathbf{Sdom}, \times \rangle$. *Suppose* $f : \mathscr{A} \to \mathscr{A}'$ *and* $g : \mathscr{B} \to \mathscr{B}'$ *are morphisms in* $\mathscr{C}$. *The following diagrams commute:*

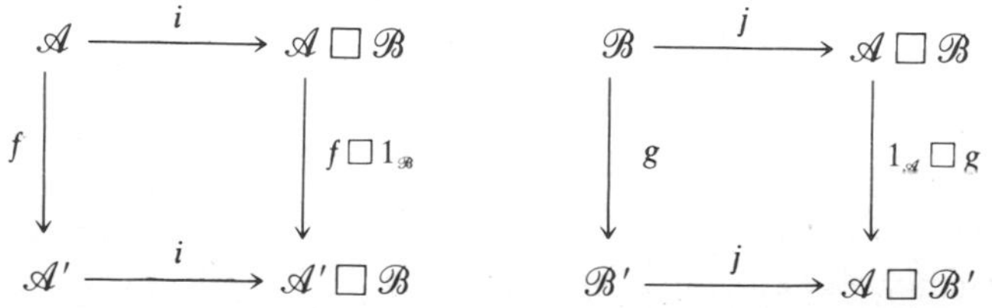

PROOF. Left to the reader.

However, in trying to extend the above proposition, $\dot\cup$ and $\times$ separate.

PROPOSITION 4.3. *In* $\langle$**Dom**$, \dot\cup\rangle$ *the following diagram commutes*:

$$\begin{array}{ccccc}
\mathcal{A} & \xrightarrow{\ i\ } & \mathcal{A}\,\dot\cup\,\mathcal{B} & \xleftarrow{\ j\ } & \mathcal{B} \\
{\scriptstyle f}\downarrow & & \downarrow{\scriptstyle f\dot\cup g} & & \downarrow{\scriptstyle g} \\
\mathcal{A}' & \xrightarrow{\ i\ } & \mathcal{A}'\,\dot\cup\,\mathcal{B}' & \xleftarrow{\ j\ } & \mathcal{B}'
\end{array}$$

PROOF. Left to the reader.

EXAMPLE. In $\langle$**Sdom**$, \times\rangle$, the following diagram does *not*, in general, commute:

$$\begin{array}{ccccc}
\mathcal{A} & \xrightarrow{\ i\ } & \mathcal{A}\times\mathcal{B} & \xleftarrow{\ j\ } & \mathcal{B} \\
{\scriptstyle f}\downarrow & & \downarrow{\scriptstyle f\times g} & & \downarrow{\scriptstyle g} \\
\mathcal{A}' & \xrightarrow{\ i\ } & \mathcal{A}'\times\mathcal{B}' & \xleftarrow{\ j\ } & \mathcal{B}'
\end{array}$$

Consider the following example. Let $\mathcal{A} = \mathcal{A}' = \{1\}$; $f = 1_{\mathcal{A}}$; $\mathcal{B} = \{2\}$; $\mathcal{B}' = \{2, 3\}$; and define $g : \mathcal{B} \to \mathcal{B}'$ by $g(2) = \{2\}$. Then

$$(f \times g)i_{\mathcal{A},\mathcal{B}}(1) = (f \times g)(\langle 1, 2\rangle) = \{\langle 1, 2\rangle\}$$

but

$$i_{\mathcal{A},\mathcal{B}'}f(1) = i_{\mathcal{A},\mathcal{B}'}(1) = \{\langle 1, 2\rangle, \langle 1, 3\rangle\}.$$

We do, however, have the following awkward substitute.

PROPOSITION 4.4. *In* $\langle$**Sdom**$, \times\rangle$:

(1) *Consider the diagram* (*which commutes by Proposition* 4.2)

$$\begin{array}{ccccc}
\mathcal{A} & \xrightarrow{\ i_{\mathcal{A},\mathcal{B}}\ } & \mathcal{A}\times\mathcal{B} & \xleftarrow{\ j_{\mathcal{A},\mathcal{B}}\ } & \mathcal{B} \\
 & & \downarrow{\scriptstyle 1_{\mathcal{A}}\times g} & & \downarrow{\scriptstyle g} \\
\mathcal{A} & \xrightarrow{\ i_{\mathcal{A},\mathcal{B}'}\ } & \mathcal{A}\times\mathcal{B}' & \xleftarrow{\ j_{\mathcal{A},\mathcal{B}'}\ } & \mathcal{B}'
\end{array}$$

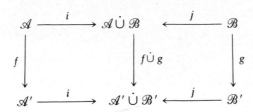

Then, for each n and for each $\mathscr{R} \subseteq \mathscr{A}^n$,

$$[(1_{\mathscr{A}} \times g)i_{\mathscr{A},\mathscr{B}}]_n(\mathscr{R}) = [j_{\mathscr{A},\mathscr{B}'}g]_n(\mathscr{B}^n) \cap [i_{\mathscr{A},\mathscr{B}'}]_n(\mathscr{R}).$$

(2) *Consider the commutative diagram*

$$\begin{array}{ccccc} \mathscr{A} & \xrightarrow{i_{\mathscr{A},\mathscr{B}}} & \mathscr{A} \times \mathscr{B} & \xleftarrow{j_{\mathscr{A},\mathscr{B}}} & \mathscr{B} \\ {\scriptstyle f}\downarrow & & \downarrow{\scriptstyle f \times 1_{\mathscr{B}}} & & \\ \mathscr{A}' & \xrightarrow{i_{\mathscr{A}',\mathscr{B}}} & \mathscr{A}' \times \mathscr{B} & \xleftarrow{j_{\mathscr{A}',\mathscr{B}}} & \mathscr{B} \end{array}$$

Then, for each n and for each $\mathscr{S} \subseteq \mathscr{B}^n$,

$$[(f \times 1_{\mathscr{B}})j_{\mathscr{A},\mathscr{B}}]_n(\mathscr{S}) = [i_{\mathscr{A}',\mathscr{B}}f]_n(\mathscr{A}^n) \cap [j_{\mathscr{A}',\mathscr{B}}]_n(\mathscr{S}).$$

PROOF. We show only (1), and to keep notation simple, suppose $n = 2$. We first note that if $\mathscr{P} \subseteq \mathscr{A}^2$,

$$(i_{\mathscr{A},\mathscr{B}})_2(\mathscr{P}) = \{\langle\langle x, y\rangle, \langle z, w\rangle\rangle \mid \langle x, z\rangle \in \mathscr{P} \text{ and } \langle y, w\rangle \in \mathscr{B}^2\}$$

and

$$(j_{\mathscr{A},\mathscr{B}})_2(\mathscr{P}) = \{\langle\langle x, y\rangle, \langle z, w\rangle\rangle \mid \langle y, w\rangle \in \mathscr{P} \text{ and } \langle x, z\rangle \in \mathscr{A}^2\}.$$

Now, for $\mathscr{R} \subseteq \mathscr{A}^2$

$$\begin{aligned} &[(1_{\mathscr{A}} \times g)i_{\mathscr{A},\mathscr{B}}]_2(\mathscr{R}) = \\ &= (1_{\mathscr{A}} \times g)_2((i_{\mathscr{A},\mathscr{B}})_2(\mathscr{R})) \\ &= (1_{\mathscr{A}} \times g)_2(\{\langle\langle x, y\rangle, \langle z, w\rangle\rangle \mid \langle x, z\rangle \in \mathscr{R} \text{ and } \langle y, w\rangle \in \mathscr{B}^2\}) \\ &= \{\langle\langle x, u\rangle, \langle z, v\rangle\rangle \mid \langle x, z\rangle \in \mathscr{R} \text{ and } \langle u, v\rangle \in g_2(\{\langle y, w\rangle\}) \\ &\qquad\qquad \text{where } \langle y, w\rangle \in \mathscr{B}^2\} \\ &= \{\langle\langle x, u\rangle, \langle z, v\rangle\rangle \mid \langle x, z\rangle \in \mathscr{R} \text{ and } \langle u, v\rangle \in g_2(\mathscr{B}^2)\} \\ &= \{\langle\langle x, u\rangle, \langle z, v\rangle\rangle \mid \langle x, z\rangle \in \mathscr{R} \text{ and } \langle u, v\rangle \in (\mathscr{B}')^2\} \\ &\quad \cap \{\langle\langle x, u\rangle, \langle z, v\rangle\rangle \mid \langle x, z\rangle \in \mathscr{A}^2 \text{ and } \langle u, v\rangle \in g_2(\mathscr{B}^2)\} \\ &= (i_{\mathscr{A},\mathscr{B}'})_2(\mathscr{R}) \cap (j_{\mathscr{A},\mathscr{B}'})_2(g_2(\mathscr{B}^2)) \\ &= [i_{\mathscr{A},\mathscr{B}'}]_2(\mathscr{R}) \cap [j_{\mathscr{A},\mathscr{B}'}g]_2(\mathscr{B}^2). \end{aligned}$$

5. Combining structures

In this section we extend our definitions of $\dot{\cup}$ and $\times$ to structures. Let $\mathscr{C}$ be one of $\langle \mathbf{Dom}, \dot{\cup} \rangle$ or $\langle \mathbf{Sdom}, \times \rangle$. As usual, we use $\Box$ for the monoidal operation of $\mathscr{C}$. Recall that for objects $\mathscr{A}$ and $\mathscr{B}$, two codings have been defined.

$$i : \mathscr{A} \to \mathscr{A} \square \mathscr{B} \quad \text{and} \quad j : \mathscr{B} \to \mathscr{A} \square \mathscr{B}.$$

DEFINITION. Let $\mathfrak{A} = \langle \mathscr{A}; \mathscr{R}_1, \ldots, \mathscr{R}_k \rangle$ and $\mathfrak{B} = \langle \mathscr{B}; \mathscr{S}_1, \ldots, \mathscr{S}_m \rangle$ be two structures, with $\mathscr{A}$ and $\mathscr{B}$ objects in $\mathscr{C}$. By $\mathfrak{A} \square \mathfrak{B}$ we mean the structure

$$\langle \mathscr{A} \square \mathscr{B}; (=_{\mathscr{A}})^i, \mathscr{R}^i_1, \ldots, \mathscr{R}^i_k, (=_{\mathscr{B}})^j, \mathscr{S}^j_1, \ldots, \mathscr{S}^j_m \rangle.$$

6. Recursion theories, for example

This section is intended to motivate what follows. We describe some very pleasant aspects of the theory assignment rec(-) and then discuss what complications must be introduced to handle other theory assignments in a similar way. The actual work involved is carried out in the next few sections.

Let $\mathscr{C}$ be one of $\langle \mathbf{Dom}, \dot{\cup} \rangle$ or $\langle \mathbf{Sdom}, \times \rangle$, with $\square$ as the monoidal operation.

The theory assignment rec(-) assigns to each structure $\mathfrak{A}$ (with its domain $\mathscr{A}$ in $\mathscr{C}$) a production system, namely the recursion theory of $\mathfrak{A}$. Now, we can define $\square$ on these recursion theories as follows: $\text{rec}(\mathfrak{A}) \square \text{rec}(\mathfrak{B})$ is to be $\text{rec}(\mathfrak{A} \square \mathfrak{B})$ where $\mathfrak{A} \square \mathfrak{B}$ was defined in the previous section. It can be shown that the $\square$ operation is well-defined in the sense that, if $\text{rec}(\mathfrak{A}) = \text{rec}(\mathfrak{A}')$ (as production systems) and $\text{rec}(\mathfrak{B}) = \text{rec}(\mathfrak{B}')$, then $\text{rec}(\mathfrak{A} \square \mathfrak{B}) = \text{rec}(\mathfrak{A}' \square \mathfrak{B}')$ also.

Suppose θ and θ' are embeddings between recursion theories, say $\theta : \text{rec}(\mathfrak{A}) \to \text{rec}(\mathfrak{B})$ and $\theta' : \text{rec}(\mathfrak{A}') \to \text{rec}(\mathfrak{B}')$. Then θ and θ' are codings, so $\theta \square \theta'$ has been defined as a coding. It can be shown that $\theta \square \theta'$ is actually an embedding, $\theta \square \theta' : \text{rec}(\mathfrak{A}) \square \text{rec}(\mathfrak{A}') \to \text{rec}(\mathfrak{B}) \square \text{rec}(\mathfrak{B}')$.

Thus we have a notion of $\square$ both for recursion theories and for embeddings. It can be shown that the category **Rec** becomes a symmetric monoidal category, using $\square$, in a natural way.

All this is a rather pleasant algebraic state of affairs. Now we ask, to what extent can this be duplicated for other theory assignments T.

In Chapter 9 we will consider theory assignments that don't make sense for *all* structures, but only for a restricted class of them. And we will not be considering *all* codings but only certain special ones. So our first modification is to work not with the objects of $\mathscr{C}$ itself, but with a subcollection $\mathscr{D}$, and likewise to restrict our attention to a subcollection $\mathscr{E}$ of all codings. We must postulate certain items about T, $\mathscr{D}$ and $\mathscr{E}$, of course, in order to get decently coherent results. In general, these postulates will be obvious for the case we begin with, rec(-) on $\mathscr{C}$ allowing all codings.

7. Combining production systems

Let $\mathscr{C}$ be one of $\langle \mathbf{Dom}, \dot{\cup} \rangle$ or $\langle \mathbf{Sdom}, \times \rangle$ and as usual $\square$ is the monoidal operation of $\mathscr{C}$. Also $\square$ is an operation on structures whose domains are in $\mathscr{C}$, as defined in Section 5. We give the following definition, which is really part of a longer definition, but this is enough to consider for the time being. The other parts will be given in the next few sections.

DEFINITION. Let $\mathscr{D}$ be a collection of structures with domains in $\mathscr{C}$, let T be a theory assignment for $\mathscr{D}$, and let $\mathscr{E}$ be a collection of codings. We say $\langle T, \mathscr{D}, \mathscr{E} \rangle$ is *object-faithful* (with respect to $\square$), if:

(1) T is elementary on $\mathscr{D}$ with respect to $\mathscr{E}$.

(2) For $\mathfrak{A}, \mathfrak{B} \in \mathscr{D}$ and for $\theta \in \mathscr{E}$, if θ is a coding from $\mathscr{A}$ to $\mathscr{B}$ then θ is a pre-embedding from $T(\mathfrak{A})$ to $T(\mathfrak{B})$.

(3) For $\mathfrak{A} \in \mathscr{D}$, $\text{inj}_{\mathscr{A}} \in \mathscr{E}$.

(4) $\mathscr{D}$ is closed under $\square$.

(5) For any $\mathfrak{A}, \mathfrak{B} \in \mathscr{D}$, $i_{\mathscr{A},\mathscr{B}}$ and $j_{\mathscr{A},\mathscr{B}}$ are in $\mathscr{E}$.

REMARK. rec(-), ω-rec(-) and f.o.(-) trivially meet these conditions if $\mathscr{D}$ is the class of all structures (all non-empty structures if $\square$ is $\times$), and $\mathscr{E}$ is the class of all codings.

LEMMA 7.1. *Let $\langle T, \mathscr{D}, \mathscr{E} \rangle$ be object faithful. Then for any $\mathfrak{A}, \mathfrak{B} \in \mathscr{D}$, both $i_{\mathscr{A},\mathscr{B}}$ and $j_{\mathscr{A},\mathscr{B}}$ are actually* embeddings,

$$i_{\mathscr{A},\mathscr{B}} : T(\mathfrak{A}) \to T(\mathfrak{A} \square \mathfrak{B}) \quad \textit{and} \quad j_{\mathscr{A},\mathscr{B}} : T(\mathfrak{B}) \to T(\mathfrak{A} \square \mathfrak{B}).$$

PROOF. $i_{\mathscr{A},\mathscr{B}}$ and $j_{\mathscr{A},\mathscr{B}}$ are in $\mathscr{E}$ by clause (5), hence they are pre-embeddings by clause (2). The result follows from the fact that T is elementary and from the definition of $\square$ on structures.

LEMMA 7.2. *Let $\langle T, \mathscr{D}, \mathscr{E} \rangle$ be object-faithful. Let $\mathfrak{A}, \mathfrak{A}', \mathfrak{B}, \mathfrak{B}' \in \mathscr{D}$. If $T(\mathfrak{A}) = T(\mathfrak{A}')$ and $T(\mathfrak{B}) = T(\mathfrak{B}')$ then $T(\mathfrak{A} \square \mathfrak{B}) = T(\mathfrak{A}' \square \mathfrak{B}')$.*

PROOF. If $T(\mathfrak{A}) = T(\mathfrak{A}')$ then $\mathscr{A} = \mathscr{A}'$. Similarly $\mathscr{B} = \mathscr{B}'$. Then $\mathscr{A} \square \mathscr{B} =$

$\mathscr{A}' \square \mathscr{B}'$. Now, let $\theta = \mathrm{inj}_{\mathscr{A}\square\mathscr{B}}$. We will show θ is an *embedding*, $\theta : T(\mathfrak{A} \square \mathfrak{B}) \to T(\mathfrak{A}' \square \mathfrak{B}')$, then since θ carries each operator to itself (and since a similar thing can be done with $\mathfrak{A}$ and $\mathfrak{A}'$, $\mathfrak{B}$ and $\mathfrak{B}'$ switched around) this will establish the lemma.

By clauses (4) and (3), $\theta \in \mathscr{E}$, and hence is a pre-embedding from $T(\mathfrak{A} \square \mathfrak{B})$ to $T(\mathfrak{A}' \square \mathfrak{B}')$ by clause (2).

Say $\mathfrak{A} = \langle \mathscr{A}; \mathscr{R}_1, \ldots, \mathscr{R}_k \rangle$ and $\mathfrak{B} = \langle \mathscr{B}; \mathscr{S}_1, \ldots, \mathscr{S}_m \rangle$. Then

$$\mathfrak{A} \square \mathfrak{B} = \langle \mathscr{A} \square \mathscr{B}; (=_{\mathscr{A}})^i, \mathscr{R}_1^i, \ldots, \mathscr{R}_k^i, (=_{\mathscr{B}})^j, \mathscr{S}_1^j, \ldots, \mathscr{S}_m^j \rangle.$$

Let $\mathscr{R}$ be one of $=_{\mathscr{A}}, \mathscr{R}_1, \ldots, \mathscr{R}_k$; we show $(\mathscr{R}^i)^\theta$ is generated in $T(\mathfrak{A}' \square \mathfrak{B}')$. A similar thing can be shown for $(=_{\mathscr{B}})^j, \mathscr{S}_1^j, \ldots, \mathscr{S}_m^j$.

Now, $\mathscr{R}$ is generated in $T(\mathfrak{A})$ (since it is a production system *on* $\mathfrak{A}$). Then $\mathscr{R}$ is generated in $T(\mathfrak{A}')$ since $T(\mathfrak{A}) = T(\mathfrak{A}')$. So by Lemma 7.1, $\mathscr{R}^i$ is generated in $T(\mathfrak{A}' \square \mathfrak{B}')$. But $(\mathscr{R}^i)^\theta = \mathscr{R}^i$.

So θ takes the "given" relations of $\mathfrak{A} \square \mathfrak{B}$ to generated relations of $T(\mathfrak{A}' \square \mathfrak{B}')$. Also θ takes $=_{\mathscr{A}\square\mathscr{B}}$ to $=_{\mathscr{A}\square\mathscr{B}}$ which is also the equality relation for $T(\mathfrak{A}' \square \mathfrak{B}')$, hence is generated. Then, since T is elementary, θ is an embedding.

DEFINITION. Let $\langle T, \mathscr{D}, \mathscr{E} \rangle$ be object-faithful. For $\mathfrak{A}, \mathfrak{B} \in \mathscr{D}$, by $T(\mathfrak{A}) \square T(\mathfrak{B})$ we mean $T(\mathfrak{A} \square \mathfrak{B})$.

We conclude with some examples. In these, $T(\text{-})$ is rec(-), $\mathscr{D}$ is the collection of all structures, and $\mathscr{E}$ is the collection of all embeddings.

EXAMPLE I. Let $\mathbb{N}'$ be the natural numbers without 0. By $\mathfrak{S}(\mathbb{N}')$ we mean $\langle \mathbb{N}', S \rangle$ where S is the successor relation on $\mathbb{N}'$. Now, $\mathrm{rec}(\mathfrak{S}(\mathbb{N})) \times \mathrm{rec}(\mathfrak{S}(\mathbb{N}'))$ is defined, and can be thought of as a recursion theory of *fractions*. The domain is the collection of ordered pairs $\langle x, y \rangle$ where x and y are natural numbers with $y \neq 0$. The "given" relations are: successor in the first component; and successor in the second component.

EXAMPLE II. $\mathfrak{S}(\mathbb{N}')$ is as in the previous example. $\mathrm{rec}(\mathfrak{S}(\mathbb{N})) \dot{\cup} \mathrm{rec}(\mathfrak{S}(\mathbb{N}'))$ can be thought of as a recursion theory on the *signed integers*.

EXAMPLE III. Let $\mathbb{R}$ be the real numbers. By $\mathfrak{S}(\mathbb{R})$ we mean the structure $\langle \mathbb{R}; +, \times, > \rangle$. Then $\mathrm{rec}(\mathfrak{S}(\mathbb{R})) \times \mathrm{rec}(\mathfrak{S}(\mathbb{R}))$ may be thought of as a recursion theory on the complex numbers.

8. Combining embeddings

We have now defined $\dot{\cup}$ and $\times$ between certain production systems. Our goal in the present section is to define related notions for embeddings. Recall that in Section 2, $\dot{\cup}$ and $\times$ were defined for *codings*; what we show now is that, at least in some circumstances, these same definitions work for *embeddings*.

Once again, $\mathscr{C}$ is one of $\langle \mathbf{Dom}, \dot{\cup} \rangle$ or $\langle \mathbf{Sdom}, \times \rangle$ and $\Box$ is $\dot{\cup}$ or $\times$ as the case may be.

DEFINITION. Let $\mathscr{D}$ be a collection of structures with domains in $\mathscr{C}$, let T be a theory assignment for $\mathscr{D}$, and let $\mathscr{E}$ be a collection of codings. We say $\langle T, \mathscr{D}, \mathscr{E} \rangle$ is *morphism faithful* (with respect to $\Box$), if:

(1) $\langle T, \mathscr{D}, \mathscr{E} \rangle$ is object-faithful.

(2) $\mathscr{E}$ is closed under $\Box$, as defined in Section 2.

(3) $\mathscr{E}$ is closed under composition.

REMARK. Again, trivially, rec(-), ω-rec(-) and f.o.(-) are all morphism-faithful using the collection of all structures (all non-empty structures if $\Box$ is $\times$) and the collection of all codings.

LEMMA 8.1. *Let $\langle T, \mathscr{D}, \mathscr{E} \rangle$ be object faithful. Let $\mathfrak{A}, \mathfrak{B}, \mathfrak{C} \in \mathscr{D}$ and suppose there is a coding $\theta : \mathscr{A} \Box \mathscr{B} \rightarrow \mathscr{C}$ with $\theta \in \mathscr{E}$. Then θ is actually an* embedding, $\theta : T(\mathfrak{A} \Box \mathfrak{B}) \rightarrow T(\mathfrak{C})$ *provided each of the following holds*:

(1) *For each relation $\mathscr{R} \in [\mathscr{A}]^n$ which is generated in $T(\mathfrak{A})$, $\theta_n(\mathscr{R}^i)$ is generated in $T(\mathfrak{C})$.*

(2) *For each relation $\mathscr{S} \in [\mathscr{B}]^n$ which is generated in $T(\mathfrak{B})$, $\theta_n(\mathscr{S}^j)$ is generated in $T(\mathfrak{C})$.*

PROOF. Suppose θ meets the conditions. First we show $\theta_2(=_{\mathscr{A}\Box\mathscr{B}})$ is generated in $T(\mathfrak{C})$.

Case 1: $\Box$ is $\dot{\cup}$. Then $=_{\mathscr{A}\Box\mathscr{B}}$ is $=_{\mathscr{A}\dot{\cup}\mathscr{B}}$ which is $(=_{\mathscr{A}})^i \cup (=_{\mathscr{B}})^j$, so by Proposition 3.2.3 $\theta_2(=_{\mathscr{A}\dot{\cup}\mathscr{B}})$ is $\theta_2((=_{\mathscr{A}})^i) \cup \theta_2((=_{\mathscr{B}})^j)$. But by hypothesis, each part of this union must be generated in $T(\mathfrak{C})$, hence the union itself is.

Case 2: $\Box$ is $\times$. Then $=_{\mathscr{A}\Box\mathscr{B}}$ is $=_{\mathscr{A}\times\mathscr{B}}$ which is $(=_{\mathscr{A}})^i \cap (=_{\mathscr{B}})^j$. Then by Proposition 3.2.3 again, $\theta_2(=_{\mathscr{A}\times\mathscr{B}})$ is $\theta_2((=_{\mathscr{A}})^i) \cap \theta_2((=_{\mathscr{B}})^j)$ which must be generated in $T(\mathfrak{C})$.

The lemma now follows from the definition of $\Box$ on structures, and the fact that T is elementary.

PROPOSITION 8.2. *Let $\langle T, \mathscr{D}, \mathscr{E}\rangle$ be morphism-faithful. Let $\mathfrak{A}, \mathfrak{A}', \mathfrak{B}, \mathfrak{B}' \in \mathscr{D}$, let $\alpha, \beta \in \mathscr{E}$, and suppose $\alpha : T(\mathfrak{A}) \rightarrow T(\mathfrak{A}')$ and $\beta : T(\mathfrak{B}) \rightarrow T(\mathfrak{B}')$ are embeddings. Then the coding $\alpha \square \beta$ is actually an embedding, $\alpha \square \beta : T(\mathfrak{A} \square \mathfrak{B}) \rightarrow T(\mathfrak{A}' \square \mathfrak{B}')$.*

PROOF. $\alpha, \beta \in \mathscr{E}$, hence $\alpha \square \beta \in \mathscr{E}$, hence $\alpha \square \beta$ is a pre-embedding. We use Lemma 8.1 to show it is an embedding. We treat $\dot{\cup}$ and $\times$ separately.

Case 1: $\square$ is $\dot{\cup}$. Let $\mathscr{R} \in [\mathfrak{A}]^n$ be generated in $T(\mathfrak{A})$; we show $(\alpha \dot{\cup} \beta)_n(\mathscr{R}^i)$ is generated in $T(\mathfrak{A}' \dot{\cup} \mathfrak{B}')$. But, by Proposition 4.3, $(\alpha \dot{\cup} \beta)_n(\mathscr{R}^i) = [i_{\mathscr{A}',\mathscr{B}'}\alpha]_n(\mathscr{R})$. Now, α is an embedding, and by Lemma 7.1, so is $i_{\mathscr{A}',\mathscr{B}'}$, hence so is the composition. But by Proposition 3.4.2, embeddings take generated relations to generated relations. Thus $(\alpha \dot{\cup} \beta)_n(\mathscr{R}^i)$ is generated in $T(\mathfrak{A}' \dot{\cup} \mathfrak{B}')$. There is a similar argument for $\mathscr{S} \in [\mathscr{B}]^n$ generated in $T(\mathfrak{B})$. Then, by Lemma 8.1, $\alpha \dot{\cup} \beta$ is an embedding.

Case 2: $\square$ is $\times$. We are to show $\alpha \times \beta$ is an embedding. By Proposition 4.1, $\alpha \times \beta = (\text{inj}_{\mathscr{A}'} \times \beta)(\alpha \times \text{inj}_{\mathscr{B}})$. If we can show each of these is an embedding, the result will follow. We do this for $\alpha \times \text{inj}_{\mathscr{B}}$, the other is similar.

As a coding, $\alpha \times \text{inj}_{\mathscr{B}} : \mathscr{A} \times \mathscr{B} \rightarrow \mathscr{A}' \times \mathscr{B}$. Suppose $\mathscr{R} \in [\mathscr{A}]^n$ is generated in $T(\mathfrak{A})$. Then by Proposition 4.2, $(\alpha \times \text{inj}_{\mathscr{B}})_n(\mathscr{R}^i) = [i_{\mathscr{A}',\mathscr{B}}\alpha]_n(\mathscr{R})$ and this must be generated in $T(\mathfrak{A}' \times \mathscr{B})$ since α and i are embeddings.

Suppose $\mathscr{S} \in [\mathscr{B}]^n$ is generated in $T(\mathfrak{B})$. This time, by Proposition 4.4 (2),

$$(\alpha \times \text{inj}_{\mathscr{B}})_n(\mathscr{S}^j) = [i_{\alpha',\mathscr{B}}\alpha]_n(\mathscr{A}^n) \cap [j_{\mathscr{A}',\mathscr{B}}]_n(\mathscr{S}).$$

Now both parts of this intersection must be generated in $T(\mathfrak{A}' \times \mathfrak{B})$ since i, j and α are embeddings, and $\mathscr{S}$ and $\mathscr{A}^n$ are generated. And generated relations are closed under $\cap$.

Now $\alpha \times \text{inj}_{\mathscr{B}}$ is an embedding, $\alpha \times \text{inj}_{\mathscr{B}} : T(\mathfrak{A} \times \mathfrak{B}) \rightarrow T(\mathfrak{A}' \times \mathfrak{B})$, by Lemma 8.1. This concludes the proof.

If $\langle T, \mathscr{D}, \mathscr{E}\rangle$ is morphism-faithful, we have now defined $\square$ both for the production systems T yields on $\mathscr{D}$, and for the embeddings between them which, as codings, are in $\mathscr{E}$. Our definitions are such that, for $\mathfrak{A}, \mathfrak{A}', \mathfrak{B}, \mathfrak{B}' \in \mathscr{D}$ and $\alpha, \beta \in \mathscr{E}$, if $\alpha : T(\mathfrak{A}) \rightarrow T(\mathfrak{A}')$ and $\beta : T(\mathfrak{B}) \rightarrow T(\mathfrak{B}')$ are embeddings, then so is

$$\alpha \square \beta : T(\mathfrak{A}) \square T(\mathfrak{B}) \rightarrow T(\mathfrak{A}') \square T(\mathfrak{B}').$$

9. Monoidal subcategories in Prod

As usual, $\mathscr{C}$ is one of $\langle \mathbf{Dom}, \dot{\cup} \rangle$ or $\langle \mathbf{Sdom}, \times \rangle$, and $\square$ is the monoidal operation. It is in this section that we "complete" the sequence of definitions started in Section 7.

Let $\langle T, \mathscr{D}, \mathscr{E} \rangle$ be morphism-faithful, as in Section 8. We define an associated category as follows.

DEFINITION. $C(T, \mathscr{D}, \mathscr{E})$ is the category with objects all production systems $T(\mathfrak{A})$ where $\mathfrak{A} \in \mathscr{D}$, and as morphisms, those embeddings between the objects which, as codings, are in $\mathscr{E}$.

It is easy to check that $C(T, \mathscr{D}, \mathscr{E})$ is a category, in fact, a subcategory of **Prod**. What we do in the present section is postulate enough about $\langle T, \mathscr{D}, \mathscr{E} \rangle$ to make $C(T, \mathscr{D}, \mathscr{E})$ into a *symmetric monoidal category.*

First, some preliminaries. Let e be the unit object of $\mathscr{C}$. That is, if $\mathscr{C}$ is $\langle \mathbf{Dom}, \dot{\cup} \rangle$, $e = \emptyset$, and if $\mathscr{C}$ is $\langle \mathbf{Sdom}, \times \rangle$, $e = \{0\}$. Then $\langle e \rangle$ is a structure which will serve as a unit structure.

Recall, there are certain natural isomorphisms in $\mathscr{C}$:

(a) For each $\mathscr{A}, \mathscr{B}, \mathscr{C} \in \mathscr{C}$ a reassociating map

$$\alpha_{\mathscr{A},\mathscr{B},\mathscr{C}} : \mathscr{A} \square (\mathscr{B} \square \mathscr{C}) \rightarrow (\mathscr{A} \square \mathscr{B}) \square \mathscr{C}.$$

(b) For each $\mathscr{A}, \mathscr{B} \in \mathscr{C}$ a commuting map

$$\gamma_{\mathscr{A},\mathscr{B}} : \mathscr{A} \square \mathscr{B} \rightarrow \mathscr{B} \square \mathscr{A}.$$

(c) For each $\mathscr{A} \in \mathscr{C}$, maps

$$\lambda_{\mathscr{A}} : e \square \mathscr{A} \rightarrow \mathscr{A} \quad \text{and} \quad \rho_{\mathscr{A}} : \mathscr{A} \square e \rightarrow \mathscr{A}.$$

DEFINITION. Let $\mathscr{D}$ be a collection of structures with domains in $\mathscr{C}$, let T be a theory assignment for $\mathscr{D}$, and let $\mathscr{E}$ be a collection of codings. We say $\langle T, \mathscr{D}, \mathscr{E} \rangle$ is *faithful*, if:

(1) $\langle T, \mathscr{D}, \mathscr{E} \rangle$ is morphism-faithful.

(2) The unit structure, $\langle e \rangle$ is in $\mathscr{D}$.

(3) For $\mathfrak{A}, \mathfrak{B}, \mathfrak{C} \in \mathscr{D}$ we have

$$\alpha_{\mathscr{A},\mathscr{B},\mathscr{C}} \in \mathscr{E}, \qquad \gamma_{\mathscr{A},\mathscr{B}} \in \mathscr{E}, \qquad \lambda_{\mathscr{A}}, \rho_{\mathscr{A}} \in \mathscr{E}.$$

THEOREM 9.1. *Let $\langle T, \mathscr{D}, \mathscr{E} \rangle$ be faithful. Then $C(T, \mathscr{D}, \mathscr{E})$ is a symmetric monoidal category.*

PROOF. Actually, all that is left to show is that α, λ, ρ and γ are embeddings, hence morphisms of the category in question. We show this for γ, and sketch it for the rest.

Let $\mathfrak{A}, \mathfrak{B} \in \mathcal{D}$. Then $\gamma_{\mathcal{A},\mathcal{B}} : \mathcal{A} \square \mathcal{B} \to \mathcal{B} \square \mathcal{A}$ is in $\mathcal{E}$. We use Lemma 8.1 to show it is an embedding.

Let $\mathcal{R} \in [\mathcal{A}]^n$ be generated in $T(\mathfrak{A})$; we must show $\gamma_n(\mathcal{R}^i)$ is generated in $T(\mathfrak{B} \square \mathfrak{A})$. Now it is easy to check that, whether $\square$ is $\dot{\cup}$ or $\times$,

$$\gamma_{\mathcal{A},\mathcal{B}} i_{\mathcal{A},\mathcal{B}} = j_{\mathcal{B},\mathcal{A}}.$$

Then $\gamma_n(\mathcal{R}^i) = \mathcal{R}^{j_{\mathcal{B},\mathcal{A}}}$. This is one of the given relations of the structure $\mathfrak{B} \square \mathfrak{A}$, hence is generated in $T(\mathfrak{B} \square \mathfrak{A})$ since this is a production system *on* the structure $\mathfrak{B} \square \mathfrak{A}$.

If $\mathcal{S} \in [\mathcal{B}]^n$ is generated in $T(\mathfrak{B})$ we proceed similarly, using

$$\gamma_{\mathcal{A},\mathcal{B}} j_{\mathcal{A},\mathcal{B}} = i_{\mathcal{B},\mathcal{A}}.$$

Then by Lemma 8.1, γ must be an embedding.

To show α is an embedding, use

$$\alpha_{\mathcal{A},\mathcal{B},\mathcal{C}} i_{\mathcal{A},\mathcal{B} \square \mathcal{C}} = i_{\mathcal{A} \square \mathcal{B},\mathcal{C}} i_{\mathcal{A},\mathcal{B}},$$

$$\alpha_{\mathcal{A},\mathcal{B},\mathcal{C}} j_{\mathcal{A},\mathcal{B} \square \mathcal{C}} i_{\mathcal{B},\mathcal{C}} = i_{\mathcal{A} \square \mathcal{B},\mathcal{C}} j_{\mathcal{A},\mathcal{B}}$$

and

$$\alpha_{\mathcal{A},\mathcal{B},\mathcal{C}} j_{\mathcal{A},\mathcal{B} \square \mathcal{C}} j_{\mathcal{B},\mathcal{C}} = j_{\mathcal{A} \square \mathcal{B},\mathcal{C}}.$$

λ is similar, but simpler. Finally, $\rho_{\mathcal{B}} = \lambda_{\mathcal{B}} \gamma_{\mathcal{B},e}$, so ρ is an embedding too.

SUMMARY. For convenience, we collect together the definition of faithful.

DEFINITION. Let $\mathcal{D}$ be a collection of structures with domains in $\mathcal{C}$, let T be a theory assignment for $\mathcal{D}$ and let $\mathcal{E}$ be a collection of codings. $\langle T, \mathcal{D}, \mathcal{E} \rangle$ is *faithful* (with respect to $\square$), if:

(1) T is elementary on $\mathcal{D}$ with respect to $\mathcal{E}$.

(2) For $\mathfrak{A}, \mathfrak{B} \in \mathcal{D}$ and $\theta \in \mathcal{E}$, if $\theta : \mathcal{A} \to \mathcal{B}$ is a coding, then θ is a pre-embedding from $T(\mathfrak{A})$ to $T(\mathfrak{B})$.

(3) For $\mathfrak{A} \in \mathcal{D}$, $\mathrm{inj}_{\mathcal{A}} \in \mathcal{E}$.

(4) $\mathcal{D}$ is closed under $\square$.

(5) For $\mathfrak{A}, \mathfrak{B} \in \mathcal{D}$, both $i_{\mathcal{A},\mathcal{B}}$ and $j_{\mathcal{A},\mathcal{B}}$ are in $\mathcal{E}$.

(6) $\mathcal{E}$ is closed under $\square$.

(7) $\mathcal{E}$ is closed under composition.

(8) The unit structure, $\langle e \rangle$, is in $\mathcal{D}$.

(9) For $\mathfrak{A}, \mathfrak{B}, \mathfrak{C} \in \mathcal{D}$

$$\alpha_{\mathcal{A},\mathcal{B},\mathcal{C}} \in \mathcal{E}, \qquad \gamma_{\mathcal{A},\mathcal{B}} \in \mathcal{E}, \qquad \lambda_{\mathcal{A}}, \rho_{\mathcal{A}} \in \mathcal{E}.$$

Then, if $\langle T, \mathcal{D}, \mathcal{E}\rangle$ is faithful, $C(T, \mathcal{D}, \mathcal{E})$ is a symmetric monoidal category.

10. Separability

Let $\mathcal{C}$ be one of $\langle \mathbf{Dom}, \dot{\cup}\rangle$ or $\langle \mathbf{Sdom}, \times\rangle$. Let $\mathcal{D}$ be a collection of structures with domains in $\mathcal{C}$, let T be a theory assignment for $\mathcal{D}$, and let $\mathcal{E}$ be a collection of codings. For this section, we suppose $\langle T, \mathcal{D}, \mathcal{E}\rangle$ is *faithful* and we investigate the symmetric monoidal category $C(T, \mathcal{D}, \mathcal{E})$.

In monoidal categories, objects called *monoids* are often singled out for study. See MacLane [1971], pp. 166–167. Here, however, that notion is too narrow to be of interest. We replace it by a weaker one, that of being a *separable* production system in $C(T, \mathcal{D}, \mathcal{E})$.

DEFINITION. Let $\mathbf{A}$ be a production system in $C(T, \mathcal{D}, \mathcal{E})$. We call f a *separation morphism* for $\mathbf{A}$, if:

(1) f is a morphism, $f : \mathbf{A} \square \mathbf{A} \to \mathbf{A}$.

(2) The relations $y \in (fi)(x)$ and $y \in (fj)(x)$ are both generated in $\mathbf{A}$, where

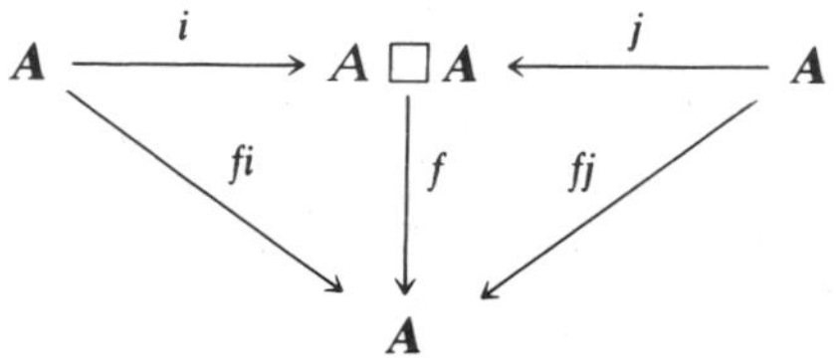

We call $\mathbf{A}$ *separable* in $C(T, \mathcal{D}, \mathcal{E})$ if $\mathbf{A}$ has a separation morphism.

In the next few sections we establish some general results about separability. Here we prove a few theorems which make it easy to give examples.

DEFINITION. $\mathbf{A}$ has an *effective pairing function* if there is a 1–1 function $J : \mathcal{A} \times \mathcal{A} \to \mathcal{A}$ which is generated in $\mathbf{A}$.

REMARK. In the proof below, and in several others in this chapter, we assume the generated relations are closed under the logical operations $\wedge$, $\vee$ and $\exists$. This follows easily from the production system axioms, or see Ch. 7, §1.

THEOREM 10.1. *Suppose* $\square$ *is* $\times$, *T assigns* total *production systems to structures, and* $\mathscr{E}$ *contains all codings. Then, for* $\mathfrak{A} \in \mathfrak{D}$, *if* $T(\mathfrak{A})$ *has an effective pairing function, then* $T(\mathfrak{A})$ *is separable.*

PROOF. Define f by $f(\langle x, y\rangle) = \{J(x, y)\}$. Since J is 1–1, f is a *coding* from $\mathscr{A} \times \mathscr{A}$ to $\mathscr{A}$. Since T assigns total production systems, f is automatically a pre-embedding from $T(\mathfrak{A} \times \mathfrak{A}) = T(\mathfrak{A}) \times T(\mathfrak{A})$ *to* $T(\mathfrak{A})$.

We next show f is an *embedding*, $f: T(\mathfrak{A} \times \mathfrak{A}) \to T(\mathfrak{A})$.

Suppose $\mathscr{R} \subseteq \mathscr{A}^n$ is generated in $T(\mathfrak{A})$; we show $f_n(\mathscr{R}^i)$ is generated in $T(\mathfrak{A})$ also. Similarly for $f_n(\mathscr{R}^j)$. Then, that f is an embedding follows from Lemma 8.1. To keep notation simple, say $n = 2$. Then

$$\mathscr{R}^i = \{\langle\langle x, y\rangle, \langle z, w\rangle\rangle \mid \langle x, z\rangle \in \mathscr{R} \text{ and } y, w \in \mathscr{A}\}.$$

So

$$\begin{aligned} f_2(\mathscr{R}^i) &= \{\langle a, b\rangle \mid a = J(x, y) \\ &\qquad b = J(z, w), \text{ and } \langle\langle x, y\rangle, \langle z, w\rangle\rangle \in \mathscr{R}^i\} \\ &= \{\langle a, b\rangle \mid a = J(x, y), \\ &\qquad b = J(z, w), \langle x, y\rangle \in \mathscr{R} \text{ and } y, w \in \mathscr{A}\}. \end{aligned}$$

Thus

$$\begin{aligned} &\langle a, b\rangle \in f_2(\mathscr{R}^i) \\ &\Leftrightarrow (\exists x, y, z, w)[\langle x, z\rangle \in \mathscr{R} \wedge a = J(x, y) \wedge b = J(z, w)]. \end{aligned}$$

Now, $\mathscr{R}$ and J are both generated in $T(\mathfrak{A})$, and it follows that this is also generated in $T(\mathfrak{A})$.

Thus f is an embedding. Finally we show fi and fj are both generated in $T(\mathfrak{A})$. We show this for fi, the other is similar. Now,

$$\begin{aligned} &y \in (fi)(x) \Leftrightarrow \\ &\Leftrightarrow y \in f(z) \quad \text{for some } z \in i(x) \\ &\Leftrightarrow y \in f(\langle x, a\rangle) \quad \text{for some } \langle x, a\rangle \in i(x) \\ &\Leftrightarrow y = J(x, a) \quad \text{for some } \langle x, a\rangle \in i(x) \\ &\Leftrightarrow y = J(x, a) \quad \text{for some } a \in \mathscr{A} \\ &\Leftrightarrow (\exists a)[y = J(x, a)]. \end{aligned}$$

But J is generated in $T(\mathfrak{A})$ and we have closure under $\exists$, that is, under projections.

REMARK. Actually, all that is needed for the above proof to work is that $f(\langle x, y\rangle) = \{J(x, y)\}$ should be in $\mathscr{E}$.

EXAMPLES. We give three examples of recursion theories having effective pairing functions, and one of an ω-recursion theory. Then, by the above theorem, these are separable production systems (under $\times$).

I. There are many well-known pairing functions on the natural numbers, $\mathbb{N}$. The oldest occurs in Georg Cantor [1895], Section 6 and is (correcting for the difference that his natural numbers start at 1 while ours start at 0)

$$J(x, y) = x + \tfrac{1}{2}(x + y + 1)(x + y).$$

It is easy to check that this is generated in $\mathrm{rec}(\mathfrak{S}(\mathbb{N}))$.

II. There are also several pairing functions on the hereditarily finite sets, L_ω. The oldest is due to Norbert Wiener in Wiener [1912], and is

$$J(x, y) = \{\{\{x\}, \emptyset\}, \{\{y\}\}\}.$$

A more common one today is due to Casimir Kuratowski in Kuratowski [1921] and is

$$J(x, y) = \{\{x\}, \{x, y\}\}.$$

Either of these is easily shown to be generated in $\mathrm{rec}(\mathfrak{S}(L_\omega))$.

III. There are also pairing functions on words, $\mathscr{W}(a_1, \ldots, a_n)$. They tend to be more complicated to state. We will develop one in Ch. 6, §6, based on Quine [1946]. Another example, coming out of computer science, may be found in Even and Rodeh [1978]. Relying on these, $\mathrm{rec}(\mathfrak{S}(a_1, \ldots, a_n))$ is also an example of a production system with an effective pairing function.

IV. Let $\mathfrak{S}(\mathbb{R})$ be the structure $\langle \mathbb{R}; +, \times, > \rangle$ where $\mathbb{R}$ is the set of real numbers and $+$, $\times$ and $>$ are the usual addition, multiplication and order for them. There are many effective pairing functions in ω-rec$(\mathfrak{S}(\mathbb{R}))$. Once again, the oldest is due to Cantor, see Cantor [1878]; his example of a map between the line and the plane, is, in fact, ω-r.e. in ω-rec$(\mathfrak{S}(\mathbb{R}))$. His map is not the one customary today; the modern one "meshes" decimal representations. It is somewhat easier to check that it, too, is effective in ω-rec$(\mathfrak{S}(\mathbb{R}))$. Finally, the various space-filling curves are generally effective pairing functions; the first such may be found in Peano [1890].

THEOREM 10.2. *Suppose* $\square$ *is* $\dot{\cup}$, *T assigns* total *production systems to structures, and* $\mathscr{C}$ *contains all codings. Then, for* $\mathfrak{A} \in \mathscr{D}$, $T(\mathfrak{A})$ *is separable if and only if there are two codings* $\mu, \nu : \mathscr{A} \to \mathscr{A}$ *with* $\mathscr{A}^\mu$ *and* $\mathscr{A}^\nu$ *disjoint, and the relations* $y \in \mu(x)$ *and* $y \in \nu(x)$ *both generated in* $T(\mathfrak{A})$.

PROOF. First, suppose $T(\mathfrak{A})$ is separable, with f as separation morphism. Set $\mu = fi$ and $\nu = fj$, and the appropriate conditions are satisfied.

Second, suppose there are codings $\mu, \nu : \mathscr{A} \to \mathscr{A}$, both generated in $T(\mathfrak{A})$, with $\mathscr{A}^{\mu}$ and $\mathscr{A}^{\nu}$ disjoint. We produce a separation morphism f for $T(\mathfrak{A})$.

Define f on $\mathscr{A} \dot{\cup} \mathscr{A}$ by $f(\langle x, 0\rangle) = \mu(x)$ and $f(\langle x, 1\rangle) = \nu(x)$. Since $\mathscr{A}^{\mu}$ and α^{ν} are disjoint, f is a coding.

By definition of f, we have $\mu(x) = fi(x)$ and $\nu(x) = fj(x)$, so both fi and fj are generated in $T(\mathfrak{A})$.

Since T assigns total production systems, f is a pre-embedding from $T(\mathfrak{A} \dot{\cup} \mathfrak{A})$ to $T(\mathfrak{A})$. All that remains is to show f is an embedding.

Suppose $\mathscr{R} \subseteq \mathscr{A}^n$ is generated in $T(\mathfrak{A})$. Then $f_n(\mathscr{R}^i) = f_n i_n(\mathscr{R}) = \mu_n(\mathscr{R})$, and this is generated in $T(\mathfrak{A})$ since $\langle y_1, \dots, y_n\rangle \in \mu_n(\mathscr{R}) \Leftrightarrow (\exists x_1, \dots, x_n)[\langle x_1, \dots, x_n\rangle \in \mathscr{R} \wedge y_1 \in \mu(x_1) \wedge \cdots \wedge y_n \in \mu(x_n)]$ and μ and $\mathscr{R}$ are generated. Similarly $f_n(\mathscr{R}^j)$ is generated in $T(\mathfrak{A})$. That $f : T(\mathfrak{A}) \dot{\cup} T(\mathfrak{A}) \to T(\mathfrak{A})$ is an embedding now follows from Lemma 8.1.

REMARK. Again, all that is needed for the above proof to work is that the map f defined from μ and ν be in $\mathscr{E}$.

EXAMPLES. The recursion theories of the following structures are separable under $\dot{\cup}$.

I. Numbers $\mathfrak{S}(\mathbb{N})$. Take $\mu(x) = \{2x\}$ and $\nu(x) = \{2x + 1\}$.

II. Sets $\mathfrak{S}(L_{\omega})$. Take $\mu(x) = \{\langle x, 0\rangle\}$ and $\nu(x) = \{\langle x, 1\rangle\}$.

III. Words, two or more letters $\mathfrak{S}(a_1, \dots, a_n)$. Take $\mu(x) = \{x * a_1\}$ and $\nu(x) = \{x * a_2\}$.

IV. Words, one letter $\mathfrak{S}(a)$. Take $\mu(x) = \{x * x\}$ and $\nu(x) = \{x * x * a\}$.

Actually, separability under $\dot{\cup}$ and $\times$ are not independent, and the above examples are consequences of the one earlier in the section, using the following.

THEOREM 10.3. *Suppose T assigns total production systems to structures, and $\mathscr{E}$ contains all codings. Then for $\mathfrak{A} \in \mathscr{E}$, if $T(\mathfrak{A})$ is separable using $\times$ then $T(\mathfrak{A})$ is separable using $\dot{\cup}$.*

PROOF. Suppose f is a separation morphism for $T(\mathfrak{A})$, under $\times$.

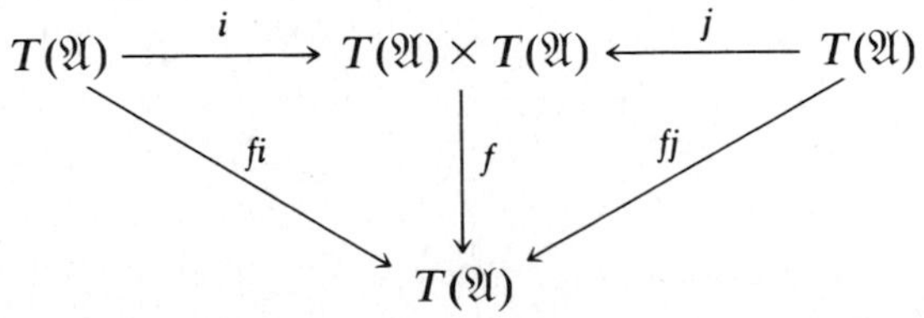

fi and fj both generated in $T(\mathfrak{A})$.

Since $f : \mathscr{A} \times \mathscr{A} \to \mathscr{A}$ is a coding, cardinality considerations give us that $\mathscr{A}$ can not be finite. Choose two members of $\mathscr{A}$, say a and b. Define maps μ and ν by

$$y \in \mu(x) \Leftrightarrow y \in (fi)(x) \quad \text{and} \quad y \in (fj)(a),$$

$$y \in \nu(x) \Leftrightarrow y \in (fi)(x) \quad \text{and} \quad y \in (fj)(b).$$

Since fi is a coding of $\mathscr{A}$ into $\mathscr{A}$, it follows that μ and ν are also codings of $\mathscr{A}$ into $\mathscr{A}$. Since fj is a coding of $\mathscr{A}$ into $\mathscr{A}$, it follows that $\mathscr{A}^{\mu}$ and $\mathscr{A}^{\nu}$ are disjoint. Finally, fi and fj are generated in $T(\mathfrak{A})$ and it follows that μ and ν are also. Now use Theorem 10.2.

11. A few results about separability

As usual, $\square$ is $\dot{\cup}$ or $\times$ and $\langle T, \mathscr{D}, \mathscr{E} \rangle$ is faithful.

PROPOSITION 11.1. *Let* $\mathbf{A}, \mathbf{B}, \mathbf{C} \in C(T, \mathscr{D}, \mathscr{E})$, *and suppose* $\mathbf{C}$ *is separable. If each of* $\mathbf{A}$ *and* $\mathbf{B}$ *has a morphism to* $\mathbf{C}$, *so does* $\mathbf{A} \square \mathbf{B}$.

PROOF. Let $f : \mathbf{C} \square \mathbf{C} \to \mathbf{C}$ be a separation morphism for $\mathbf{C}$, and let $g : \mathbf{A} \to \mathbf{C}$ and $h : \mathbf{B} \to \mathbf{C}$ be embeddings. Then $f(g \square h) : \mathbf{A} \square \mathbf{B} \to \mathbf{C}$ is an embedding.

PROPOSITION 11.2. *Let* $\mathbf{A}, \mathbf{B} \in C(T, \mathscr{D}, \mathscr{E})$. *If* $\mathbf{A}$ *and* $\mathbf{B}$ *are both separable, so is* $\mathbf{A} \square \mathbf{B}$.

PROOF. Suppose f is a separation morphism for $\mathbf{A}$, and g for $\mathbf{B}$. Then we have

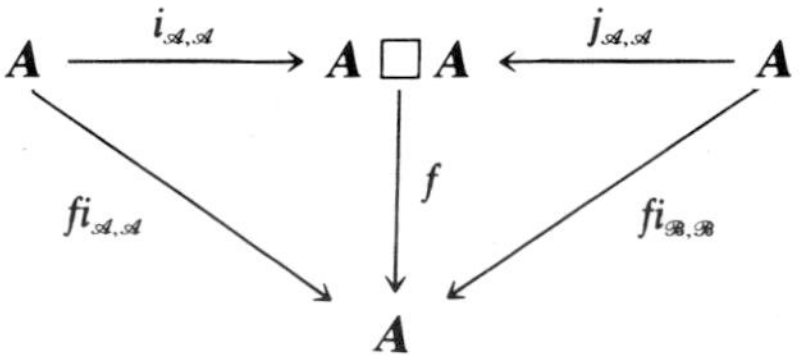

$fi_{\mathscr{A},\mathscr{A}}$ and $fj_{\mathscr{B},\mathscr{B}}$ both generated in $\mathbf{A}$, and

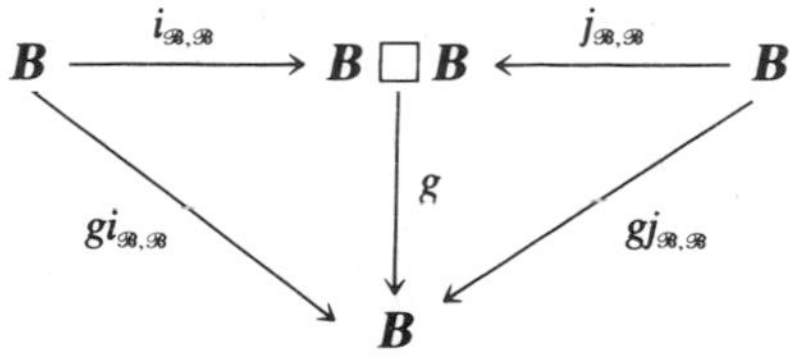

$gi_{\mathscr{B},\mathscr{B}}$ and $gj_{\mathscr{B},\mathscr{B}}$ both generated in $\boldsymbol{B}$. We *want* an appropriate embedding h:

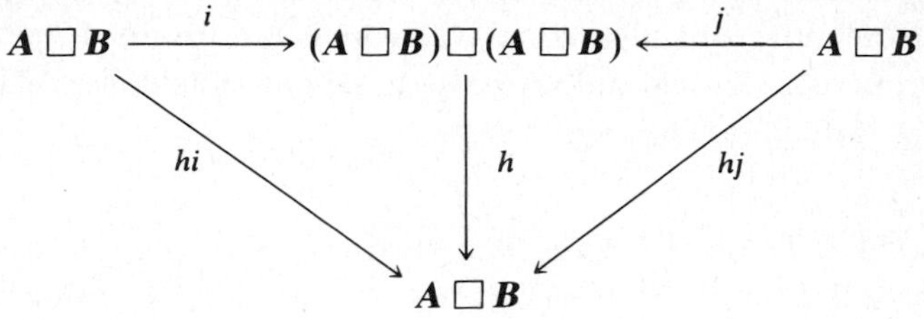

(where we omit subscripts on i and j, which are horrifying).

Let π be the obvious isomorphism

$$\pi : (\mathscr{A} \,\square\, \mathscr{B}) \,\square\, (\mathscr{C} \,\square\, \mathscr{D}) \cong (\mathscr{A} \,\square\, \mathscr{C}) \,\square\, (\mathscr{B} \,\square\, \mathscr{D})$$

compounded appropriately out of the reassociating maps and the commuting maps. Now set

$$h = (f \,\square\, g)\pi.$$

Then h is a composition of embeddings from $\mathscr{E}$ (Theorem 9.1 enters here); hence h is an embedding from $\mathscr{E}$. We must show hi and hj are generated. But it is easy to see that

$$i = \pi(i_{\mathscr{A},\mathscr{A}} \,\square\, i_{\mathscr{B},\mathscr{B}}) \quad \text{and} \quad \pi\pi = 1.$$

So

$$hi = (f \,\square\, g)\pi\pi(i_{\mathscr{A},\mathscr{A}} \,\square\, i_{\mathscr{B},\mathscr{B}})$$
$$= (f \,\square\, g)(i_{\mathscr{A},\mathscr{A}} \,\square\, i_{\mathscr{B},\mathscr{B}}) = (fi_{\mathscr{A},\mathscr{A}}) \,\square\, (gi_{\mathscr{B},\mathscr{B}}). \qquad (*)$$

Now, $fi_{\mathscr{A},\mathscr{A}}$ is generated in $\boldsymbol{A}$, that is, the relation

$$\langle y, x\rangle \in (fi_{\mathscr{A},\mathscr{A}}) \;\Leftrightarrow\; y \subset (fi_{\mathscr{A},\mathscr{A}})(x)$$

is generated in $\boldsymbol{A}$. Similarly for $gi_{\mathscr{B},\mathscr{B}}$ in $\boldsymbol{B}$. We leave it to the reader to verify, using (*), that

(1) if $\square$ is $\dot{\cup}$, $y \in (hi)(x) \Leftrightarrow \langle y, x\rangle \in (fi_{\mathscr{A},\mathscr{A}})^{i_{\mathscr{A},\mathscr{B}}} \cup (gi_{\mathscr{B},\mathscr{B}})^{j_{\mathscr{A},\mathscr{B}}}$,
(2) if $\square$ is $\times$, $y \in (hi)(x) \Leftrightarrow \langle y, x\rangle \in (fi_{\mathscr{A},\mathscr{A}})^{i_{\mathscr{A},\mathscr{B}}} \cap (gi_{\mathscr{B},\mathscr{B}})^{j_{\mathscr{A},\mathscr{B}}}$.

Now, $(fi_{\mathscr{A},\mathscr{A}})^{i_{\mathscr{A},\mathscr{B}}}$ and $(gi_{\mathscr{B},\mathscr{B}})^{j_{\mathscr{A},\mathscr{B}}}$ are both generated in $\boldsymbol{A} \,\square\, \boldsymbol{B}$ by Lemma 7.1, and the generated relations are closed under $\cap$ and $\cup$.

Thus hi is generated, as a relation, in $\boldsymbol{A} \,\square\, \boldsymbol{B}$. Similarly for hj, so h is a separation morphism.

12. Injections and co-embeddings

We follow our usual conventions: $\square$ is one of $\dot{\cup}$ or $\times$, and $\langle T, \mathscr{D}, \mathscr{E} \rangle$ is faithful. The entire of this section is devoted to a proof of the following theorem, our first result about the existence of co-embeddings. This will be developed further in Theorem 5.3.4.

THEOREM 12.1. *Let* $\mathbf{A}, \mathbf{B} \in C(T, \mathscr{D}, \mathscr{E})$, *suppose* $\mathbf{B}$ *is separable, and* $\mathbf{A}$ *has a morphism to* $\mathbf{B}$. *Let* $j_{\mathscr{A},\mathscr{B}} : \mathbf{B} \rightarrow \mathbf{A} \square \mathbf{B}$ *be the usual injection embedding. Then* $(j_{\mathscr{A},\mathscr{B}})^{-1}$ *is a co-embedding.*

The proof of this theorem is different if $\square$ is $\dot{\cup}$ or $\times$ (because of the failure of the analog of Proposition 4.3 for $\times$). We begin with a useful result, independent of $\square$, then turn to $\dot{\cup}$ and $\times$ separately.

PROPOSITION 12.2. *Let* $\delta : \mathscr{A} \rightarrow \mathscr{A}$ *be a coding such that the relation* $y \in \delta(x)$ *is generated in* $\mathbf{A}$. *Then, for each* n, *both* δ_n *and* δ_n^{-1} *are operators of order* $\langle n, n \rangle$ *in* $\mathbf{A}$.

PROOF. We show this for δ_n, the proof for δ_n^{-1} is similar.

Since $y \in \delta(x)$ is generated in $\mathbf{A}$, so is the relation

$$\mathscr{R}(x_1, \ldots, x_n, y_1, \ldots, y_n) \Leftrightarrow y_1 \in \delta(x_1) \wedge \cdots \wedge y_n \in \delta(x_n).$$

But then, for $\mathscr{P} \in [\mathscr{A}]^n$, $\delta_n(\mathscr{P}) = \mathscr{R}''(\mathscr{P})$, which is an operator in $\mathbf{A}$ by Corollary 2.5.2.

Now we prove a sequence of lemmas which establish Theorem 12.1.

LEMMA 12.3. *For each* n, *and for any coding* $f : \mathscr{A} \rightarrow \mathscr{B}$, $(j_{\mathscr{A},\mathscr{B}})_n^{-1} = (j_{\mathscr{B},\mathscr{B}})_n^{-1}(f \square \mathrm{inj}_{\mathscr{B}})_n$, *where*

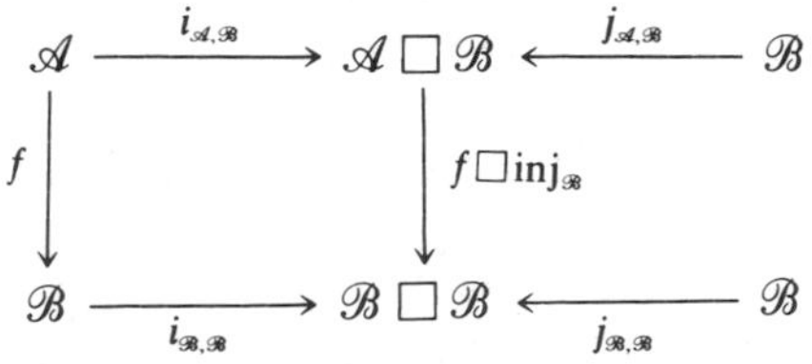

PROOF. To keep the notation simple, we take $n = 2$. We give separate arguments for $\dot{\cup}$ and $\times$.

(1) $\square$ is $\dot{\cup}$. Let $\mathscr{P} \subseteq (\mathscr{A} \dot{\cup} \mathscr{B})^2$. Then

$$\begin{aligned}
&\langle x, y\rangle \in (j_{\mathscr{B},\mathscr{B}})_2^{-1}(f \dot{\cup} \mathrm{inj}_{\mathscr{B}})_2(\mathscr{P}) \Leftrightarrow \\
&\Leftrightarrow \langle\langle x, 1\rangle, \langle y, 1\rangle\rangle \in (f \dot{\cup} \mathrm{inj}_{\mathscr{B}})_2(\mathscr{P}) \\
&\Leftrightarrow \langle\langle x, 1\rangle, \langle y, 1\rangle\rangle \in \mathscr{P} \\
&\Leftrightarrow \langle x, y\rangle \in (j_{\mathscr{A},\mathscr{B}})_2^{-1}(\mathscr{P}).
\end{aligned}$$

(2) $\square$ is $\times$. Let $\mathscr{P} \subseteq (\mathscr{A} \times \mathscr{B})^2$. Then

$$\begin{aligned}
&\langle x, y\rangle \in (j_{\mathscr{B},\mathscr{B}})_2^{-1}(f \times \mathrm{inj}_{\mathscr{B}})_2(\mathscr{P}) \Leftrightarrow \\
&\Leftrightarrow \langle\langle a, x\rangle, \langle b, y\rangle\rangle \in (f \times \mathrm{inj}_{\mathscr{B}})_2(\mathscr{P}) \quad \text{for some } a, b \in \mathscr{B}, \\
&\Leftrightarrow \langle\langle c, x\rangle, \langle d, y\rangle\rangle \in \mathscr{P} \quad \text{for some } c, d \in \mathscr{A}, \\
&\Leftrightarrow \langle x, y\rangle \in (j_{\mathscr{A},\mathscr{B}})_2^{-1}(\mathscr{P}).
\end{aligned}$$

LEMMA 12.4. *For each n, and for any coding* $f : \mathscr{A} \to \mathscr{B}$,

$$(j_{\mathscr{A},\mathscr{B}})_n = (f \dot{\cup} \mathrm{inj}_{\mathscr{B}})_n^{-1}(j_{\mathscr{B},\mathscr{B}})_n.$$

PROOF. By Proposition 4.3, the following commutes,

$$\begin{array}{ccccc}
\mathscr{A} & \xrightarrow{\;i_{\mathscr{A},\mathscr{B}}\;} & \mathscr{A} \dot{\cup} \mathscr{B} & \xleftarrow{\;i_{\mathscr{A},\mathscr{B}}\;} & \mathscr{B} \\
{\scriptstyle f}\big\downarrow & & \big\downarrow{\scriptstyle f \dot{\cup} \mathrm{inj}_{\mathscr{B}}} & & \big\downarrow{\scriptstyle \mathrm{inj}_{\mathscr{B}}} \\
\mathscr{B} & \xrightarrow{\;i_{\mathscr{B},\mathscr{B}}\;} & \mathscr{B} \dot{\cup} \mathscr{B} & \xleftarrow{\;j_{\mathscr{B},\mathscr{B}}\;} & \mathscr{B}
\end{array}$$

Hence $(f \dot{\cup} \mathrm{inj}_{\mathscr{B}})(j_{\mathscr{A},\mathscr{B}}) = j_{\mathscr{B},\mathscr{B}}$. Then, using Proposition 3.2.3,

$$(f \dot{\cup} \mathrm{inj}_{\mathscr{B}})_n^{-1}(f \dot{\cup} \mathrm{inj}_{\mathscr{B}})_n (j_{\mathscr{A},\mathscr{B}})_n = (f \dot{\cup} \mathrm{inj}_{\mathscr{B}})_n^{-1}(j_{\mathscr{B},\mathscr{B}})_n,$$

$$(j_{\mathscr{A},\mathscr{B}})_n = (f \dot{\cup} \mathrm{inj}_{\mathscr{B}})_n^{-1}(j_{\mathscr{B},\mathscr{B}})_n.$$

PROPOSITION 12.5 (Theorem 12.1 for $\dot{\cup}$). *Suppose* $\square$ *is* $\dot{\cup}$, $\mathbf{A}, \mathbf{B} \in C(T, \mathscr{D}, \mathscr{E})$, $\mathbf{B}$ *is separable, and* $f : \mathbf{A} \to \mathbf{B}$ *is a morphism. Then* $(j_{\mathscr{A},\mathscr{B}})^{-1}$ *is a co-embedding.*

PROOF. Since $\langle T, \mathscr{D}, \mathscr{E}\rangle$ is faithful, $j_{\mathscr{A},\mathscr{B}}$ is an embedding. We show $(j_{\mathscr{A},\mathscr{B}})^{-1}$ is a co-embedding using the definition of co-embedding directly (Ch. 3, §4).

Let g be a separation morphism for $\mathbf{B}$. Then, using Proposition 4.3, the following commutes.

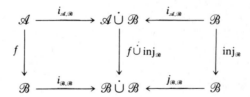

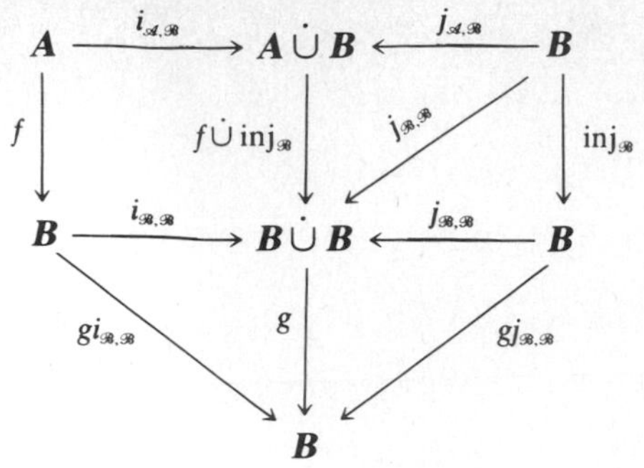

Also, set $h = g(f \dot\cup \mathrm{inj}_{\mathcal{B}}) : \boldsymbol{A} \dot\cup \boldsymbol{B} \to \boldsymbol{B}$.

Now, let Φ be an operator in $\boldsymbol{A} \dot\cup \boldsymbol{B}$ of order $\langle n, m\rangle$. We show $\Phi^{(j_{\mathcal{A},\mathcal{B}})^{-1}}$ is an operator in $\boldsymbol{B}$, which establishes that $(j_{\mathcal{A},\mathcal{B}})^{-1}$ is a co-embedding.

Set $\Psi = (\Phi^h)^{(gj_{\mathcal{B},\mathcal{B}})^{-1}}$. We first show that Ψ is an operator in $\boldsymbol{B}$, and second, that $\Psi = \Phi^{(j_{\mathcal{A},\mathcal{B}})^{-1}}$.

(1) Φ is an operator in $\boldsymbol{A} \dot\cup \boldsymbol{B}$ and h is the composition of two embeddings, hence an embedding, so Φ^h is an operator in $\boldsymbol{B}$. g is a separation morphism, so $gj_{\mathcal{B},\mathcal{B}}$ is generated in $\boldsymbol{B}$. Then by Proposition 12.2, both $(gj_{\mathcal{B},\mathcal{B}})_n$ and $(gj_{\mathcal{B},\mathcal{B}})_m^{-1}$ are operators in $\boldsymbol{B}$. Then the following is an operator in $\boldsymbol{B}$:

$$(gj_{\mathcal{B},\mathcal{B}})_m^{-1}\Phi^h(gj_{\mathcal{B},\mathcal{B}})_n = (\Phi^h)^{(gj_{\mathcal{B},\mathcal{B}})^{-1}} = \Psi.$$

(2) To see that $\Psi = \Phi^{(j_{\mathcal{A},\mathcal{B}})^{-1}}$, we make the following computation.

$$\begin{aligned}
\Psi &= (\Phi^h)^{(gj_{\mathcal{B},\mathcal{B}})^{-1}} \\
&= (gj_{\mathcal{B},\mathcal{B}})_m^{-1}(\Phi^h)(gj_{\mathcal{B},\mathcal{B}})_n \\
&= (gj_{\mathcal{B},\mathcal{B}})_m^{-1} h_m \Phi h_n^{-1}(gj_{\mathcal{B},\mathcal{B}})_n \\
&= (gj_{\mathcal{B},\mathcal{B}})_m^{-1}[g(f \dot\cup \mathrm{inj}_{\mathcal{B}})]_m \Phi [g(f \dot\cup \mathrm{inj}_{\mathcal{B}})]_n^{-1}(gj_{\mathcal{B},\mathcal{B}})_n \\
&= (j_{\mathcal{B},\mathcal{B}})_m^{-1} g_m^{-1} g_m (f \dot\cup \mathrm{inj}_{\mathcal{B}})_m \Phi (f \dot\cup \mathrm{inj}_{\mathcal{B}})_n^{-1} g_n^{-1} g_n (j_{\mathcal{B},\mathcal{B}})_n \\
&= (j_{\mathcal{B},\mathcal{B}})_m^{-1}(f \dot\cup \mathrm{inj}_{\mathcal{B}})_m \Phi (f \dot\cup \mathrm{inj}_{\mathcal{B}})_n^{-1}(j_{\mathcal{B},\mathcal{B}})_n \\
&= (j_{\mathcal{A},\mathcal{B}})_m^{-1}\Phi(j_{\mathcal{A},\mathcal{B}})_n \quad \text{(by Lemmas 12.3 and 12.4)} \\
&= \Phi^{(j_{\mathcal{A},\mathcal{B}})^{-1}}.
\end{aligned}$$

This completes the proof for $\dot\cup$.

LEMMA 12.6. *Suppose* $f : \mathcal{A} \to \mathcal{B}$ *is a coding, so the following commutes (by Proposition* 4.2):

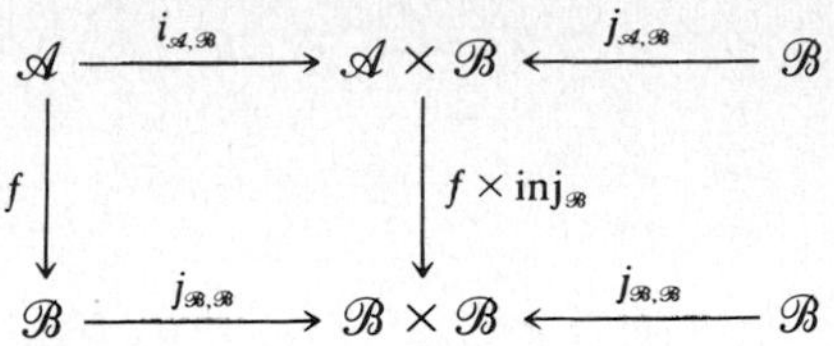

Then (1) *for* $\mathscr{P} \subseteq \mathscr{B}^n$,

$$(j_{\mathscr{A},\mathscr{B}})_n(\mathscr{P}) = (f \times \mathrm{inj}_{\mathscr{B}})_n^{-1}[(i_{\mathscr{B},\mathscr{B}}f)_n(\mathscr{A}^n) \cap (j_{\mathscr{B},\mathscr{B}})_n(\mathscr{P})].$$

(2) *For* $\mathscr{Q} \subseteq (\mathscr{A} \times \mathscr{B})^m$,

$$(j_{\mathscr{A},\mathscr{B}})_m^{-1}(\mathscr{Q}) = (j_{\mathscr{B},\mathscr{B}})_m^{-1}[(i_{\mathscr{B},\mathscr{B}}f)_m(\mathscr{A}^m) \cap (f \times \mathrm{inj}_{\mathscr{B}})_m(\mathscr{Q})].$$

PROOF. (1) Use Proposition 4.4 (2), and take $(f \times \mathrm{inj}_{\mathscr{B}})_n^{-1}$ of both sides.

(2) By Proposition 4.2

$$\begin{aligned}
&(j_{\mathscr{B},\mathscr{B}})_m^{-1}[(i_{\mathscr{B},\mathscr{B}}f)_m(\mathscr{A}^m) \cap (f \times \mathrm{inj}_{\mathscr{B}})_m(\mathscr{Q})] \\
&= (j_{\mathscr{B},\mathscr{B}})_m^{-1}[(f \times \mathrm{inj}_{\mathscr{B}})_m(i_{\mathscr{A},\mathscr{B}})_m(\mathscr{A}^m) \cap (f \times \mathrm{inj}_{\mathscr{B}})_m(\mathscr{Q})] \\
&= (j_{\mathscr{B},\mathscr{B}})_m^{-1}(f \times \mathrm{inj}_{\mathscr{B}})_m[(i_{\mathscr{A},\mathscr{B}})_m(\mathscr{A}^m) \cap \mathscr{Q}] \\
&= (j_{\mathscr{B},\mathscr{B}})_m^{-1}(f \times \mathrm{inj}_{\mathscr{B}})_m(\mathscr{Q}) \\
&= (j_{\mathscr{A},\mathscr{B}})_m^{-1}(\mathscr{Q}) \quad \text{(by Lemma 12.3).}
\end{aligned}$$

PROPOSITION 12.7 (Theorem 12.1 for $\times$). *Suppose* $\square$ *is* $\times$, $\mathbf{A}, \mathbf{B} \in C(T, \mathscr{D}, \mathscr{E})$, $\mathbf{B}$ *is separable, and* $f : \mathbf{A} \to \mathbf{B}$ *is a morphism. Then* $(j_{\mathscr{A},\mathscr{B}})^{-1}$ *is a co-embedding.*

PROOF. Again, since $\langle T, \mathscr{D}, \mathscr{E} \rangle$ is faithful, $j_{\mathscr{A},\mathscr{B}}$ is an embedding.

Let g be a separation morphism for $\mathbf{B}$. Then, by Proposition 4.2, the following commutes.

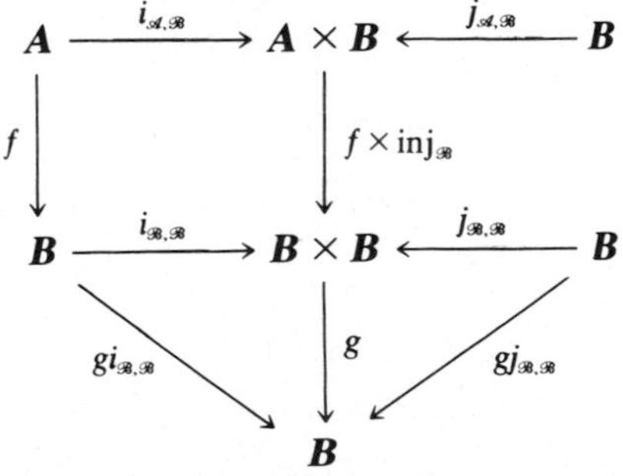

Also, set $h = g(f \times \mathrm{inj}_{\mathscr{B}})$, so h is an embedding, $h : \mathbf{A} \times \mathbf{B} \to \mathbf{B}$.

Let Φ be an operator in $\boldsymbol{A} \times \boldsymbol{B}$ of order $\langle n, m \rangle$. We show $\Phi^{(j_{\mathscr{A},\mathscr{B}})^{-1}}$ is an operator in $\boldsymbol{B}$.

Let $\mathscr{A}' = (gi_{\mathscr{B},\mathscr{B}}f)_m(\mathscr{A}^m)$ and $\mathscr{A}'' = (gi_{\mathscr{B},\mathscr{B}}f)_n(\mathscr{A}^n)$. Define a map Ψ of order $\langle n, m \rangle$ by: for $\mathscr{P} \subseteq \mathscr{B}^n$,

$$\Psi(\mathscr{P}) = (gj_{\mathscr{B},\mathscr{B}})_m^{-1}[\mathscr{A}' \cap \Phi^h(\mathscr{A}'' \cap (gj_{\mathscr{B},\mathscr{B}})_n(\mathscr{P}))].$$

Claim 1. *Ψ is an operator in $\boldsymbol{B}$.* This follows since $\mathscr{A}'$ and $\mathscr{A}''$ are generated in $\boldsymbol{B}$, $(gj_{\mathscr{B},\mathscr{B}})_n$ and $(gj_{\mathscr{B},\mathscr{B}})_m^{-1}$ are operators in $\boldsymbol{B}$, Φ^h is an operator in $\boldsymbol{B}$, and production systems have appropriate closure properties.

Claim 2. $\Psi = \Phi^{(j_{\mathscr{A},\mathscr{B}})^{-1}}$. One may show, for $\mathscr{P} \subseteq \mathscr{B}^n$, that

$$\Psi(\mathscr{P}) = \Phi^{(j_{\mathscr{A},\mathscr{B}})^{-1}}(\mathscr{P})$$

by a computation similar to that in the proof of Proposition 12.5, but using Lemma 12.6 in place of Lemma 12.4. We omit the steps.

CHAPTER FIVE

EFFECTIVE EMBEDDINGS

1. Introduction

In proving the incompleteness of elementary arithmetic, Gödel introduced the device now called Gödel numbering. It can be looked at as a coding, of words by numbers, and is, essentially, Example III in Ch. 3, §6. Often the "effectiveness" of the Gödel numbering procedure is emphasized, for example, Beth [1964], p. 297 says, "It will be clear that the Gödel number $g(U)$ of any expression U is uniquely defined and effectively calculable; and conversely, any natural number n can be the Gödel number of at most one expression and if $g(U)$ is given the U can be effectively determined." But then Beth continues with the following interesting statement. "So we have established, by a recursive definition, a one-to-one correspondence between the expressions U of sentential logic and the elements $g(U)$ of a certain set G of natural numbers; this set G also admits of a certain recursive definition."

Now when Beth says the Gödel numbering is given by a recursive definition he is speaking informally, since he only discusses recursive definitions *in* arithmetic, while the Gödel numbering g is *between* words and numbers. Nevertheless it is clear that his statement is meaningful (and true) and it is worthwhile to introduce the technical machinery to deal with such concepts.

Prof. Smullyan informs us that in that lost paper of 1957 (see Ch. 1, §14) he gave a technical definition of an effective Gödel numbering along the following lines. He worked in a recursion theory of words, and identified the number n with the word (numeral) consisting of n 1's; there were other letters present besides 1, and hence other words besides numerals. Then, a Gödel numbering, being a function g from words to numbers, could be thought of as a function from words to numerals, and as such could be dealt with by the recursion theory of words. Prof. Smullyan called g an effective Gödel numbering if, in his recursion theory of words, g was a recursive function.

For us, Gödel numbering is one kind of embedding, and we would like to say, in general, when an embedding is effective. For Gödel numbering, Prof. Smullyan's approach was available because there is a single recursion theory that "discusses" both words and numerals. Now, in general, if $\theta : \text{rec}(\mathfrak{A}) \to \text{rec}(\mathfrak{B})$ is an embedding, we must have a recursion theory that can "discuss" things pertinent to $\mathfrak{A}$ and also to $\mathfrak{B}$, if we are to give a meaning to "effectiveness" for θ. Well, $\text{rec}(\mathfrak{A}) \square \text{rec}(\mathfrak{B})$ is precisely such a recursion theory (for either $\dot{\cup}$ or $\times$). In brief, we will say θ is effective (relative to $\square$) if θ is generated in $\text{rec}(\mathfrak{A}) \square \text{rec}(\mathfrak{B})$. The only, minor, difficulty is that θ itself is not quite the right thing to look at. We replace θ by $\theta^{\square}$, where, if $y \in \theta(x)$, then $y' \in \theta^{\square}(x')$ where $x' \in i(x)$ and $y' \in j(y)$. That is, $\theta^{\square}$ behaves like θ, but on the "copies" of x and y that are present in $\mathscr{A} \square \mathscr{B}$. Then, θ is effective if $\theta^{\square}$ is generated.

In this chapter we develop basic properties of effective embeddings, not only for recursion theories, but more generally.

Let $\langle T, \mathscr{D}, \mathscr{E} \rangle$ be faithful, and $\boldsymbol{A} = T(\mathfrak{A})$ for $\mathfrak{A} \in \mathscr{D}$. Now, $\text{inj}_{\mathscr{A}} : \boldsymbol{A} \to \boldsymbol{A}$ is a trivial example of an embedding, but curiously enough, it need not be effective. We call $\boldsymbol{A}$ *reflexive* (relative to $\square$) if $\text{inj}_{\mathscr{A}}$ is an effective embedding. Reflexivity seems to capture, at least in part, the idea that a production system domain should be capable of being "built up" using the given machinery. We do not investigate this connection however.

Ordinary recursion theory is reflexive, as are many others. We investigate the properties of production systems that are both reflexive and separable. They are quite pleasant. We conclude with some general theorems that imply that all the embeddings of Ch. 3, §6 can be replaced by isomorphisms. It should be understood that when $T(\text{-})$ is rec(-) or ω-sec(-) we automatically take $\mathscr{E}$ to be all codings and $\mathscr{D}$ to be all structures (all non-empty structures, if $\square$ is $\times$). Thus mention of $\mathscr{D}$ and $\mathscr{E}$ is suppressed in the examples of this chapter.

For the entire of this chapter, $\square$ is either $\dot{\cup}$ or $\times$, and $\langle T, \mathscr{D}, \mathscr{E} \rangle$ is faithful.

2. Effective embeddings

DEFINITION. Let $\theta : \mathscr{A} \to \mathscr{B}$ be a coding. By $\theta^{\square}$ we mean the binary relation on $\mathscr{A} \square \mathscr{B}$ given by $\langle x', y' \rangle \in \theta^{\square} \Leftrightarrow$ for some x, y, $y \in \theta(x)$, $x' \in i_{\mathscr{A},\mathscr{B}}(x)$ and $y' \in j_{\mathscr{A},\mathscr{B}}(y)$.

PROPOSITION 2.1. *Let $\theta : \mathscr{A} \to \mathscr{B}$ be a coding. Suppose $x' \in i_{\mathscr{A},\mathscr{B}}(x)$ and $y' \in j_{\mathscr{A},\mathscr{B}}(y)$. Then $\langle x', y' \rangle \in \theta^{\square} \Leftrightarrow y \in \theta(x)$.*

PROOF. If $y \in \theta(x)$ then $\langle x', y'\rangle \in \theta^{\square}$ by definition.

Suppose $\langle x', y'\rangle \in \theta^{\square}$. Then by definition there are a, b, with $b \in \theta(a)$, $x' \in i_{\mathcal{A},\mathcal{B}}(a)$ and $y' \in j_{\mathcal{A},\mathcal{B}}(b)$. But i is a coding, so if $x' \in i_{\mathcal{A},\mathcal{B}}(a)$ and $x' \in i_{\mathcal{A},\mathcal{B}}(x)$ then $a = x$. Similarly $b = y$, so $y \in \theta(x)$ since $b \in \theta(a)$.

REMARK. Let $\theta : \mathcal{A} \to \mathcal{B}$ be a coding. Then the above gives that

$$\theta^{\dot{\cup}} = \{\langle\langle x, 0\rangle, \langle y, 1\rangle\rangle \mid y \in \theta(x)\},$$

$$\theta^{\times} = \{\langle\langle x, b\rangle, \langle a, y\rangle\rangle \mid y \in \theta(x), a \in \mathcal{A}, b \in \mathcal{B}\}.$$

DEFINITION. Let $\mathfrak{A}, \mathfrak{B} \in \mathcal{D}$. If $\theta : T(\mathfrak{A}) \to T(\mathfrak{B})$ is an embedding, we say θ is an *effective embedding* (relative to $\square$) if the relation $\theta^{\square}$ *is generated in* $T(\mathfrak{A}) \square T(\mathfrak{B})$.

EXAMPLES. In Ch. 3, §6 we defined (Example I) a coding $\theta : \mathsf{N} \to \mathrm{L}_\omega$ by: $\theta(\mathsf{N}) = \{n^{\#}\}$ where $n^{\#}$ is the Von Neumann ordinal for n. And we showed it gave an embedding of the corresponding recursion theories, $\theta : \mathrm{rec}(\mathfrak{S}(\mathsf{N})) \to \mathrm{rec}(\mathfrak{S}(\mathrm{L}_\omega))$. In fact, θ is effective (for both $\dot{\cup}$ and $\times$). We show it is effective relative to $\dot{\cup}$; the proof that it is effective relative to $\times$ is quite similar, or see Theorem 3.5 below.

$$\mathrm{rec}(\mathfrak{S}(\mathsf{N}))\, \dot{\cup}\, \mathrm{rec}(\mathfrak{S}(\mathrm{L}_\omega)) = \mathrm{rec}(\langle \mathsf{N} \dot{\cup} \mathrm{L}_\omega\, ; (=_{\mathsf{N}})^i, A^i, (=_{\mathrm{L}_\omega})^j, B^j\rangle)$$

where A is the successor relation on N, $A(x, y) \Leftrightarrow y = x^+$; and B is the relation on L_ω, $B(x, y, z) \Leftrightarrow x \cup \{y\} = z$. Now, for this recursion theory, $\theta^{\dot{\cup}}$ has the following elementary formal system axioms: (we use $\theta^{\dot{\cup}}$ as a predicate symbol too, for convenience)

$$\theta^{\dot{\cup}}(\langle 0, 0\rangle, \langle \emptyset, 1\rangle),$$

$$\theta^{\dot{\cup}}(x, y) \to A^i(x, z) \to B^j(y, y, w) \to \theta^{\dot{\cup}}(z, w).$$

As a matter of fact, *every* embedding given in Ch. 3, §6 is effective, both for $\dot{\cup}$ and $\times$. This is important, and the reader should convince himself that it is true.

3. Results about effective embeddings

THEOREM 3.1. *Let $\mathbf{A}, \mathbf{B}, \mathbf{C} \in C(T, \mathcal{D}, \mathcal{E})$ and suppose $\mathbf{A}$ and $\mathbf{C}$ are separable. If $f : \mathbf{A} \to \mathbf{B}$ and $g : \mathbf{B} \to \mathbf{C}$ are morphisms of $C(T, \mathcal{D}, \mathcal{E})$ which are effective embeddings, then $(gf) : \mathbf{A} \to \mathbf{C}$ is also an effective embedding (and of course a morphism).*

PROOF. By Proposition 3.4.1, gf is an embedding. We show it is effective, that is, $(gf)^{\square}$ is generated in $\boldsymbol{A} \square \boldsymbol{C}$.

$f: \boldsymbol{A} \to \boldsymbol{B}$ is effective so $f^{\square}$ is generated in $\boldsymbol{A} \square \boldsymbol{B}$. Also $g: \boldsymbol{B} \to \boldsymbol{C}$ is effective so $g^{\square}$ is generated in $\boldsymbol{B} \square \boldsymbol{C}$. Now $C(T, \mathcal{D}, \mathcal{E})$ is a symmetric monoidal category (Theorem 4.9.1) so we have

$$\boldsymbol{A} \square \boldsymbol{B} \xrightarrow{i} (\boldsymbol{A} \square \boldsymbol{B}) \square \boldsymbol{C} \xrightarrow{\gamma \square 1_{\mathscr{C}}} (\boldsymbol{B} \square \boldsymbol{A}) \square \boldsymbol{C} \xrightarrow{\alpha^{-1}} \boldsymbol{B} \square (\boldsymbol{A} \square \boldsymbol{C})$$

and

$$\boldsymbol{B} \square \boldsymbol{C} \xrightarrow{j} \boldsymbol{A} \square (\boldsymbol{B} \square \boldsymbol{C}) \xrightarrow{\alpha^{-1}} (\boldsymbol{A} \square \boldsymbol{B}) \square \boldsymbol{C} \xrightarrow{\gamma \square 1_{\mathscr{C}}}$$
$$(\boldsymbol{B} \square \boldsymbol{A}) \square \boldsymbol{C} \xrightarrow{\alpha} \boldsymbol{B} \square (\boldsymbol{A} \square \boldsymbol{C}).$$

So, let

$$\mu = \alpha^{-1}(\gamma \square 1_{\mathscr{C}})i, \qquad \nu = \alpha(\gamma \square 1_{\mathscr{C}})\alpha^{-1}j.$$

Then these are both embeddings,

$$\mu : \boldsymbol{A} \square \boldsymbol{B} \to \boldsymbol{B} \square (\boldsymbol{A} \square \boldsymbol{C}),$$
$$\nu : \boldsymbol{B} \square \boldsymbol{C} \to \boldsymbol{B} \square (\boldsymbol{A} \square \boldsymbol{C}).$$

Then, since embeddings take generated relations to generated relations, both $(f^{\square})^{\mu}$ and $(g^{\square})^{\nu}$ are generated in $\boldsymbol{B} \square (\boldsymbol{A} \square \boldsymbol{C})$.

Now define a relation h in $\boldsymbol{B} \square (\boldsymbol{A} \square \boldsymbol{C})$ by:

$$\langle x, y \rangle \in h \iff \text{for some } z,\ \langle x, z \rangle \in (f^{\square})^{\mu} \quad \text{and} \quad \langle z, y \rangle \in (g^{\square})^{\nu}.$$

Using the closure properties of production systems, h is generated in $\boldsymbol{B} \square (\boldsymbol{A} \square \boldsymbol{C})$. We leave it to the reader to verify (treating $\dot{\cup}$ and $\times$ separately is easiest) that

$$h = ((gf)^{\square})^{j}$$

where $j = j_{\mathscr{B}, \mathscr{A} \square \mathscr{C}} : \mathscr{A} \square \mathscr{C} \to \mathscr{B} \square (\mathscr{A} \square \mathscr{C})$.

Thus $((gf)^{\square})^{j}$ is generated in $\boldsymbol{B} \square (\boldsymbol{A} \square \boldsymbol{C})$. Now, both $\boldsymbol{A}$ and $\boldsymbol{C}$ are separable, hence so is $\boldsymbol{A} \square \boldsymbol{C}$ by Proposition 4.11.2. Also $\boldsymbol{B}$ has a morphism to $\boldsymbol{C}$ (namely g) and $\boldsymbol{C}$ has a morphism to $\boldsymbol{A} \square \boldsymbol{C}$ (injection), so $\boldsymbol{B}$ has a morphism to $\boldsymbol{A} \square \boldsymbol{C}$. Then by Theorem 4.12.1, j^{-1} is a co-embedding. But co-embeddings also take generated relations to generated relations, hence $(((gf)^{\square})^{j})^{j^{-1}}$ is generated in $\boldsymbol{A} \square \boldsymbol{C}$. But this is $j_2^{-1} j_2((gf)^{\square}) = (gf)^{\square}$. And thus gf is *effective.*

DEFINITION. Let $\theta : \mathscr{B} \to \mathscr{A}$ be a coding. For each n we define maps $(\theta^{\square})_n$ and $(\theta^{\square})_n^{-1}$ by: for $\mathscr{P} \subseteq (\mathscr{B} \square \mathscr{A})^n$

(1) $\langle y_1, \ldots, y_n \rangle \in (\theta^\square)_n(\mathcal{P}) \Leftrightarrow$ for some $\langle x_1, \ldots, x_n \rangle \in \mathcal{P}$, $\langle x_1, y_1 \rangle \in \theta^\square \wedge \cdots \wedge \langle x_n, y_n \rangle \in \theta^\square$.

(2) $\langle x_1, \ldots, x_n \rangle \in (\theta^\square)_n^{-1}(\mathcal{P}) \Leftrightarrow$ for some $\langle y_1, \ldots, y_n \rangle \in \mathcal{P}$, $\langle x_1, y_1 \rangle \in \theta^\square \wedge \cdots \wedge \langle x_n, y_n \rangle \in \theta^\square$.

LEMMA 3.2. *Let $f : \mathcal{B} \to \mathcal{A}$ be a coding. For each n,*

$$f_n = (i_{\mathcal{A},\mathcal{B}})_n^{-1}(f^\square)_n (j_{\mathcal{A},\mathcal{B}})_n,$$

$$f_n^{-1} = (j_{\mathcal{A},\mathcal{B}})_n^{-1}(f^\square)_n^{-1}(i_{\mathcal{A},\mathcal{B}})_n.$$

PROOF. Left to the reader.

LEMMA 3.3. *Let $f : \boldsymbol{B} \to \boldsymbol{A}$ be an effective embedding. Then for each n, $(f^\square)_n$ and $(f^\square)_n^{-1}$ are operators in $\boldsymbol{A} \square \boldsymbol{B}$ (note order of $\boldsymbol{A}$ and $\boldsymbol{B}$).*

PROOF. f is effective, so $f^\square$ is generated in $\boldsymbol{B} \square \boldsymbol{A}$. But there is an isomorphism $\gamma : \boldsymbol{B} \square \boldsymbol{A} \to \boldsymbol{A} \square \boldsymbol{B}$, so $\gamma_2(f^\square)$ is generated in $\boldsymbol{A} \square \boldsymbol{B}$. Define a relation R by

$$R(x_1, \ldots, x_n, y_1, \ldots, y_n) \Leftrightarrow \langle y_1, x_1 \rangle \in \gamma_2(f^\square) \wedge \cdots \wedge \langle y_n, x_n \rangle \in \gamma_2(f^\square).$$

Then R is also generated in $\boldsymbol{A} \square \boldsymbol{B}$. Also, for $\mathcal{P} \subseteq (\mathcal{A} \square \mathcal{B})^n$, $(f^\square)_n(\mathcal{P}) = R''(\mathcal{P})$, and we have Corollary 2.5.2. Similarly for $(f^\square)_n^{-1}$.

THEOREM 3.4. *Suppose, in $C(T, \mathcal{D}, \mathcal{E})$,*
(1) $\boldsymbol{A}$ has a morphism to $\boldsymbol{B}$,
(2) $\boldsymbol{B}$ is separable,
(3) $f : \boldsymbol{B} \to \boldsymbol{A}$ is an effective embedding.
Then f^{-1} is a co-embedding.

PROOF. Let Φ be an operator in $\boldsymbol{A}$ of order $\langle n, m \rangle$. We show $\Phi^{f^{-1}}$ is an operator in $\boldsymbol{B}$. But, by Lemma 3.2,

$$\begin{aligned}\Phi^{f^{-1}} &= f_m^{-1}\Phi f_n \\ &= (j_{\mathcal{A},\mathcal{B}})_m^{-1}(f^\square)_m^{-1}(i_{\mathcal{A},\mathcal{B}})_m\Phi(i_{\mathcal{A},\mathcal{B}})_n^{-1}(f^\square)_n(j_{\mathcal{A},\mathcal{B}})_n \\ &= (j_{\mathcal{A},\mathcal{B}})_m^{-1}(f^\square)_m^{-1}\Phi^{i_{\mathcal{A},\mathcal{B}}}(f^\square)_n(j_{\mathcal{A},\mathcal{B}})_n.\end{aligned}$$

Set $\Psi = (f^\square)_m^{-1}\Phi^{i_{\mathcal{A},\mathcal{B}}}(f^\square)_n$. Since Φ is an operator in $\boldsymbol{A}$, $\Phi^{i_{\mathcal{A},\mathcal{B}}}$ is an operator in $\boldsymbol{A} \square \boldsymbol{B}$. Then, using Lemma 3.3, so is Ψ. Then, continuing the above,

$$\Phi^{f^{-1}} = (j_{\mathcal{A},\mathcal{B}})_m^{-1}\Psi(j_{\mathcal{A},\mathcal{B}})_n = \Psi^{(j_{\mathcal{A},\mathcal{B}})^{-1}}.$$

But this is an operator in $\boldsymbol{B}$ since j is a co-embedding by Theorem 4.12.1.

REMARKS. In Ch. 3, §6 we gave an example of an embedding from rec($\mathfrak{S}(\mathbb{N})$) to rec($\mathfrak{S}(\mathrm{L}_\omega)$) [Example I] and the other way around [Example VII]. Both are effective embeddings, and both recursion theories are separable. Hence both are also co-embeddings, by the above theorem. It follows that a set of numbers is r.e. in rec($\mathfrak{S}(\mathbb{N})$) iff the corresponding set of ordinals is r.e. in rec($\mathfrak{S}(\mathrm{L}_\omega)$). Similarly, all the other embeddings of Ch. 3, §6 are also co-embeddings.

Actually we have been discussing two notions of effective embeddings, depending on whether $\square$ is $\dot\cup$ or $\times$. The following relates them somewhat.

THEOREM 3.5. *Suppose T assigns total production systems to structures, and $\mathscr{E}$ contains all codings. Let $\boldsymbol{A}, \boldsymbol{B} \in C(T, \mathscr{D}, \mathscr{E})$. Suppose $\boldsymbol{A}$ and $\boldsymbol{B}$ are both separable using $\times$. If $f : \boldsymbol{A} \to \boldsymbol{B}$ is effective relative to $\dot\cup$, it is also effective relative to $\times$.*

PROOF. In this proof we need to work with both $\dot\cup$ and $\times$. To keep things straight, we use i, j for the injection maps where $\dot\cup$ is concerned, and I, J for the injection maps where $\times$ is concerned.

$\boldsymbol{A}$ and $\boldsymbol{B}$ are both separable using $\times$; then by Proposition 4.11.2, $\boldsymbol{A} \times \boldsymbol{B}$ is also separable using $\times$; and then by Theorem 4.10.3, $\boldsymbol{A} \times \boldsymbol{B}$ is separable using $\dot\cup$. Let g be a separation morphism for it. Then (using Proposition 4.4.3) the following commutes.

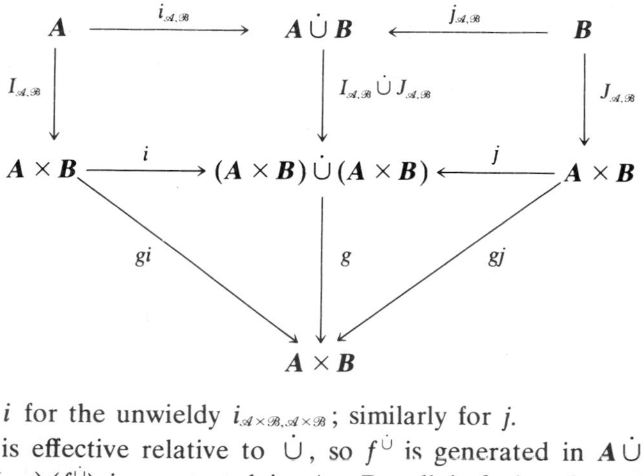

We write i for the unwieldy $i_{\mathscr{A}\times\mathscr{B},\mathscr{A}\times\mathscr{B}}$; similarly for j.

Now f is effective relative to $\dot\cup$, so $f^{\dot\cup}$ is generated in $\boldsymbol{A} \dot\cup \boldsymbol{B}$. Then $g_2(I_{\mathscr{A},\mathscr{B}} \dot\cup J_{\mathscr{A},\mathscr{B}})_2(f^{\dot\cup})$ is generated in $\boldsymbol{A} \times \boldsymbol{B}$, call it h for short. Define a relation R on $\mathscr{A} \times \mathscr{B}$ by

$$\langle x, y\rangle \in R \;\Leftrightarrow\; \text{for some } u, v, \langle u, v\rangle \in h,\ u \in (gi)(x) \text{ and } v \in (gj)(y).$$

Since gi and gj are generated in $\boldsymbol{A} \times \boldsymbol{B}$, so is R. Finally, we claim

$$\langle x, y\rangle \in R \;\Leftrightarrow\; \langle x, y\rangle \in f^{\times}$$

which says that f is effective relative to $\times$.

(1) $f^{\times} \subseteq R$.

Suppose $\langle x, y\rangle \in f^{\times}$. Then for some z, w, $w \in f(z)$, $x \in I_{\mathcal{A},\mathcal{B}}(z)$ and $y \in J_{\mathcal{A},\mathcal{B}}(w)$. Choose any u, v with $u \in (gi)(x)$ and $v \in (gj)(y)$. Then $u \in (giI_{\mathcal{A},\mathcal{B}})(z)$ and $v \in (gjJ_{\mathcal{A},\mathcal{B}})(w)$. By the commutativity of the above diagram, this gives us

$$u \in g(I_{\mathcal{A},\mathcal{B}} \dot{\cup} J_{\mathcal{A},\mathcal{B}})i_{\mathcal{A},\mathcal{B}}(z) \quad \text{and} \quad v \in g(I_{\mathcal{A},\mathcal{B}} \dot{\cup} J_{\mathcal{A},\mathcal{B}})j_{\mathcal{A},\mathcal{B}}(w).$$

Since $w \in f(z)$, this says $\langle u, v\rangle \in g_2(I_{\mathcal{A},\mathcal{B}} \dot{\cup} J_{\mathcal{A},\mathcal{B}})_2(f^{\dot{\cup}})$ or $\langle u, v\rangle \in h$. Since $u \in (gi)(x)$ and $v \in (gj)(y)$, $\langle x, y\rangle \in R$.

(2) $R \subseteq f^{\times}$.

Suppose $\langle x, y\rangle \in R$. Then for some $\langle u, v\rangle \in h$, $u \in (gi)(x)$ and $v \in (gj)(y)$. $\langle u, v\rangle \in h$ so $\langle u, v\rangle \in g_2(I_{\mathcal{A},\mathcal{B}} \dot{\cup} J_{\mathcal{A},\mathcal{B}})_2(f^{\dot{\cup}})$. Then for some $\langle a, b\rangle \in (I_{\mathcal{A},\mathcal{B}} \dot{\cup} J_{\mathcal{A},\mathcal{B}})_2(f^{\dot{\cup}})$, $u \in g(a)$ and $v \in g(b)$. But then, for some $\langle c, d\rangle \in f^{\dot{\cup}}$, $\langle a, b\rangle \in (I_{\mathcal{A},\mathcal{B}} \dot{\cup} J_{\mathcal{A},\mathcal{B}})_2(\langle c, d\rangle)$. If $\langle c, d\rangle \in f^{\dot{\cup}}$, for some z, w, $w \in f(z)$, $c \in i_{\mathcal{A},\mathcal{B}}(z)$ and $d \in j_{\mathcal{A},\mathcal{B}}(w)$. Now $a \in (I_{\mathcal{A},\mathcal{B}} \dot{\cup} J_{\mathcal{A},\mathcal{B}})(c)$ and $c \in i_{\mathcal{A},\mathcal{B}}(z)$ so $a \in (I_{\mathcal{A},\mathcal{B}} \dot{\cup} J_{\mathcal{A},\mathcal{B}})i_{\mathcal{A},\mathcal{B}}(z)$. Using the commutativity of the above diagram, $a \in iI_{\mathcal{A},\mathcal{B}}(z)$. Similarly, $b \in jJ_{\mathcal{A},\mathcal{B}}(w)$. Further $u \in g(a)$, so $u \in (giI_{\mathcal{A},\mathcal{B}})(z)$. Similarly $v \in (gjJ_{\mathcal{A},\mathcal{B}})(w)$. Now $u \in (gi)(x)$ and $u \in (giI_{\mathcal{A},\mathcal{B}})(z)$, so by the "disjointness" feature of codings we must have $x \in I_{\mathcal{A},\mathcal{B}}(z)$. Similarly $y \in J_{\mathcal{A},\mathcal{B}}(w)$. Finally, since $w \in f(z)$, we have $\langle x, y\rangle \in f^{\times}$.

In Theorem 3.1 we considered the closure of effective embeddings under composition. We conclude this section with a companion result concerning closure under $\Box$ itself.

THEOREM 3.6. *Let $\boldsymbol{A}, \boldsymbol{A}', \boldsymbol{B}, \boldsymbol{B}' \in C(T, \mathcal{D}, \mathcal{E})$ and suppose $\alpha : \boldsymbol{A} \to \boldsymbol{A}'$ and $\beta : \boldsymbol{B} \to \boldsymbol{B}'$ are both effective embeddings. Then $\alpha \Box \beta : \boldsymbol{A} \Box \boldsymbol{B} \to \boldsymbol{A}' \Box \boldsymbol{B}'$ is also effective.*

PROOF. Let π be the obvious isomorphism

$$\pi : (\mathcal{A} \Box \mathcal{A}') \Box (\mathcal{B} \Box \mathcal{B}') \cong (\mathcal{A} \Box \mathcal{B}) \Box (\mathcal{A}' \Box \mathcal{B}')$$

compounded out of the reassociating maps and the commuting maps. (We used this earlier in the proof of Proposition 4.11.2.) Also for convenience in notation we write i and j for

$$i = i_{\mathcal{A}\Box\mathcal{A}',\mathcal{B}\Box\mathcal{B}'} : \mathcal{A}\Box\mathcal{A}' \to (\mathcal{A}\Box\mathcal{A}')\Box(\mathcal{B}\Box\mathcal{B}')$$

and

$$j = j_{\mathcal{A}\Box\mathcal{A}',\mathcal{B}\Box\mathcal{B}'} : \mathcal{B}\Box\mathcal{B}' \to (\mathcal{A}\Box\mathcal{A}')\Box(\mathcal{B}\Box\mathcal{B}').$$

We claim

$$(\alpha\Box\beta)^{\Box} = \begin{cases} ((\alpha^{\dot\cup})^i)^\pi \cup ((\beta^{\dot\cup})^j)^\pi & \text{if } \Box = \dot\cup, \\ ((\alpha^{\times})^i)^\pi \cap ((\beta^{\times})^j)^\pi & \text{if } \Box = \times. \end{cases}$$

This has an essentially computational verification, which we leave to the reader.

Then, since i, j and π are embeddings, they take generated relations to generated relations. And generated relations are closed under $\cup$ and $\cap$, which completes the proof.

4. Reflexivity

Let $\boldsymbol{A} \in C(T, \mathcal{D}, \mathcal{E})$. $\boldsymbol{A}$ has a morphism to itself, the identity one, $\text{inj}_{\mathcal{A}} : \boldsymbol{A} \to \boldsymbol{A}$. But, this embedding need not be effective! For example, let $\mathfrak{A} = \langle \mathbb{N} \rangle$ be the structure with the natural numbers as domain, but no relations given. $\text{inj}_{\mathbb{N}} : \text{rec}(\mathfrak{A}) \to \text{rec}(\mathfrak{A})$ is an embedding, but not effective since in $\text{rec}(\mathfrak{A})\Box\text{rec}(\mathfrak{A})$ we simply haven't enough machinery to show $(\text{inj}_{\mathbb{N}})^{\Box}$ is r.e. (whether $\Box$ is $\dot\cup$ or $\times$).

DEFINITION. Let $\boldsymbol{A} \in C(T, \mathcal{D}, \mathcal{E})$. We say $\boldsymbol{A}$ is *reflexive* (relative to $\Box$) if $\text{inj}_{\mathcal{A}}$ is an effective embedding of $\boldsymbol{A}$ to $\boldsymbol{A}$.

REMARKS. There is the feeling that the proper subject matter of generalized recursion theory is those structures whose relations can be used to "build up" the domain. Without going into details, it should be clear that if a structure does allow such "building up" in a systematic way, its recursion theory will be reflexive, for in $\boldsymbol{A}\Box\boldsymbol{A}$ we can then build up, side by side, two copies of $\mathcal{A}$, which amounts to showing $(\text{inj}_{\mathcal{A}})^{\Box}$ is r.e. in $\boldsymbol{A}\Box\boldsymbol{A}$.

Similar remarks apply to ω-recursion theories, of course. But, it is likely that for first order theories, the only ones that are reflexive are those on finite domains.

EXAMPLES. I. $\text{rec}(\mathfrak{S}(\mathbb{N}))$, $\text{rec}(\mathfrak{S}(L_\omega))$ and $\text{rec}(\mathfrak{S}(a_1, \ldots, a_n))$ are all reflexive. We leave the verification to the reader.

II. Let $\mathbb{R}$ be the real numbers, and let $\mathfrak{S}(\mathbb{R})$ be the structure

$\langle \mathbb{R}, +, \times, > \rangle$. It would surprise us if rec($\mathfrak{S}(\mathbb{R})$) turned out to be reflexive. But ω-rec($\mathfrak{S}(\mathbb{N})$) is reflexive relative to $\dot{\cup}$. We sketch the idea and leave it to the reader to fill in the details.

In ω-rec($\mathfrak{S}(\mathbb{R})$) $\dot{\cup}$ ω-rec($\mathfrak{S}(\mathbb{R})$), first show $(\mathrm{inj}_{\mathbb{R}})^{\dot{\cup}}$ *on the integers* is ω-r.e. (it can be done much as in rec($\mathfrak{S}(\mathbb{N})$); the ω-rule will not be needed). Next show $(\mathrm{inj}_{\mathbb{R}})^{\dot{\cup}}$ *on the rationals* is ω-r.e. (represent rationals by fractions, which involves integers; again the ω-rule will not be needed). Finally show $(\mathrm{inj}_{\mathbb{R}})^{\dot{\cup}}$, in full, is ω-r.e. (use the fact that a real r is determined by the set of rationals $< r$; here the ω-rule will be needed).

PROPOSITION 4.1. *Let* $\boldsymbol{A}, \boldsymbol{B} \in C(T, \mathscr{D}, \mathscr{E})$. *If* $\boldsymbol{A}$ *is reflexive, then* $i_{\mathscr{A},\mathscr{B}} : \boldsymbol{A} \to \boldsymbol{A} \,\square\, \boldsymbol{B}$ *is an effective embedding. (Similarly for* $j_{\mathscr{A},\mathscr{B}}$.)

PROOF. Since $C(T, \mathscr{D}, \mathscr{E})$ is a symmetric monoidal category, we have the natural isomorphism

$$\alpha^{-1} : (\boldsymbol{A} \,\square\, \boldsymbol{A}) \,\square\, \boldsymbol{B} \to \boldsymbol{A} \,\square\, (\boldsymbol{A} \,\square\, \boldsymbol{B}).$$

Now, $\mathrm{inj}_{\mathscr{A}}$ is effective, so $(\mathrm{inj}_{\mathscr{A}})^{\square}$ is generated in $\boldsymbol{A} \,\square\, \boldsymbol{A}$. Then $((\mathrm{inj}_{\mathscr{A}})^{\square})^{i_{\mathscr{A}\square\mathscr{A},\mathscr{B}}}$ is generated in $(\boldsymbol{A} \,\square\, \boldsymbol{A}) \,\square\, \boldsymbol{B}$, and so $(((\mathrm{inj}_{\mathscr{A}})^{\square})^{i_{\mathscr{A}\square\mathscr{A},\mathscr{B}}})^{\alpha^{-1}}$ is generated in $\boldsymbol{A} \,\square\, (\boldsymbol{A} \,\square\, \boldsymbol{B})$. We leave it to the reader to show this is $(i_{\mathscr{A},\mathscr{B}})^{\square}$.

PROPOSITION 4.2. *Suppose* $\boldsymbol{A} \in C(T, \mathscr{D}, \mathscr{E})$ *is reflexive, and* $h : \boldsymbol{A} \to \boldsymbol{A}$ *is an embedding. If the relation* $y \in h(x)$ *is generated in* $\boldsymbol{A}$, *then* h *is effective. Further, if* $\boldsymbol{A}$ *is also separable, the converse holds.*

PROOF. If the relation $y \in h(x)$ is generated in $\boldsymbol{A}$, then $h^{i_{\mathscr{A},\mathscr{A}}}$ is generated in $\boldsymbol{A} \,\square\, \boldsymbol{A}$. If also $\boldsymbol{A}$ is reflexive, then $(\mathrm{inj}_{\mathscr{A}})^{\square}$ is generated in $\boldsymbol{A} \,\square\, \boldsymbol{A}$. Define a relation H by

$$H(x, y) \Leftrightarrow (\exists z)[z \in h^{i_{\mathscr{A},\mathscr{A}}}(x) \text{ and } \langle z, y \rangle \in (\mathrm{inj}_{\mathscr{A}})^{\square}].$$

Then H must be generated in $\boldsymbol{A} \,\square\, \boldsymbol{A}$, and $H = h^{\square}$ is effective.

For the converse, suppose h is effective and $\boldsymbol{A}$ is separable and reflexive. Both $h^{\square}$ and $(\mathrm{inj}_{\mathscr{A}})^{\square}$ are generated in $\boldsymbol{A} \,\square\, \boldsymbol{A}$, hence so is K where $K(x, y)$ $\Leftrightarrow (\exists z)[\langle x, z \rangle \in h^{\square}$ and $\langle y, z \rangle \in (\mathrm{inj}_{\mathscr{A}})^{\square}]$. Now $h = K^{i^{-1}_{\mathscr{A},\mathscr{A}}}$ and the result follows by Theorem 4.12.1.

PROPOSITION 4.3. *Suppose* $\boldsymbol{A}, \boldsymbol{B} \in C(T, \mathscr{D}, \mathscr{E})$ *are both reflexive. Then so is* $\boldsymbol{A} \,\square\, \boldsymbol{B}$.

PROOF. If $\boldsymbol{A}$ and $\boldsymbol{B}$ are reflexive, each of $\mathrm{inj}_{\mathscr{A}}$ and $\mathrm{inj}_{\mathscr{B}}$ are effective. But $\mathrm{inj}_{\mathscr{A}\square\mathscr{B}} = \mathrm{inj}_{\mathscr{A}} \,\square\, \mathrm{inj}_{\mathscr{B}}$, and this is effective by Theorem 3.6.

5. An "effective" symmetric monoidal category

Consider the symmetric monoidal category in which $\square$ is one of $\dot{\cup}$ or $\times$, the objects are all recursion theories (all non-empty ones, if $\square$ is $\times$) and the morphisms are all embeddings. Cut it down to a "more effective" category as follows:

(1) Keep only the reflexive, separable recursion theories.

(2) Keep only the effective embeddings between those recursion theories.

Theorem 3.1 says the remaining morphisms are closed under composition. Since our remaining objects are all reflexive, we still have identity morphisms. It follows that we have a sub-category of the category of all recursion theories.

Proposition 4.11.2 and Proposition 4.3 say the remaining objects are closed under $\square$, while Theorem 3.6 says the remaining morphisms are closed under $\square$ too.

Proposition 4.1 says we still have the i and j embeddings. Similar arguments show that, in fact, we have all the appropriate symmetric monoidal machinery. For example, let γ be the "commuting map"

$$\gamma_{\mathscr{A},\mathscr{B}} : \mathscr{A} \square \mathscr{B} \to \mathscr{B} \square \mathscr{A}.$$

Suppose $\boldsymbol{A}$ and $\boldsymbol{B}$ are reflexive and separable, we show $\gamma_{\mathscr{A},\mathscr{B}}$ is an effective embedding

$$\gamma_{\mathscr{A},\mathscr{B}} : \boldsymbol{A} \square \boldsymbol{B} \to \boldsymbol{B} \square \boldsymbol{A}.$$

In fact, as the reader may compute,

$$(\gamma_{\mathscr{A},\mathscr{B}})^{\square} = ((\mathrm{inj}_{\mathscr{A} \square \mathscr{B}})^{\square})^{\pi}$$

where $\pi = \mathrm{inj}_{\mathscr{A} \square \mathscr{B}} \square \gamma_{\mathscr{A},\mathscr{B}}$. Now, since $\boldsymbol{A}$ and $\boldsymbol{B}$ are reflexive, so is $\boldsymbol{A} \square \boldsymbol{B}$ (Proposition 4.3), hence $(\mathrm{inj}_{\mathscr{A} \square \mathscr{B}})^{\square}$ is generated in $(\boldsymbol{A} \square \boldsymbol{B}) \square (\boldsymbol{A} \square \boldsymbol{B})$. π is an embedding, and so takes generated relations to generated relations. Thus $(\gamma_{\mathscr{A},\mathscr{B}})^{\square}$ is generated, so $\gamma_{\mathscr{A},\mathscr{B}}$ is effective.

The embeddings α, λ and ρ may be handled similarly.

Thus we have yet another symmetric monoidal catgegory. And it has a rather "constructive" air to it. All objects are reflexive recursion theories, and we conjecture that these are the ones whose domains can be "effectively" built up, using the given relations. Also all morphisms are effective embeddings, so means exist for calculating them.

The extra restriction to separable recursion theories is, technically, so that Theorem 3.1 can be applied. But it is no serious drawback. By Theorems 4.10.1 and 4.10.3, every recursion theory with an effective pairing function is separable (under either $\dot{\cup}$ or $\times$). And in the next chapter we will show how to extend any recursion theory to one that has an effective pairing function. It is really the restriction to reflexive recursion theories that is significant.

Of course, everything we have said above applies equally well to ω-recursion theories too.

6. Extensions

In rec($\mathfrak{A}$) we may not have enough machinery for a well-behaved theory. One possibility is to "extend" the structure $\mathfrak{A}$ to another structure $\mathfrak{A}'$, and work with rec($\mathfrak{A}'$) instead. This will be a common occurrence in Chapter 6 where we replace a structure $\mathfrak{A}$ by a structure $\mathfrak{A}'$ whose domain consists of all *words* whose letters are members of the domain of $\mathfrak{A}$. By identifying a one-letter word with the letter itself, we have $\mathscr{A} \subseteq \mathscr{A}'$. The "given" relations of $\mathfrak{A}'$ will be those of $\mathfrak{A}$ (on letters) together with concatenation. rec($\mathfrak{A}'$) is generally a much nicer production system to work with than rec($\mathfrak{A}$). It is an example of an *extension* of the production system rec($\mathfrak{A}$).

In this section we define the notion of extension in general, and derive some theorems that will be important for the development in Chapter 6.

DEFINITION. Let $\boldsymbol{A}, \boldsymbol{B} \in C(T, \mathscr{D}, \mathscr{E})$, with $\mathscr{A} \subseteq \mathscr{B}$. Define $\mathrm{inj}_{\mathscr{A},\mathscr{B}} : \mathscr{A} \to \mathscr{B}$ by $\mathrm{inj}_{\mathscr{A},\mathscr{B}}(x) = \{x\}$ for $x \in \mathscr{A}$. If $\mathrm{inj}_{\mathscr{A},\mathscr{B}} : \boldsymbol{A} \to \boldsymbol{B}$ is a morphism of $C(T, \mathscr{D}, \mathscr{E})$ we call $\boldsymbol{B}$ *an extension* of $\boldsymbol{A}$ [in $C(T, \mathscr{D}, \mathscr{E})$].

PROPOSITION 6.1. *Let* $\mathbf{A}, \boldsymbol{B} \in C(T, \mathscr{D}, \mathscr{E})$ *and suppose* $\boldsymbol{B}$ *is an extension of* $\mathbf{A}$. *If a relation* $\mathscr{R}$ *is generated in* $\mathbf{A}$, *then* $\mathscr{R}$ *is generated in* $\mathbf{B}$.

PROOF. By Proposition 3.4.2.

DEFINITION. Suppose $\boldsymbol{B}$ is an extension of $\mathbf{A}$. If $(\mathrm{inj}_{\mathscr{A},\mathscr{B}})^{-1}$ is a co-embedding, we call $\boldsymbol{B}$ a *conservative extension* of $\mathbf{A}$.

PROPOSITION 6.2. *Suppose* $\boldsymbol{B}$ *is a conservative extension of* $\mathbf{A}$ *and* $\mathscr{R}$ *is a relation on* $\mathscr{A}$. $\mathscr{R}$ *is generated in* $\mathbf{A}$ *iff* $\mathscr{R}$ *is generated in* $\mathbf{B}$.

PROOF. By Proposition 3.4.2 again.

LEMMA 6.3. *Let* $\mathbf{A}, \boldsymbol{B} \in C(T, \mathscr{D}, \mathscr{E})$ *and suppose* $\boldsymbol{B}$ *is an extension of* $\mathbf{A}$ *where* $\mathbf{A}$ *is reflexive. Then* $\mathrm{inj}_{\mathscr{A},\mathscr{B}}$ *is an* effective *embedding of* $\mathbf{A}$ *in* $\mathbf{B}$.

PROOF. Using Proposition 4.4.2, the following diagram commutes.

$$\begin{array}{ccc} \boldsymbol{A} & \xrightarrow{\quad j \quad} & \boldsymbol{A} \,\square\, \boldsymbol{A} \\ \Big\downarrow{\scriptstyle \mathrm{inj}_{\mathscr{A},\mathscr{B}}} & & \Big\downarrow{\scriptstyle 1_{\mathscr{A}} \square\, \mathrm{inj}_{\mathscr{A},\mathscr{B}} = \theta} \\ \boldsymbol{B} & \xrightarrow{\quad j \quad} & \boldsymbol{A} \,\square\, \boldsymbol{B} \end{array}$$

Now, $\boldsymbol{A}$ is reflexive, so $(\mathrm{inj}_{\mathscr{A}})^{\square}$ is generated in $\boldsymbol{A} \,\square\, \boldsymbol{A}$. Then $((\mathrm{inj}_{\mathscr{A}})^{\square})^{\theta}$ is generated in $A \,\square\, \boldsymbol{B}$. It is $(\mathrm{inj}_{\mathscr{A},\mathscr{B}})^{\square}$.

THEOREM 6.4. *Let* $\boldsymbol{A}, \boldsymbol{B} \in C(T, \mathscr{D}, \mathscr{E})$. *Suppose* $\boldsymbol{B}$ *is an extension of* $\boldsymbol{A}$, $\boldsymbol{A}$ *is reflexive and separable, and* $\boldsymbol{B}$ *has some morphism to* $\boldsymbol{A}$. *Then* $\boldsymbol{B}$ *is a* conservative *extension of* $\boldsymbol{A}$.

PROOF. By the lemma, $\mathrm{inj}_{\mathscr{A},\mathscr{B}}$ is effective. The result now follows by Theorem 3.4.

Suppose $\boldsymbol{B}$ is an extension of $\boldsymbol{A}$ and $h : \boldsymbol{B} \to \boldsymbol{A}$ is a morphism of $C(T, \mathscr{D}, \mathscr{E})$. Let $h \restriction \mathscr{A}$ be the coding h restricted to members of $\mathscr{A}$ (recall, if $\boldsymbol{B}$ extends $\boldsymbol{A}$, $\mathscr{A} \subseteq \mathscr{B}$). Clearly $h \restriction \mathscr{A}$ is a coding of $\mathscr{A}$ to $\mathscr{A}$. But also, it is easy to see that $h \restriction \mathscr{A} = h\ \mathrm{inj}_{\mathscr{A},\mathscr{B}}$, hence it is a morphism of $\boldsymbol{A}$ to $\boldsymbol{A}$.

DEFINITION. Let $\boldsymbol{B}$ be an extension of $\boldsymbol{A}$ and $h : \boldsymbol{B} \to \boldsymbol{A}$ be a morphism. If the relation $y \in (h \restriction \mathscr{A})(x)$ is generated in $\boldsymbol{A}$ we call $\boldsymbol{B}$ an *inessential extension* of $\boldsymbol{A}$ (via h).

THEOREM 6.5. *If* $\boldsymbol{B}$ *is an inessential extension of* $\boldsymbol{A}$ *then* $\boldsymbol{B}$ *is a conservative extension of* $\boldsymbol{A}$.

PROOF. Say $\boldsymbol{B}$ is an inessential extension of $\boldsymbol{A}$ via h. Let Φ be an operator in $\boldsymbol{B}$ of order $\langle n, m\rangle$. We show $\Phi^{\mathrm{inj}^{-1}}$ is an operator in $\boldsymbol{A}$ to establish directly that inj^{-1} is a co-embedding (we are writing inj for $\mathrm{inj}_{\mathscr{A},\mathscr{B}}$).

$h : \boldsymbol{B} \to \boldsymbol{A}$ is a morphism, so Φ^h is an operator in $\boldsymbol{A}$. Also $y \in (h \restriction \mathscr{A})$ is generated in $\boldsymbol{A}$ so by Proposition 4.12.2, both $(h \restriction \mathscr{A})_n$ and $(h \restriction \mathscr{A})_m^{-1}$ are operators in $\boldsymbol{A}$. Then the following is an operator in $\boldsymbol{A}$:

$$\begin{aligned} (h \restriction \mathscr{A})_m^{-1} \Phi^h (h \restriction \mathscr{A})_n &= (h\ \mathrm{inj})_m^{-1} h_m \Phi h_n^{-1} (h\ \mathrm{inj})_n \\ &= \mathrm{inj}_m^{-1} h_m^{-1} h_m \Phi h_n^{-1} h_n\ \mathrm{inj}_n \\ &= \mathrm{inj}_m^{-1} \Phi\ \mathrm{inj}_n = \Phi^{\mathrm{inj}^{-1}}. \end{aligned}$$

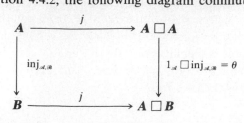

7. Isomorphisms

Let $\boldsymbol{A}$ be a production system. It can be viewed as a category, with objects the various $[\mathscr{A}]^n$, and with operators as morphisms. It is the 1–1, onto embeddings that can serve as *functors* between such categories (see Propositions 3.3.3 and 3.3.4) so it is natural to ask when 1–1, onto embeddings exist, and when they are unique. We begin with uniqueness. We will see roughly that *effective* 1–1, onto embeddings between *reflexive*, *separable* production systems are unique up to natural equivalence. Then we give some conditions for existence, which are enough to guarantee that there are 1–1, onto, effective embeddings between rec($\mathfrak{S}(\mathbb{N})$), rec($\mathfrak{S}(\mathrm{L}_\omega)$) and rec($\mathfrak{S}(a_1, \ldots, a_n)$), our "standard" recursion theories.

DEFINITION. Let $\boldsymbol{A}, \boldsymbol{B} \in C(T, \mathscr{D}, \mathscr{E})$, and let $f : \boldsymbol{A} \to \boldsymbol{B}$ be a morphism. If f is 1–1 and onto, we call f an *isomorphism*. If f is also effective, we call it an *effective isomorphism*.

REMARKS. If $f : \boldsymbol{A} \to \boldsymbol{B}$ is an isomorphism, then as a coding, it has an inverse. The symbol f^{-1} is already in use for the co-embedding, so we use f' for the inverse coding. Specifically, $f : \mathscr{B} \to \mathscr{A}$ is given by

$$y \in f'(x) \Leftrightarrow x \in f(y).$$

Clearly $ff' = 1_{\mathscr{B}}$ and $f'f = 1_{\mathscr{A}}$. Also f' and f^{-1} are closely related: if Φ is an operator in $\boldsymbol{B}$, $\Phi^{f'} = \Phi^{f^{-1}}$.

DEFINITION. We call $\langle T, \mathscr{D}, \mathscr{E} \rangle$ *full*, relative to $\square$, if, for any objects $\boldsymbol{A}, \boldsymbol{B}$ of $C(T, \mathscr{D}, \mathscr{E})$, and embedding $\theta : \boldsymbol{A} \to \boldsymbol{B}$, if θ is effective (relative to $\square$), then θ is a morphism of $C(T, \mathscr{D}, \mathscr{E})$.

PROPOSITION 7.1. *Suppose $\langle T, \mathscr{D}, \mathscr{E} \rangle$ is full, $\boldsymbol{A}, \boldsymbol{B} \in C(T, \mathscr{D}, \mathscr{E})$, $f : \boldsymbol{A} \to \boldsymbol{B}$ is an effective isomorphism, $\boldsymbol{A}$ is separable, and $\boldsymbol{B}$ has some morphism to $\boldsymbol{A}$. Then $f' : \boldsymbol{B} \to \boldsymbol{A}$ is also an effective isomorphism.*

PROOF. By Theorem 3.4, f^{-1} is a co-embedding. For each operator $\Phi \in \boldsymbol{B}$, $\Phi^{f'} = \Phi^{f^{-1}}$ and it follows that f' is an embedding. It is trivially 1–1 and onto.

Let L be the logical operator in $\boldsymbol{B} \square \boldsymbol{A}$ such that $L : \langle u, v \rangle \to \langle v, u \rangle$. Also let γ be the "commuting map" of $C(T, \mathscr{D}, \mathscr{E})$. Then f' is effective, since

$$(f')^{\square} = L((f^{\square})^{\gamma}).$$

Since $\langle T, \mathscr{D}, \mathscr{E} \rangle$ is full, f' is a morphism.

COROLLARY 7.2. *Suppose* $\langle T, \mathcal{D}, \mathcal{E}\rangle$ *is full,* $h : \boldsymbol{B} \to \boldsymbol{B}$ *is an effective isomorphism and* $\boldsymbol{B}$ *is separable. Then* h' *is also an effective isomorphism.*

PROOF. Take $\boldsymbol{A} = \boldsymbol{B}$ above.

THEOREM 7.3. *Let* $\boldsymbol{A}, \boldsymbol{B} \in C(T, \mathcal{D}, \mathcal{E})$ *where* $\langle T, \mathcal{D}, \mathcal{E}\rangle$ *is full. Suppose* $\boldsymbol{B}$ *is separable and reflexive. Let* $f, g : \boldsymbol{A} \to \boldsymbol{B}$ *be two effective isomorphisms. Then there is a coding* $h : \mathcal{B} \to \mathcal{B}$ *such that*:

(1) *The relation* $y \in h(x)$ *is generated in* $\boldsymbol{B}$.
(2) $h : \boldsymbol{B} \to \boldsymbol{B}$ *is an effective isomorphism.*
(3) $g = hf$.

PROOF. f and g are effective, so both $f^{\square}$ and $g^{\square}$ are generated in $\boldsymbol{A} \square \boldsymbol{B}$. Define a relation R by

$$R(u, v) \iff (\exists x)[\langle x, u\rangle \in f^{\square} \text{ and } \langle x, v\rangle \in g^{\square}].$$

Then R is also generated in $\boldsymbol{A} \square \boldsymbol{B}$. By Theorem 4.12.1, $R^{(j_{\mathcal{A},\mathcal{B}})^{-1}} = S$ is generated in $\boldsymbol{B}$. And it is easy to see that

$$S(q, r) \iff q \in f(x) \text{ and } r \in g(x) \text{ for some } x \in \mathcal{A}.$$

Now define a mapping h by

$$r \in h(q) \iff S(q, r).$$

Since f and g are 1–1 and onto, it follows that $h : \mathcal{B} \to \mathcal{B}$ is a 1–1, onto coding. And the relation $y \in h(x)$ is generated in $\boldsymbol{B}$.

By Proposition 4.12.2, for each n, both h_n and h_n^{-1} are operators in $\boldsymbol{B}$. Since operators take allowable relations to allowable relations, h is a pre-embedding.

Let Φ be an operator in $\boldsymbol{B}$ of order $\langle n, m\rangle$. Then $\Phi^h = h_m \Phi h_n^{-1}$. But this is the composition of three operators in $\boldsymbol{B}$, hence is an operator in $\boldsymbol{B}$. Thus h is an embedding.

h is effective by Proposition 4.2, hence also an isomorphism since $\langle T, \mathcal{D}, \mathcal{E}\rangle$ is full.

Finally, that $g = hf$ is a straightforward calculation.

COROLLARY 7.4. *Let* $\boldsymbol{A}, \boldsymbol{B} \in C(T, \mathcal{D}, \mathcal{E})$ *where* $\langle T, \mathcal{D}, \mathcal{E}\rangle$ *is full. Suppose* $\boldsymbol{B}$ *is separable and reflexive. Any two effective isomorphisms from* $\boldsymbol{A}$ *to* $\boldsymbol{B}$ *are naturally equivalent* (*thought of as functors*).

PROOF. Let $f, g : \boldsymbol{A} \to \boldsymbol{B}$ both be effective isomorphisms. Let h be as in Theorem 7.3. We use it to define a natural transformation

$$\tau : f \dot{\to} g.$$

Since $y \in h(x)$ is generated in $\boldsymbol{B}$, for each n, h_n is an operator in $\boldsymbol{B}$. Let $[\mathscr{A}]^n$ be an object in $\boldsymbol{A}$; we let τ assign to it the morphism h_n in $\boldsymbol{B}$.

Now, let $\Phi : [\mathscr{A}]^n \to [\mathscr{A}]^m$ be an operator in $\boldsymbol{A}$. We show the following diagram commutes, which says τ is a natural transformation.

$$\begin{array}{ccc} [\mathscr{B}]^n & \xrightarrow{\quad \tau([\mathscr{A}]^n) \quad} & [\mathscr{B}]^n \\ \Big\downarrow \Phi^f & & \Big\downarrow \Phi^g \\ [\mathscr{B}]^m & \xrightarrow{\quad \tau([\mathscr{A}]^m) \quad} & [\mathscr{B}]^m \end{array}$$

That is, we must show $\Phi^g h_n = h_m \Phi^f$ or

$$g_m \Phi g_n^{-1} h_n = h_m f_m \Phi f_n^{-1}. \tag{$*$}$$

Now, $hf = g$, so $h_n f_n = g_n$, so

$$g_n^{-1} h_n f_n = 1.$$

And since f is 1–1 and onto,

$$g_n^{-1} h_n = f_n^{-1}.$$

Also $h_m f_m = g_m$, and these facts give ($*$).

Thus τ is a natural transformation. And each component of τ is invertible; specifically, the inverse of h_n is h_n^{-1}, also an operator in $\boldsymbol{B}$. Thus we have a natural equivalence.

Now we turn to the question of the *existence* of effective isomorphisms. We begin with replacing embeddings by 1–1 embeddings.

THEOREM 7.5. *Let* $\boldsymbol{A}, \boldsymbol{B} \in C(T, \mathscr{D}, \mathscr{E})$ *and let* $f : \boldsymbol{A} \to \boldsymbol{B}$ *be an embedding. Suppose the following conditions hold*:

(1) $f_2(=_{\mathscr{A}})$ *is bi-generated in* $\boldsymbol{B}$.

(2) *The domain of* $\boldsymbol{B}$ *is an ordinal.*

(3) *If* $R(x, y)$ *and its complement are both generated in* $\boldsymbol{B}$, *then* $S(z, y)$ *is generated where*

$$S(z, y) \iff z = \mu x R(x, y) \iff z = \text{the least } x \text{ such that } R(x, y).$$

Then there is a 1–1 *embedding* $g : \boldsymbol{A} \to \boldsymbol{B}$.

Further, if f *is effective, so is* g.

PROOF. f may not be 1–1. We define a new coding g so that $g(x)=\{y\}$ where y is the *least* member of $f(x)$. Then we show g is an embedding.

Let w be some member of $f(x)$. Then in fact,

$$f(x)=\{z \mid w(=_{\mathscr{A}})^f z\}.$$

So, our definition of g is:

$$y \in g(x) \Leftrightarrow (\exists w)[w \in f(x) \wedge y = \mu z(w(=_{\mathscr{A}})^f z)].$$

It should be clear that g is a 1–1 coding, $g: \mathscr{A} \to \mathscr{B}$.

$(=_{\mathscr{A}})^f$ is bi-generated in $\boldsymbol{B}$, so $S(y, w) \Leftrightarrow y = \mu z[w(=_{\mathscr{A}})^f z]$ is generated in $\boldsymbol{B}$. For each n, we define relations S_n and S_n^{-1} by

$$S_n(y_1, \ldots, y_n, w_1, \ldots, w_n) \Leftrightarrow S(y_1, w_1) \wedge \cdots \wedge S(y_n, w_n),$$

$$S_n^{-1}(\mathbf{w}, \mathbf{y}) \Leftrightarrow S_n(\mathbf{y}, \mathbf{w}).$$

Then S_n and S_n^{-1} are also generated in $\boldsymbol{B}$.

For $\mathscr{P} \subseteq \mathscr{B}^n$, $S_n''(\mathscr{P})$ is the result of enlarging $\mathscr{P}$ by throwing in all f-codes for members of $\mathscr{A}^n$ which have minimal f-codes in $\mathscr{P}$, while $(S_n^{-1})''(\mathscr{P})$ cuts things down to just minimal f-codes. It is not hard to see, then, that

$$g_n = (S_n^{-1})'' f_n \quad \text{and} \quad g_n^{-1} = f_n^{-1} S_n''.$$

Further, by Corollary 2.5.2, both S_n'' and $(S_n^{-1})''$ are operators in $\boldsymbol{B}$.

Now, if $\mathscr{P} \subseteq \mathscr{A}^n$ is allowable in $\boldsymbol{A}$, $g_n(\mathscr{P}) = (S_n^{-1})'' f_n(\mathscr{P})$, which is allowable in $\boldsymbol{B}$ since f is an embedding and $(S_n^{-1})''$ is an operator. Similarly for g_n^{-1}. Thus g is a pre-embedding.

Let $\Phi \in \boldsymbol{A}$ be of order $\langle n, m\rangle$. Then

$$\begin{aligned}\Phi^g &= g_m \Phi g_n^{-1} = (S_m^{-1})'' f_m \Phi f_n^{-1} S_n'' \\ &= (S_m^{-1})'' \Phi^f S_n''.\end{aligned}$$

Since f is an embedding, $\Phi^f \in \boldsymbol{B}$, hence $\Phi^g \in \boldsymbol{B}$ since $\boldsymbol{B}$ is closed under composition. Thus g is an embedding.

Now suppose f is effective; we show g also is. S is generated in $\boldsymbol{B}$, so S^j is generated in $\boldsymbol{A} \square \boldsymbol{B}$. Then Q is also generated in $\boldsymbol{A} \square \boldsymbol{B}$ where

$$Q(x, y) \Leftrightarrow (\exists w)[\langle x, w\rangle \in f^{\square} \text{ and } S^j(y, w)].$$

We leave it to the reader to verify that

$$\langle x, y\rangle \in g^{\square} \Leftrightarrow Q(x, y)$$

which concludes the proof.

REMARKS. $\boldsymbol{B} = \mathrm{rec}(\mathfrak{S}(\mathbb{N}))$ meets conditions (2) and (3) above. (2) is obvious, while (3) is a standard result of ordinary recursion theory. Now, in Ch. 3, §6 we gave embeddings of rec($\mathfrak{S}(L_\omega)$) [VIIa] and rec($\mathfrak{S}(a_1, \ldots, a_n)$) [III] into rec($\mathfrak{S}(\mathbb{N})$). In the present chapter we remarked that these embeddings are effective. It is easy to check that they carry the respective equality relations to recursive relations on $\mathbb{N}$. Then, by the above theorem, there are effective 1–1 embeddings of rec($\mathfrak{S}(L_\omega)$) and rec($\mathfrak{S}(a_1, \ldots, a_n)$) into rec($\mathfrak{S}(\mathbb{N})$).

Next we turn to the existence of *onto* effective embeddings.

THEOREM 7.6. *Let $\boldsymbol{A}, \boldsymbol{B} \in C(T, \mathcal{D}, \mathcal{E})$ where $\langle T, \mathcal{D}, \mathcal{E} \rangle$ is full, and suppose each has an effective* 1–1 *embedding in the other. Suppose the following additional conditions hold*:

(1) ***B** is reflexive.*

(2) ***B** is separable.*

(3) *The domain of **B** is an ordinal, β.*

(4) *Every generated set in **B** is the range of a* 1–1 *generated function whose domain is an initial segment of β.*

(5) *There is no* 1–1 *generated function from β to a proper initial segment. Then **A** and **B** are effectively isomorphic.*

PROOF. Let $f : \boldsymbol{A} \to \boldsymbol{B}$ and $g : \boldsymbol{B} \to \boldsymbol{A}$ be effective, 1–1 embeddings, hence morphisms. $\mathcal{A}$ is generated in **A**, so $\mathcal{A}^f$ is generated in **B**. Then there must be a 1–1 generated function, say h, whose domain is an initial segment of β, say it is the ordinal $\alpha \leq \beta$, and whose range is $\mathcal{A}^f$. We first show $\alpha = \beta$.

Both f and g are effective, hence the composition $fg : \boldsymbol{B} \to \boldsymbol{B}$ is also effective, by Theorem 3.1, and trivially 1–1. By Proposition 4.2, the relation $y \in (fg)(x)$ is generated in **B**. Now define a map k by

$$y = k(x) \Leftrightarrow (\exists z)[h(y) = z \text{ and } z \in (fg)(x)].$$

k is well-defined, is 1–1, and $k : \beta \to \alpha$. Further, k is generated in **B**. Then by condition (5), $\alpha = \beta$. So $h : \beta \to \mathcal{A}^f$.

Now we define a coding $p : \mathcal{A} \to \mathcal{B}$ as follows.

$$y \in p(x) \Leftrightarrow (\exists z)[z \in f(x) \text{ and } y = h^{-1}(z)].$$

It should be clear that p is a 1–1 onto coding. That p is an embedding, and is effective, are both routine.

REMARKS. $\boldsymbol{B}$ = rec($\mathfrak{S}(\mathbb{N})$) meets the five conditions of the above theorem. The only one that is not immediate at this point is (4), which is a standard result of ordinary recursion theory not proved here.

In Ch. 3, §6 we gave 1–1 embeddings of rec($\mathfrak{S}(\mathbb{N})$) into rec($\mathfrak{S}(L_\omega)$) [I] and rec($\mathfrak{S}(a_1, \ldots, a_n)$) [IV]. These embeddings are effective. By the remarks above, there are effective 1–1 embeddings the other way around. It follows that rec($\mathfrak{S}(\mathbb{N})$), rec($\mathfrak{S}(L_\omega)$) and rec($\mathfrak{S}(a_1, \ldots, a_n)$) are all effectively isomorphic.

In Example 7 of Ch. 3, §6, two embeddings from rec($\mathfrak{S}(L_\omega)$) to rec($\mathfrak{S}(\mathbb{N})$) were given. This gives two effective isomorphisms between these recursion theories. They are not the same, but they are naturally equivalent.

CHAPTER SIX

INDEXED RECURSION AND ω-RECURSION THEORIES

1. Introduction

In any recursion theory the r.e. sets are the ones that have elementary formal systems for generating them. In Ch. 1, §10 we proved a number of facts about r.e. sets by manipulating elementary formal systems. Now, in *ordinary* recursion theory it is possible to "code" elementary formal systems by natural numbers in a way that captures all the pertinent information. An *index* of an r.e. set $\mathcal{R}$ is a natural number that thus "codes" instructions for generating $\mathcal{R}$. In ordinary recursion theory every r.e. relation has an index. (Kleene's Enumeration Theorem.) These indexes, being natural numbers, are subjects to be worked on *within* ordinary recursion theory. It turns out that the indexing can be done in such a way that the manipulations *we* do on the r.e. sets (forming $\cap$, $\cup$, etc.) correspond to recursive functions on their indexes. In effect, ordinary recursion theory can "talk about itself" via this indexing.

All this is equally true of hyperarithmetic theory, but it is not the case for every recursion and ω-recursion theory. In this chapter we investigate which ones can be indexed in a way that makes "talking about itself" possible. Then in the next two chapters we examine the consequences of this ability.

Our plan of procedure is quite direct. We "talk about" the r.e. sets of rec($\mathfrak{A}$) by using elementary formal systems. An elementary formal system axiom is a word, whose letters are members of $\mathcal{A}$ and various formal symbols. So we set up another recursion theory, rec($\mathfrak{A}'$), which has these words as members of its domain. We construct it in such a way that the sorts of things we do with the elementary formal systems of rec($\mathfrak{A}$), from the outside, can be done *within* rec($\mathfrak{A}'$), and in pretty much the way we were doing them. Thus for every recursion theory rec($\mathfrak{A}$), there is another recursion theory, rec($\mathfrak{A}'$), that can "talk about it" in essentially the same way we can. (The same is true for ω-recursion theories as well.)

Then we show that if rec($\mathfrak{A}$) has the right properties, rec($\mathfrak{A}'$) will be an *inessential extension* of it (Ch. 5, §6), and this has the consequence that rec($\mathfrak{A}$) can "talk about" itself.

As described above, for each $\mathfrak{A}$, rec($\mathfrak{A}'$) is a recursion theory of *words*, and may be thought of as generalizing rec($\mathfrak{S}(a_1, \ldots, a_n)$). It is possible to create a recursion theory of *sets* that also can "talk about" rec($\mathfrak{A}$). It may be thought of as generalizing rec($\mathfrak{S}(L_\omega)$). We find the word approach more natural, but we define the basics of the set approach as well, partly because it allows us to make connections between our work and other notions of generalized recursion theory in the literature, notably search computability. See Moschovakis [1969].

Finally, by techniques similar to those described above, we show that the *finite* sets can often be given codes, so that they can be "discussed" within our theory. This is important for recursion theories, at least, since enumeration operators are compact, and so are completely determined by their behavior on the finite sets. It is not the case for ω-enumeration operators, however. In Chapter 8, we will produce sets that play the role, for ω-enumeration operators, that the finite sets play for enumeration operators. But the situation is much more complicated, and we will find *partial* production systems emerging then.

IMPORTANT NOTE. For the entire of this chapter, $T(\text{-})$ is either rec(-) or ω-rec(-). No other theory assignments are considered.

2. Elaborating a structure

Let $\mathfrak{A} = \langle \mathcal{A}; \mathcal{R}_1, \ldots, \mathcal{R}_k \rangle$ be a structure. It may happen that $\mathfrak{A}$ itself is not rich enough for our purposes, and so we extend $\mathfrak{A}$ to a structure $\mathfrak{A}^*$ which is more elaborate, and we do our work there instead. For example, in Montague [1968], $\mathcal{A}$ is enlarged to a set closed under the formation of subsets bounded in size by a given cardinal. In Moschovakis [1969], closure under a pairing function is similarly introduced.

In this section we describe three methods of elaborating a structure: (1) closing it under a pairing function, as in Moschovakis [1969]; (2) closing it under the formation of finite words; and (3) closing it under the formation of finite subsets, which is essentially the $\aleph_0$ case of Montague [1968]. In the next few sections we develop methods (2) and (3) in some detail, then concentrate on method (2) for the rest of the chapter. Any of the three methods can be used to derive similar results; we simply find the word approach most congenial.

(1) *Closure under pairs*; Moschovakis [1969].

Choose some object not in $\mathscr{A}$, call it 0. Let $\mathscr{A}^0 = \mathscr{A} \cup \{0\}$. Choose some pairing function J having the property that no member of $\mathscr{A}^0$ is a pair under J. Form the least set, $\mathscr{A}^P$, which

(a) extends $\mathscr{A}^0$, and

(b) is closed under J.

Now, by $\mathfrak{A}^P$ we mean the structure $\langle \mathscr{A}^P; \mathscr{A}, J \restriction \mathscr{A}^P, \mathscr{R}_1, \ldots, \mathscr{R}_k \rangle$.

REMARKS. It is essential that no member of $\mathscr{A}^0$ be a pair under J. We often wish to extend a function f from $\mathscr{A}^0$ to $\mathscr{A}^P$, and we do this inductively; we define f on $J(x, y)$ in terms of its values on x and y. If some member a of $\mathscr{A}^0$ were itself a pair under J, an ambiguity could arise, since f has its original value on a, and may get a different value under the process of extending f to pairs.

It is an important point where the pairing function J is to come from. We assume that all the mathematics in this book is being carried out in the framework of set theory as a metatheory (ZFC if you like). In set theory it is easy to construct many examples of pairing functions, and to produce one, J, meeting the condition that no member of $\mathscr{A}^0$ is a pair under J. (As one example, let $J(x, y) = \langle\langle x, \mathscr{A}^0\rangle, \langle y, \mathscr{A}^0\rangle\rangle$ where $\langle\ ,\ \rangle$ is the usual Kuratowski pairing function. If some member of $\mathscr{A}^0$ were a pair under J the axiom of regularity would be contradicted.) This section is one of the few places we need to be conscious of our metatheory.

(2) *Closure under finite words.*

Choose some notion of finite sequence (word, string) having the property that no member of $\mathscr{A}$ is a finite sequence. By $\mathscr{A}^w$ we mean the collection of all words over $\mathscr{A}$. We identify a word of length one with its only letter, so $\mathscr{A} \subseteq \mathscr{A}^w$. We allow the empty word, of length 0.

Let $*$ be a notion of concatenation appropriate to the notion of finite sequence chosen.

By $\mathfrak{A}^w$ we mean the structure $\langle \mathscr{A}^w; \mathscr{A}, * \restriction \mathscr{A}^w, \mathscr{R}_1, \ldots, \mathscr{R}_k \rangle$. We call $\mathfrak{A}^w$ the *word structure over* $\mathfrak{A}$.

Using the notation of Ch. 1, §2, $\{a_1, \ldots, a_n\}^w$ and $\mathscr{W}(a_1, \ldots, a_n)$ are essentially the same, so any results we derive about our present word structures apply to $\mathfrak{S}(a_1, \ldots, a_n)$. If $\mathscr{A}$ is infinite, $\mathfrak{A}^w$ can be thought of as a structure of words on an infinite alphabet, with some relations given on the letters.

REMARKS. Again it is essential that a notion of finite sequence is chosen so that no member of $\mathcal{A}$ is also a finite sequence. In the framework of our set-metatheory there are many ways this can be done.

(3) *Closure under finite subsets*; Montague [1968].

Choose some notion of "set", a corresponding notion, "$\in$", of "member-of", so that no member of $\mathcal{A}$ is a "set". Define a sequence by

$$\mathcal{A}^0 = \emptyset,$$

$$\mathcal{A}^{n+1} = \text{the collection of all finite "subsets" of } \mathcal{A}^n \cup \mathcal{A}.$$

Let $\mathcal{A}^\omega = \bigcup_{n\in\omega} \mathcal{A}^n$.

The members of $\mathcal{A}$ are called *urelements* and the members of $\mathcal{A}^\omega$ are called *hereditarily finite sets over* $\mathcal{A}$. Let $\mathcal{A}^s = \mathcal{A}^\omega \cup \mathcal{A}$. Let us denote by $x \cup \{y\} = z$ the relation: "set" z is "set" x with y added as "member".

By $\mathfrak{A}^s$ we mean the structure $\langle \mathcal{A}^s; \mathcal{A}, (x \cup \{y\} = z) \restriction \mathcal{A}^s, \mathcal{R}_1, \ldots, \mathcal{R}_k \rangle$. We call $\mathfrak{A}^s$ the *set structure over* $\mathfrak{A}$. We note that it is an obvious generalization of $\mathfrak{S}(L_\omega)$ of Ch. 1, §2.

REMARKS. It is essential that no member of $\mathcal{A}$ be a "set", just as above, no member of $\mathcal{A}$ was to be a pair, or a finite sequence. One way to do this is as follows. Call something a "set" if it is of the form $\langle x, \mathcal{A}\rangle$ [$\langle\ ,\ \rangle$ is the usual pairing function of set theory]. Say $\langle x, \mathcal{A}\rangle$ "$\in$" $\langle y, \mathcal{A}\rangle$ if $x \in y$. This produces, within set theory, an isomorphic copy of the universe of sets, but in which no member of $\mathcal{A}$ is a "set".

One might feel a certain discomfort with this third method. The problem is psychological, rather than logical, but nonetheless real. Within our metatheory, which is set theory, one is shifting the meaning of set. There is no comparable problem with finite sequence, or ordered pair, since these are not primitives of set theory. And there would be no psychological difficulty at all if our metatheory were something other than set theory. The unfortunate fact is that for most of us, set theory, as a foundation for all mathematics, is the only game in town. This, surely, is an accident of history, and will not continue. On the other hand, whatever foundation of mathematics is accepted, a similar sense of uneasiness may arise when its primitives must be shifted.

3. Recursion and ω-recursion theories

Recall the definition of *extension*, in Ch. 5, §6. Also recall, $T(\text{-})$ is one of rec(-) or ω-rec(-) for this chapter.

PROPOSITION 3.1. *$T(\mathfrak{A}^p)$, $T(\mathfrak{A}^w)$ and $T(\mathfrak{A}^s)$ are all extensions of $T(\mathfrak{A})$.*

PROOF. We state the proof for $T(\mathfrak{A}^p)$; the others are similar. Define $\text{inj}(v) = \{v\}$ for $v \in \mathcal{A}$. We must show $\text{inj} : T(\mathfrak{A}) \to T(\mathfrak{A}^p)$ is an embedding.

Now, inj is a pre-embedding since T assigns total production systems to structures. Observe that $(=_{\mathcal{A}})^{\text{inj}}$ is simply the equality relation on $\mathcal{A}^p$, intersected with $\mathcal{A} \times \mathcal{A}$. Since $\mathcal{A}$ is "given" in $\mathfrak{A}^p$, it follows from the closure features of production systems that $(=_{\mathcal{A}})^{\text{inj}}$ is generated in $T(\mathfrak{A}^p)$. The proposition now follows by the definition of $\mathfrak{A}^p$ and the fact that T is an elementary theory assignment.

The question of when $T(\mathfrak{A}^p)$ or $T(\mathfrak{A}^w)$ or $T(\mathfrak{A}^s)$ is an *inessential* extension of $T(\mathfrak{A})$ is an important one, and we will turn to it in Section 8. We devote the rest of this section to extending embeddings to the elaborated structures we have introduced.

DEFINITION. Let $\mathcal{A}$ and $\mathcal{B}$ be sets and consider $\mathcal{A}^p$ and $\mathcal{B}^p$. We denote the extra object added to $\mathcal{A}$ by $0_{\mathcal{A}}$, and the pairing function for $\mathcal{A}^p$ by $J_{\mathcal{A}}$. Similarly for $\mathcal{B}$.

Let $\theta : \mathcal{A} \to \mathcal{B}$ be a coding. We extend it to a map θ^p on $\mathcal{A}^p$ as follows:

(1) if $a \in \mathcal{A}$, $\theta^p(a) = \theta(a)$,

(2) $\theta^p(0_{\mathcal{A}}) = \{0_{\mathcal{B}}\}$,

(3) $\theta^p(J_{\mathcal{A}}(x, y)) = \{J_{\mathcal{B}}(u, v) \mid u \in \theta^p(x) \text{ and } v \in \theta^p(y)\}$.

It should be clear that θ^p is also a *coding*, $\theta^p : \mathcal{A}^p \to \mathcal{B}^p$, that agrees with θ on $\mathcal{A}$.

DEFINITION. Again let $\mathcal{A}$ and $\mathcal{B}$ be sets, and consider $\mathcal{A}^w$ and $\mathcal{B}^w$. We denote by juxtaposition the concatenation operations of each.

Let $\theta : \mathcal{A} \to \mathcal{B}$ be a coding. We extend it to a map θ^w on $\mathcal{A}^w$ as follows. Let $a_1 a_2 \cdots a_n$ be a word over $\mathcal{A}$, where each a_i is a letter (member of $\mathcal{A}$). By $\theta^w(a_1 a_2 \cdots a_n)$ we mean

$$\{b_1 b_2 \cdots b_n \mid b_1 \in \theta(a_1) \wedge \cdots \wedge b_n \in \theta(a_n)\}.$$

Again, it should be clear that θ^w is a *coding*, $\theta^w : \mathcal{A}^w \to \mathcal{B}^w$, that agrees with θ on $\mathcal{A}$.

DEFINITION. Finally, let $\mathcal{A}$ and $\mathcal{B}$ be sets again, and consider $\mathcal{A}^s$ and $\mathcal{B}^s$. We use $\{x_1, \ldots, x_n\}$ to indicate the "set" with $x_1, \ldots, x_n$ as "members", both in $\mathcal{A}^s$ and $\mathcal{B}^s$.

Let $\theta : \mathcal{A} \to \mathcal{B}$ be a coding. We extend it to a map θ^s on $\mathcal{A}^s$ as follows.

(1) if $a \in \mathcal{A}$, $\theta^s(a) = \theta(a)$,

(2) if $\{a_1, \ldots, a_n\}$ is a "set" in $\mathcal{A}^s$,

$$\theta^s(\{a_1, \ldots, a_n\}) = \{\{b_1, \ldots, b_n\} \in \mathcal{B}^s \mid b_1 \in \theta^s(a_1) \wedge \cdots \wedge b_n \in \theta^s(a_n)\}.$$

Once again, it should be clear that θ^s is a *coding*, $\theta^s : \mathcal{A}^s \to \mathcal{B}^s$, that agrees with θ on $\mathcal{A}$.

PROPOSITION 3.2. *Let $*$ be either p or w or s. If $\theta : T(\mathfrak{A}) \to T(\mathfrak{B})$ is an embedding, then $\theta^* : T(\mathfrak{A}^*) \to T(\mathfrak{B}^*)$ is also an embedding.*

PROOF. Left to the reader.

4. Σ completeness

This section is the first of several in which we study word and set structures in some detail, dropping the pairing function elaboration. In word structures we use $*$ for concatenation. In set structures we use the customary $\in$ symbol for membership. No confusion should result.

We define two notions of Σ formula, one for set structures, one for word structures. We call production systems Σ complete if every relation definable by the appropriate version of Σ formula is generated. In the next section we investigate the Σ completeness of recursion and ω-recursion theories.

Let $\mathfrak{A} = \langle \mathcal{A}; \mathcal{R}_1, \ldots, \mathcal{R}_k \rangle$ be a structure. By an *atomic set formula* over $\mathfrak{A}$ we mean one of $(x = y)$, $(x \neq y)$, $(x \in y)$, $(x \notin y)$, $R_1\boldsymbol{x}, \ldots, R_k\boldsymbol{z}$, where $x, y, \boldsymbol{x}, \ldots, \boldsymbol{z}$ are variables or members of $\mathcal{A}^s$.

By a *Δ_0^s formula* over $\mathfrak{A}$ we mean any formula built up from atomic set formulas over $\mathfrak{A}$ using: conjunction, $\wedge$, disjunction, $\vee$, and bounded quantification, $(\forall x \in y)$ and $(\exists x \in y)$, where y is a variable or a member of $\mathcal{A}^s$. (This notion comes from Levy [1965].)

By a *Σ^s formula* over $\mathfrak{A}$ we mean a formula built up from atomic set formulas over $\mathfrak{A}$, using the machinery of Δ_0^s formulas, and also unbounded existential quantification, $(\exists x)$.

When we talk about the *truth* of a Σ^s formula over $\mathfrak{A}$, we mean, in the structure $\mathfrak{A}^s$, where $\mathcal{R}_i$ interprets R_i, and $\in$ and $=$ are given the obvious interpretations.

We call a relation $\mathcal{R}$ on $\mathcal{A}^s$ a *Σ^s relation* (a *Δ_0^s relation*) over $\mathfrak{A}^s$ if there is some formula $\varphi(\boldsymbol{x})$ which is Σ^s over $\mathfrak{A}$ (Δ_0^s over $\mathfrak{A}$) and

$$\mathcal{R}(\boldsymbol{x}) \Leftrightarrow \varphi(\boldsymbol{x}) \text{ is true.}$$

Let $\boldsymbol{A}$ be a production system on $\mathfrak{A}^s$. We call $\boldsymbol{A}$ Σ^s *complete* (with respect to $\mathfrak{A}^s$) if every relation which is Σ^s over $\mathfrak{A}^s$ is generated in $\boldsymbol{A}$.

Next we set up similar notions for word structures. By an *atomic word formula* over $\mathfrak{A}$ we mean one of $(x = y)$, $(x \neq y)$, $(x * y = z)$, $(x * y \neq z)$, $R_1\boldsymbol{x}, \ldots, R_k\boldsymbol{z}$, where $x, y, z, \boldsymbol{x}, \ldots, \boldsymbol{z}$ are variables or members of $\mathcal{A}^w$.

If u and v are words over $\mathcal{A}$ we say u is *part of* v if there are words x and y (possibly empty) so that $x * u * y = v$. We write $u \text{ pt } v$ to indicate that u is part of v.

By a Δ_0^w *formula* over $\mathfrak{A}$ we mean any formula built up from atomic word formulas over $\mathfrak{A}$ using: conjunction, $\wedge$, disjunction, $\vee$, and bounded quantification, $(\forall x \text{ pt } y)$ and $(\exists x \text{ pt } y)$, where y is a variable or a member of $\mathcal{A}^w$.

By a Σ^w *formula* over $\mathfrak{A}$ we mean any formula built up from atomic word formulas over $\mathfrak{A}$, using the machinery of Δ_0^w formulas, and also unbounded existential quantification, $(\exists x)$.

Truth of a Σ^w formula over $\mathfrak{A}$ means, in the structure $\mathfrak{A}^w$, in the obvious way.

A relation $\mathcal{R}$ on $\mathcal{A}^w$ is a Σ^w *relation* (a Δ_0^w *relation*) if there is some formula $\varphi(\boldsymbol{x})$ which is Σ^w over $\mathfrak{A}$ (Δ_0^w over $\mathfrak{A}$) and

$$\mathcal{R}(\boldsymbol{x}) \Leftrightarrow \varphi(\boldsymbol{x}) \text{ is true.}$$

NOTE. Δ_0^w relations are esentially a generalization of the *rudimentary* relations of Smullyan [1961].

Let $\boldsymbol{A}$ be a production system over $\mathfrak{A}^w$. We call $\boldsymbol{A}$ Σ^w *complete* (with respect to $\mathfrak{A}^w$) if every relation which is Σ^w over $\mathfrak{A}^w$ is generated in $\boldsymbol{A}$.

5. Elementary formal systems and Σ completeness

In this section we show that both recursion and ω-recursion theories over word or set structures have the appropriate version of Σ completeness (provided equality is well behaved). Also, for recursion theories, we sketch a converse result, that allows us to relate our notion of recursion theory with certain other generalizations in the literature.

LEMMA 5.1. *Suppose* rec($\mathfrak{A}$) (ω-rec($\mathfrak{A}$)) *is a theory with equality. Then*

(1) *In* rec($\mathfrak{A}^s$) (*in* ω-rec($\mathfrak{A}^s$)) *the following are r.e.* (ω*-r.e.*): $=, \neq, \in, \notin$.

(2) *In* rec($\mathfrak{A}^w$) (*in* ω-rec($\mathfrak{A}^w$)) *the following are r.e.* (ω*-r.e.*): $=, \neq$, pt.

PROOF. (1) $=$ is always generated. The following axioms generate, in both rec($\mathfrak{A}^s$) and ω-rec($\mathfrak{A}^s$), $\bar{\mathscr{A}} = \mathscr{A}^\omega$, $\in$, $\notin$ and $\neq$:

$$\text{axioms for } \neq_{\mathscr{A}},$$
$$x \cup \{y\} = z \rightarrow y \in z,$$
$$\bar{\mathscr{A}}\emptyset,$$
$$y \in x \rightarrow \bar{\mathscr{A}}x,$$
$$\mathscr{A}x \rightarrow \mathscr{A}y \rightarrow x \neq_{\mathscr{A}} y \rightarrow x \neq y,$$
$$\mathscr{A}x \rightarrow \bar{\mathscr{A}}y \rightarrow x \neq y,$$
$$\bar{\mathscr{A}}x \rightarrow \bar{\mathscr{A}}y \rightarrow z \in x \rightarrow z \notin y \rightarrow x \neq y,$$
$$x \neq y \rightarrow y \neq x,$$
$$\mathscr{A}x \rightarrow y \notin x,$$
$$y \notin \emptyset,$$
$$x \notin y \rightarrow x \neq w \rightarrow y \cup \{w\} = z \rightarrow x \notin z$$

[here we used $\emptyset$ for the "empty set" of $\mathscr{A}^s$].

(2) Again $=$ is generated. The following axioms generate, in both rec($\mathfrak{A}^w$) and ω-rec($\mathfrak{A}^w$), the relation $\neq$ (we begin with an iterated concatenation, for convenience):

$$x * a = b \rightarrow b * u = y \rightarrow x * a * u = y,$$
$$\text{axioms for } \neq_{\mathscr{A}},$$
$$\mathscr{A}c \rightarrow x * c = y \rightarrow \text{nonempty } y,$$
$$\text{nonempty } y \rightarrow x * y = z \rightarrow x \neq z,$$
$$\text{nonempty } y \rightarrow x * y = z \rightarrow z \neq x,$$
$$a \neq_{\mathscr{A}} b \rightarrow x * a * u = q \rightarrow x * b * v \doteq r \rightarrow q \neq r.$$

And the following generate the relation pt:

$$x * u * y = v \rightarrow u \text{ pt } v.$$

LEMMA 5.2. *Suppose* rec($\mathfrak{A}$) (ω-rec($\mathfrak{A}$)) *is a theory with equality. Then*

(1) *the* r.e. *relations* (*the* ω-r.e. *relations*) *of* rec($\mathfrak{A}^s$) (*of* ω-rec($\mathfrak{A}^s$)) *are closed under bounded quantification,* $(\forall x \in y)$ *and* $(\exists x \in y)$;

(2) *the* r.e. *relations* (*the* ω-r.e. *relations*) *of* rec($\mathfrak{A}^w$) (*of* ω-rec($\mathfrak{A}^w$)) *are closed under bounded quantification,* $(\forall x \text{ pt } y)$ *and* $(\exists x \text{ pt } y)$.

PROOF. (1) We have closure under $\exists$ and $\wedge$, and $\in$ is generated by the previous lemma, thus we have closure under $(\exists x \in y)$ since

$$(\exists x \in y)Rx, z \iff (\exists x)[x \in y \wedge Rx, z].$$

We show closure under $(\forall x \in y)$.

Suppose we have axioms for Rx, z. To get axioms for $(\forall x \in y)Rx, z$ add the following:

$$\mathscr{A}y \to (\forall x \in y)Rx, z,$$

$$(\forall x \in \emptyset)Rx, z,$$

$$(\forall x \in a)Rx, z \to Rb, z \to a \cup \{b\} = c \to (\forall x \in c)Rx, z.$$

(2) We have closure under $(\exists x \text{ pt } y)$, using Lemma 5.1, since

$$(\exists x \text{ pt } y)\, Rx, z \iff (\exists x)[x \text{ pt } y \wedge Rx, z].$$

Suppose we have axioms for Rx, z. To get axioms for $(\forall x \text{ pt } y)Rx, z$ add the following (where 0 is the empty word):

$$R0, z \to (\forall x \text{ pt } 0)Rx, z,$$

$$\mathscr{A}m \to R0, z \to Rm, z \to (\forall x \text{ pt } m)Rx, z,$$

$$\mathscr{A}m \to \mathscr{A}n \to m * v = b \to v * n = e \to b * n = w$$

$$\to (\forall x \text{ pt } b)Rx, z \to (\forall x \text{ pt } e)Rx, z$$

$$\to Rw, z \to (\forall x \text{ pt } w)Rx, z.$$

In following the last axiom the reader should have in mind the picture

$$\underbrace{m * \overbrace{v}^{}}_{b}\overbrace{{}* n}^{e} = w$$

where m and n are letters, i.e. members of $\mathscr{A}$.

THEOREM 5.3. *If* rec($\mathfrak{A}$) *is a theory with equality, then* rec($\mathfrak{A}^w$) *is* Σ^w *complete, and* rec($\mathfrak{A}^s$) *is* Σ^s *complete. Similarly for* ω-rec(-).

PROOF. Immediate from the above lemmas and the closure properties of production systems.

In a sense, recursion theories are minimal among Σ complete production systems. More precisely,

THEOREM 5.4. *If* rec($\mathfrak{A}$) *is a theory with equality, then*

(1) *If* $\mathcal{R}$ *is a relation on* $\mathcal{A}^s$, $\mathcal{R}$ *is* Σ^s *over* $\mathfrak{A}^s$ *if and only if* $\mathcal{R}$ *is r.e. in* rec($\mathfrak{A}^s$).

(2) *If* $\mathcal{R}$ *is a relation on* $\mathcal{A}^w$, $\mathcal{R}$ *is* Σ^w *over* $\mathfrak{A}^w$ *if and only if* $\mathcal{R}$ *is r.e. in* rec($\mathfrak{A}^w$).

We do not need this result in this book, so we merely suggest the proof, leaving the details to the reader.

Half of (1) is from the previous theorem. For the other half, (if $\mathcal{R}$ is r.e., then $\mathcal{R}$ is Σ^s), elementary formal systems for $\mathcal{R}$ may be coded into $\mathfrak{A}^s$, and derivations from them shown to be Σ^s (indeed Δ_0^s) using the chart on p. 14 in Barwise [1975]. Then $x \in \mathcal{R} \Leftrightarrow (\exists D)\,[D$ is a derivation of $Rx]$ makes $\mathcal{R}$ Σ^s. (2) has a parallel development.

In Montague [1968], in effect, a relation $\mathcal{R}$ on $\mathcal{A}$ is called $\aleph_0$ *recursively enumerable* if $\mathcal{R}$ is Σ^s over $\mathfrak{A}^s$ (this is not his terminology). By the above theorem, then, if rec($\mathfrak{A}$) is a theory with equality, $\mathcal{R}$ is $\aleph_0$ recursively enumerable over $\mathfrak{A}$ if and only if $\mathcal{R}$ is r.e., in our sense, in rec($\mathfrak{A}^s$). Indeed, we will see later in the Chapter, that if rec($\mathfrak{A}$) has an effective pairing function, this is further equivalent to: $\mathcal{R}$ is r.e. in rec($\mathfrak{A}$) in our sense.

In Moschovakis [1969] a definition of *search computability* is given, by using an inductive definition on $\mathfrak{A}^p$, involving indexes. This is yet another generalization of ordinary recursion theory, and what he calls the σ_0^1 relations are intended to be generalizations of the r.e. relations.

Gordon [1970] shows that the σ_0^1 relations on a structure $\langle \mathcal{A}; \mathcal{R}_1, \ldots, \mathcal{R}_k \rangle$ are exactly the $\aleph_0$ recursively enumerable relations on the structure $\langle \mathcal{A}; \mathcal{R}_1, \ldots, \mathcal{R}_k, \bar{\mathcal{R}}_1, \ldots, \bar{\mathcal{R}}_k, =_{\mathcal{A}}, \neq_{\mathcal{A}} \rangle$.

Further, in Moschovakis [1969A], the search computability approach is shown to be equivalent to two other generalizations of ordinary recursion theory, due to Fraïsse [1961] and Lacombe [1964], [1964A].

In Ch. 1, §13,14 we also established equivalences with inductive definability without universal quantifiers, and to R-definability. Clearly the notions involved possess great stability.

6. Separability

The entire of this section is devoted to a proof of the following.

THEOREM 6.1. *If* $T(\mathfrak{A})$ *is a theory with equality, then both* $T(\mathfrak{A}^s)$ *and* $T(\mathfrak{A}^w)$ *have effective pairing functions, and are separable with respect to both* $\times$ *and* $\dot{\cup}$.

PROOF. If we show the existence of an effective pairing function, then separability under $\times$ follows by Theorem 4.10.1, and separability under $\dot{\cup}$ by Theorem 4.10.3. We show the existence of an effective pairing function by producing an appropriate Σ formula and using Theorem 5.3.

Case 1: sets.

By $\langle x, y\rangle$ we mean the usual $\{\{x\},\{x, y\}\}$. The relation $z = \langle x, y\rangle$ is Σ^s (indeed Δ_0^s) over $\mathfrak{A}^s$ by the following.

$$w = \{x, y\} \Leftrightarrow x \in w \wedge y \in w \wedge (\forall z \in w)(z = x \vee z = y),$$

$$z = \langle x, y\rangle \Leftrightarrow (\exists u \in z)(\exists v \in z)(u = \{x, x\} \wedge v = \{x, y\} \wedge z = \{u, v\}).$$

Then $J(x, y) = \langle x, y\rangle$ will serve as an effective pairing function for both $\text{rec}(\mathfrak{A}^s)$ and $\omega\text{-rec}(\mathfrak{A}^s)$.

Case 2: words.

Actually, we must consider two cases depending on whether $\mathscr{A}$ has one member or more than one. Suppose $\mathscr{A}$ has one member, say $\mathscr{A} = \{1\}$. We can "identify" a string of n 1's with the number n and it is easy to see that $\mathfrak{A}^w$ is rather like $\mathfrak{S}(\mathbb{N})$ in disguise. Then any of the pairing functions from ordinary recursion theory (say Example I in Ch. 4, §10) can be "transferred" to both $\text{rec}(\mathfrak{A}^w)$ and $\omega\text{-rec}(\mathfrak{A}^w)$. We leave details to the reader.

Now suppose $\mathscr{A}$ has more than one member. We produce a pairing function for $\mathscr{A}^w$, which is Σ^w, which completes the proof. The method we give is from Quine [1946]. See Even and Rodeh [1978] for another.

Say $1, 2 \in \mathscr{A}$ (we use these without numerical significance). By a *tally* we mean a non-empty string of 1's. If π is a tally, let $\mathring{\pi} = 2 * \pi * 2$. If π is the shortest tally not occurring as part of x or y, set

$$J(x, y) = \mathring{\pi} * x * \mathring{\pi} * y * \mathring{\pi}.$$

We leave it to the reader to verify that this serves as a pairing function for $\mathscr{A}^w$. We show it is Σ^w (in fact, Δ_0^w). Below is a list of Δ_0^w relations, ending with $z = J(x, y)$.

(1) For each n, the relation

$$y = x_1 * \cdots * x_n \Leftrightarrow (\exists y_1 \,\text{pt}\, y) \cdots (\exists y_{n-2} \,\text{pt}\, y) [y_1 = x_1 * x_2 \wedge y_2 = y_1 * x_3 \wedge \cdots \wedge y = y_{n-2} * x_n]$$

(2) $y \,\text{pt}\, x \Leftrightarrow (\exists u \,\text{pt}\, x)(\exists v \,\text{pt}\, x)[x = u * y * v]$.

(3) $\sigma x \Leftrightarrow x$ is a tally $\Leftrightarrow x \neq 0 \wedge (\forall y \,\text{pt}\, x)[y = 0 \vee 1 \,\text{pt}\, y]$ (here 0 is the empty word).

(4) $vMy \Leftrightarrow v$ is the longest tally in word y
$\Leftrightarrow \sigma v \wedge v \text{ pt } y \wedge (\forall x \text{ pt } y)[x \neq v * 1]$.

(5) $vBx, y \Leftrightarrow v$ is the longer of tallys x and y
$\Leftrightarrow \sigma x \wedge \sigma y \wedge [(x \text{ pt } y \wedge v = y) \vee (y \text{ pt } x \wedge v = x)]$.

(6) $z = J(x, y) \Leftrightarrow (\exists v \text{ pt } z)(\exists q \text{ pt } z)(\exists r \text{ pt } z)(\exists \pi \text{ pt } z)(\exists \mathring{\pi} \text{ pt } z)$
$[qMx \wedge rMy \wedge vBq, r \wedge \pi = v * 1 \wedge \mathring{\pi} = 2 * \pi * 2$
$\wedge z = \mathring{\pi} * x * \mathring{\pi} * y * \mathring{\pi}]$.

7. Reflexivity

Again the entire section is devoted to a single result.

THEOREM 7.1. *If $T(\mathfrak{A})$ is reflexive with respect to $\dot{\cup}$, so are $T(\mathfrak{A}^s)$ and $T(\mathfrak{A}^w)$.*

PROOF. We give the proof for rec(-); that for ω-rec(-) is similar.

Suppose rec($\mathfrak{A}$) is reflexive with respect to $\dot{\cup}$. As usual, let inj : $\mathcal{A} \to \mathcal{A}$ be defined by $\text{inj}(x) = \{x\}$. Then by definition, $\text{inj}^{\dot{\cup}}$ is r.e. in rec($\mathfrak{A}$) $\dot{\cup}$ rec($\mathfrak{A}$). It is immediate that $\text{inj}^{\dot{\cup}}$ is also r.e. in rec($\mathfrak{A}^s$) $\dot{\cup}$ rec($\mathfrak{A}^s$) and in rec($\mathfrak{A}^w$) $\dot{\cup}$ rec($\mathfrak{A}^w$). Now we treat these cases separately.

Case 1: sets.

Define $f : \mathcal{A}^s \to \mathcal{A}^s$ by $f(x) = \{x\}$. We show $f^{\dot{\cup}}$ is r.e. in rec($\mathfrak{A}^s$) $\dot{\cup}$ rec($\mathfrak{A}^s$), which establishes reflexivity of rec($\mathfrak{A}^s$).

Now, $u = v \cup \{w\}$ is r.e. in rec($\mathfrak{A}^s$) hence $[u = v \cup \{w\}]^i$ and $[u = v \cup \{w\}]^j$ are r.e. in rec($\mathfrak{A}^s$) $\dot{\cup}$ rec($\mathfrak{A}^s$). Then, as elementary formal system axioms for $f^{\dot{\cup}}$, the following will serve.

$$y \in \text{inj}^{\dot{\cup}} x \to y \in f^{\dot{\cup}} x,$$

$$y \in f^{\dot{\cup}} x \to b \in f^{\dot{\cup}} a \to [y' = y \cup \{b\}]^j$$
$$\to [x' = x \cup \{a\}]^i \to y' = f^{\dot{\cup}} x'.$$

Case 2: words.

Define $g : \mathcal{A}^w \to \mathcal{A}^w$ by $g(x) = \{x\}$. We show $g^{\dot{\cup}}$ is r.e. in rec($\mathfrak{A}^w$) $\dot{\cup}$ rec($\mathfrak{A}^w$). Now, $u = v * w$ is r.e. in rec($\mathfrak{A}^w$), hence $[u = v * w]$ and $[u = v * w]^j$ are all r.e. in rec($\mathfrak{A}^w$) $\dot{\cup}$ rec($\mathfrak{A}^w$). As axioms for $g^{\dot{\cup}}$ we take

$$y \in \text{inj}^{\dot{\cup}} x \to y \in g^{\dot{\cup}} x,$$

$$y \in g^{\dot{\cup}} x \to b \in \text{inj}^{\dot{\cup}} a \to [y' = y * b]^j$$
$$\to [x' = x * a]^i \to y' \in g^{\dot{\cup}} x'.$$

8. Sequence codings

We have carried our parallel discussion of set structures and word structures far enough. From now on we concentrate on the details of word structures only. This will lead to some general results that incidentally give us information about set structures, however. Alternately, the entire development from here on could be modified to fit set structures (or pairing function structures) directly. We leave that to the dedicated reader.

Since we are concentrating on word structures, this section concerns itself with the possibility of "coding" words over a structure back into the original structure. It is the first of a series of sections ending in Section 14, which combine to show there are many recursion and ω-recursion theories that can "talk about" themselves.

DEFINITION. We say $T(\mathfrak{A})$ has a *sequence coding* if there is a way of assigning to every word x in $\mathcal{A}^w$, some distinct *sequence code* $x\#$ for it, where $x\# \in \mathcal{A}$, so that

(1) the function c, given by $c(u\#, v\#) = (u * v)\#$, is generated in $T(\mathfrak{A})$;

(2) the function d, given by $d(a) = a\#$ for all $a \in \mathcal{A}$, is generated in $T(\mathfrak{A})$.

Primarily, in this section, we show $T(\mathfrak{A})$ has a sequence coding if and only if $T(\mathfrak{A})$ has an effective pairing function.

THEOREM 8.1. *If $T(\mathfrak{A})$ has a sequence coding, then $T(\mathfrak{A})$ has an effective pairing function (and hence is separable with respect to both $\times$ and $\dot{\cup}$).*

PROOF. Suppose $T(\mathfrak{A})$ has a sequence coding. For $x, y \in \mathcal{A}$, set

$$J(x, y) = (x * y)\# = c(d(x), d(y)).$$

This is generated in $T(\mathfrak{A})$ since

$$J(x, y) = z \iff (\exists q)(\exists r)[q = d(x) \wedge r = d(y) \wedge z = c(q, r)]$$

and c and d are generated. And it is trivial that J serves as a pairing function.

THEOREM 8.2. *If $T(\mathfrak{A})$ has a sequence coding, then $T(\mathfrak{A}^w)$ is an inessential extension of $T(\mathfrak{A})$, via a 1–1 embedding.*

PROOF. Let $T(\mathfrak{A})$ have a sequence coding. Define h by: for a word $x \in \mathcal{A}^w$,

$h(x)=\{x\#\}$. h is a coding of $\mathscr{A}^w$ in $\mathscr{A}$, and is 1–1. It is trivially a pre-embedding of $T(\mathfrak{A}^w)$ in $T(\mathfrak{A})$; we use the fact that T is an elementary theory assignment to show h is an embedding.

Let $\mathfrak{A}=\langle\mathscr{A};\mathscr{R}_1,\ldots,\mathscr{R}_k\rangle$. Then $\mathfrak{A}^w=\langle\mathscr{A}^w;\mathscr{A},*,\mathscr{R}_1,\ldots,\mathscr{R}_k\rangle$.

First, $\mathscr{A}^h=\{y\mid(\exists x)d(x)=y\}$ and hence is generated in $T(\mathfrak{A})$.

Next, $(\mathscr{A}^w)^h=\{y\mid y\in\mathscr{A}^h\vee(\exists q)(\exists r)c(q,r)=y\}$, so it is generated in $T(\mathfrak{A})$. Then $(=_{\mathscr{A}^w})^h$ is generated, since it is $(=_{\mathscr{A}})\cap[(\mathscr{A}^w)^h\times(\mathscr{A}^w)^h]$.

$*^h$ is simply the function c, which is generated.

For each $i=1,2,\ldots,k$, $(\mathscr{R}_i)^h$ is $\{\langle y_1,\ldots,y_{n_i}\rangle\mid(\exists x_1)\cdots(\exists x_{n_i})[\langle x_1,\ldots,x_{n_i}\rangle\in\mathscr{R}_i\wedge y_1=d(x_1)\wedge\cdots\wedge y_{n_i}=d(x_{n_i})]\}$. Thus it is generated in $T(\mathfrak{A})$.

Since T is elementary, h is an embedding. Finally, to show $T(\mathfrak{A}^w)$ is an inessential extension of $T(\mathfrak{A})$ we need that $y\in(h\restriction\mathscr{A})(x)$ is generated in $T(\mathfrak{A})$. But in fact, $y\in(h\restriction\mathscr{A})(x)\Leftrightarrow y=d(x)$. This completes the proof.

PROPOSITION 8.3. *Suppose $\mathscr{A}$ is infinite and $T(\mathfrak{A})$ has an effective pairing function. Then* rec($\mathfrak{S}$(N)) *has a* 1–1 *embedding into $T(\mathfrak{A})$.*

REMARK. If $T(\mathfrak{A})$ has a pairing function at all, cardinality considerations show $\mathscr{A}$ must have 0 or 1 members, or be infinite. Thus the hypotheses are stated in a stronger form than strictly necessary. The conclusion is, of course, that a copy of ordinary recursion theory will be present.

PROOF. Let J be an effective pairing function in $T(\mathfrak{A})$. Pick some $x_0\in\mathscr{A}$ and define $J'(x,y)=J(J(x_0,x),J(x_0,y))$. Then J' is also an effective pairing function, and its range omits infinitely many members of $\mathscr{A}$ (in particular, if $x_1\neq x_0$, $J(J(x_1,x),J(x_1,y))$ is not in the range of J').

Now pick $a,b\in\mathscr{A}$ with $a\neq b$, and a and b not in the range of J'. For each natural number n we define a corresponding member $n'\in\mathscr{A}$ as follows. $0'=a$ and $1'=b$. Now suppose $0',1',\ldots,(n-1)',n'$ have been defined, and are all distinct. Set $(n+1)'=J'((n-1)',n')$. With this definition, $(n+1)'$ must be new, for the following reasons. $(n+1)'$ cannot be $0'$ or $1'$ since a and b are not in the range of J'. And if $(n+1)'=k'$ where $2\leq k\leq n$, then $J'((n-1)',n')=J'((k-2)',(k-1'))$, and so $n'=(k-1)'$ contradicting the supposition that all of $0',1',\ldots,n'$ were distinct.

Now define a coding $\theta:\mathsf{N}\to\mathscr{A}$ by $\Phi(n)=\{n'\}$. We omit the routine verification that $\theta:\text{rec}(\mathfrak{S}(\mathsf{N}))\to T(\mathfrak{A})$ is an embedding.

THEOREM 8.4. *Suppose $\mathscr{A}$ is infinite and $T(\mathfrak{A})$ has an effective pairing function. Then $T(\mathfrak{A})$ has a sequence coding.*

PROOF. Let J be an effective pairing function for $T(\mathfrak{A})$. We write $y = \langle x_1, x_2, x_3\rangle$ for $y = J(x_1, J(x_2, x_3))$. Then this too is generated; an effective tripling function. By the previous theorem there is a 1–1 embedding $\theta : \text{rec}(\mathfrak{S}(\mathbf{N})) \to T(\mathfrak{A})$. For simple notation, we write n for the only member of $\theta(n)$, thus identifying some members of $\mathcal{A}$ with the natural numbers. Also we write $+$ for $+^\theta$, which is generated in $T(\mathfrak{A})$ of course.

Now, the idea is to code the empty sequence by $\langle 0,0,0\rangle$, and the sequence $abcd$, for example, by $\langle a, 4, \langle b, 3, \langle c, 2, \langle d, 1, \langle 0,0,0\rangle\rangle\rangle\rangle\rangle$. We give the following elementary formal systems which establish that appropriate c and d functions are generated in $T(\mathfrak{A})$.

Axioms for the d function:

$$y = \langle x, 1, \langle 0,0,0\rangle\rangle \to y = d(x).$$

Let L be the relation $Lx, a, y \Leftrightarrow x$ is a sequence code for, say, x_0, and y is a sequence code for $x_0 * a$, where $a \in \mathcal{A}$. L has axioms:

$$y = d(a) \to L\langle 0,0,0\rangle, a, y,$$

$$Lx, a, y \to x = \langle u, v, w\rangle \to v' = v + 1 \to x' = \langle q, v', x\rangle$$

$$\to y = \langle e, f, g\rangle \to f' = f + 1 \to y' = \langle q, f', y\rangle \to Lx', a, y'.$$

$Cx \Leftrightarrow x$ is a sequence code. C has axioms

$$C\langle 0,0,0\rangle,$$

$$Cx \to Lx, a, y \to Cy.$$

Finally, axioms for the c function

$$Cx \to x = c(x, \langle 0,0,0\rangle)$$

$$y = c(u, v) \to Lv, a, v' \to Ly, a, y' \to y' = c(u, v').$$

COROLLARY 8.5. *Suppose $\mathcal{A}$ is infinite and $T(\mathfrak{A})$ has an effective pairing function. Then $T(\mathfrak{A}^w)$ is an inessential extension of $T(\mathfrak{A})$, via a* 1–1 *embedding.*

PROOF. Theorems 8.2 and 8.4.

9. Adding letters

An elementary formal system axiom over $\mathfrak{A}$ is a word, made up of letters which are members of $\mathcal{A}$ and various formal symbols (predicate letters, arrow, comma). We will be treating such an axiom as a word *in* some word structure. Now, $\mathfrak{A}^w$ would do if we only needed letters which are members of $\mathscr{A}$, but we also need some letters to play the role of the formal symbols. Additional letters are called for.

In this section we show that $\mathfrak{A}$ can be expanded by adding what amounts to countably many new letters, so that the (ω) recursion theory of the word structure over this expansion is an inessential extension of the (ω) recursion theory of $\mathfrak{A}^w$. (Ch. 5, §6.)

Recall, $\mathfrak{S}(\mathbb{N})$ is the structure of arithmetic, $\langle \mathbb{N}; y = x^+\rangle$. We use $\emptyset$ for the trivial, empty structure.

THEOREM 9.1. *Let $\mathfrak{A}$ be a structure such that $T(\mathfrak{A})$ is a theory with equality. Then $T([\mathfrak{A} \dot\cup \mathfrak{S}(\mathbb{N})]^w)$ is an inessential extension of $T([\mathfrak{A} \dot\cup \emptyset]^w)$ via a* 1–1 *embedding.*

REMARKS. We are forced to use $\mathfrak{A} \dot\cup \emptyset$ rather than $\mathfrak{A}$ itself for the technical reason that $\mathscr{A} \dot\cup \emptyset \subseteq \mathscr{A} \dot\cup \mathbb{N}$ but not $\mathscr{A} \subseteq \mathscr{A} \dot\cup \mathbb{N}$. We have more to say about this in the next section.

By Theorem 5.6.5, inessential extensions are also conservative ones, so no new sets are generated on the common domain.

PROOF. The plan of the proof is as follows. In part I we define a 1–1 coding $\theta : [\mathscr{A} \dot\cup \mathbb{N}]^w \to [\mathscr{A} \dot\cup \emptyset]^w$. In part II we show θ is an embedding. Finally in part III we show the relation $y \in [\theta \restriction (\mathscr{A} \dot\cup \emptyset)^w](x)$ is generated in $T([\mathfrak{A} \dot\cup \emptyset]^w)$.

Part I: If $x \in \mathscr{A}$, then $\langle x, 0\rangle \in \mathscr{A} \dot\cup \mathbb{N}$ and $\langle x, 0\rangle \in \mathscr{A} \dot\cup \emptyset$. We write x^i for $\langle x, 0\rangle$. Similarly, if $n \in \mathbb{N}$ then $\langle n, 1\rangle \in \mathscr{A} \dot\cup \mathbb{N}$. We write n^j for $\langle n, 1\rangle$.

$T(\mathfrak{A})$ is a theory with equality so trivially $T(\mathfrak{A} \dot\cup \emptyset)$ also is. Then by Theorem 6.1, $T([\mathfrak{A} \dot\cup \emptyset]^w)$ has an effective pairing function J, and then by Theorem 8.4 it also has a sequence coding $\#$, with generated c and d functions as in Section 8. We make use of this machinery to define our coding θ.

Choose some member $c \in \mathscr{A}$, fixed for the rest of this section.

We first define an auxiliary map, on letters, $\circ : \mathscr{A} \dot\cup \mathbb{N} \to (\mathscr{A} \dot\cup \emptyset)^w$ as follows.

(1) If $x \in \mathscr{A}$, by $(x^i)^\circ$ we mean $J(x^i, c^i)$.

(2) If $n \in \mathbb{N}$, by $(n^j)^\circ$ we mean $J(\underbrace{c^i c^i \cdots c^i}_{n}, c^i c^i)$.

(Here juxtaposition denotes concatenation in $(\mathscr{A} \cup \emptyset)^w$.)

Then z° has been defined on every member of $\mathscr{A} \dot{\cup} \mathbb{N}$, and clearly the mapping is 1–1.

Now, let s be a word in $(\mathscr{A} \dot{\cup} \mathbb{N})^w$.

(1) If s is the empty word of $(\mathscr{A} \dot{\cup} \mathbb{N})^w$, s is also the empty word of $(\mathscr{A} \dot{\cup} \emptyset)^w$. Set $\theta(s) = \{s\#\}$. Briefly, $\theta(\emptyset) = \{\emptyset\#\}$.

(2) If $x = z_1 z_2 \cdots z_n$, where each z_i is a letter, i.e. a member of $\mathscr{A} \dot{\cup} \mathbb{N}$, then for each i, z_i° is a member of $(\mathscr{A} \dot{\cup} \emptyset)^w$. Then $z_1^\circ * z_2^\circ * \cdots * z_n^\circ$ is a sequence over $(\mathscr{A} \dot{\cup} \emptyset)^w$ (caution: over, not in). Set

$$\theta(s) = \{(z_1^\circ * z_2^\circ * \cdots * z_n^\circ)\#\}.$$

It is straightforward that θ is a 1–1 coding, $\theta : (\mathscr{A} \dot{\cup} \mathbb{N})^w \to (\mathscr{A} \dot{\cup} \emptyset)^w$. This ends part I.

Part II: We show θ is an embedding, $\theta : T([\mathfrak{A} \dot{\cup} \mathfrak{S}(\mathbb{N})]^w) \to T([\mathfrak{A} \dot{\cup} \emptyset]^w)$ by making use of the fact that T is an elementary theory assignment. Since T(-) assigns total production systems to structures, θ is automatically a pre-embedding, so we need only show that certain relations are taken over to generated relations under θ. This part of the proof is largely bookkeeping. We begin by carefully saying what relations we must be concerned with.

Let $\mathfrak{A} = \langle \mathscr{A}; \mathscr{R}_1, \ldots, \mathscr{R}_k \rangle$. Then

$$\mathfrak{A} \dot{\cup} \mathfrak{S}(\mathbb{N}) = \langle \mathscr{A} \dot{\cup} \mathbb{N}; (=_{\mathscr{A}})^i, \mathscr{R}_1^i, \ldots, \mathscr{R}_k^i, (=_{\mathbb{N}})^j, (y = x^+)^j \rangle$$

so

$$[\mathfrak{A} \dot{\cup} \mathfrak{S}(\mathbb{N})]^w$$
$$= \langle (\mathscr{A} \dot{\cup} \mathbb{N})^w; \mathscr{A} \dot{\cup} \mathbb{N}, *, (=_{\mathscr{A}})^i, \mathscr{R}_1^i, \ldots, \mathscr{R}_k^i, (=_{\mathbb{N}})^j, (y = x^+)^j \rangle.$$

Likewise, but slightly simpler since $\mathscr{A} \dot{\cup} \emptyset = \mathscr{A}^i$,

$$[\mathfrak{A} \dot{\cup} \emptyset]^w = \langle (\mathscr{A}^i)^w; \mathscr{A}^i, *, (=_{\mathscr{A}})^i, \mathscr{R}_1^i, \ldots, \mathscr{R}_k^i \rangle.$$

We have denoted concatenation by $*$ in both structures.

Now, to show θ is an embedding, we must show θ takes each of the following relations to a relation generated in $T([\mathfrak{A} \dot{\cup} \emptyset]^w)$ (see Ch. 3, §5):

$$=_{(\mathscr{A} \dot{\cup} \mathbb{N})^w}, \quad \mathscr{A} \dot{\cup} \mathbb{N}, \quad *, \quad (=_{\mathscr{A}})^i, \quad \mathscr{R}_1^i, \ldots, \mathscr{R}_k^i, \quad (=_{\mathbb{N}})^j, \quad (y = x^+)^j.$$

We first show some auxilary relations are generated, then proceed to these.

(1) $(\mathscr{A}^i)^\theta$ is generated in $T([\mathfrak{A} \dot{\cup} \emptyset]^w)$:

By definition of θ,

$$(\mathscr{A}^i)^\theta = \{t \mid \text{for some } z \in \mathscr{A}^i,\ t = (z^0)\#\}$$
$$= \{t \mid \text{for some } z \in \mathscr{A}^i,\ t = (J(z, c^i))\#\}$$
$$= \{t \mid \text{for some } z \in \mathscr{A}^i,\ t = d(J(z, c^i))\}.$$

Now, $\mathscr{A}^i$ is a "given" relation of $T([\mathfrak{A}\,\dot{\cup}\,\emptyset]^w)$, and d and J are generated in it. It follows from the closure properties of production systems that $(\mathscr{A}^i)^\theta$ is generated in $T([\mathfrak{A}\,\dot{\cup}\,\emptyset]^w)$.

(2) Let A be the set of words in $(\mathscr{A}\,\dot{\cup}\,\emptyset)^w$ of the form $c^ic^i\cdots c^i$ (including the empty word $\emptyset$). A is generated in $T([\mathfrak{A}\,\dot{\cup}\,\emptyset]^w)$. Indeed, by the elementary formal system

$$A\emptyset,$$

$$Ax \rightarrow y = xc^i \rightarrow Ay$$

(juxtaposition denotes concatenation).

(3) $(\mathsf{N}^j)^\theta$ is generated in $T([\mathfrak{A}\,\dot{\cup}\,\emptyset]^w)$:

As in item (1), $(\mathsf{N}^j)^\theta = \{t \mid \text{for some } z \in A,\ t = d(J(z, c^ic^i))\}$ and hence is generated since A, t and d are.

Now we turn to some of the relations in our list earlier.

(4) $(\mathscr{A}\,\dot{\cup}\,\mathsf{N})^\theta$ is generated in $T([\mathfrak{A}\,\dot{\cup}\,\emptyset]^w)$, since it is in fact, $(\mathscr{A}^i)^\theta \cup (\mathsf{N}^j)^\theta$ and we have closure under unions, and items (1) and (3).

(5) $((=_{\mathscr{A}})^i)^\theta$ is generated in $T([\mathfrak{A}\,\dot{\cup}\,\emptyset]^w)$. Since θ is 1–1, this relation is simply the equality relation of $(\mathfrak{A}\,\dot{\cup}\,\emptyset)^w$ restricted to $(\mathscr{A}^i)^\theta$, that is,

$$((=_{\mathscr{A}})^i)^\theta \quad \text{is} \quad =_{\mathscr{A}\dot{\cup}\emptyset} \cap [(\mathscr{A}^i)^\theta \times (\mathscr{A}^i)^\theta].$$

This is generated by item (1) and the properties of production systems.

(6) $((=_{\mathsf{N}})^j)^\theta$ is generated in $T([\mathfrak{A}\,\dot{\cup}\,\emptyset]^w)$. This is similar to item (5).

(7) Each of $(\mathscr{R}^i_1)^\theta, \ldots, (\mathscr{R}^i_k)^\theta$ is generated in $T(\lfloor\mathfrak{A}\,\dot{\cup}\,\emptyset\rfloor^w)$. This is like item (1), taking into account the fact that $\mathscr{R}_p$ may be more than one-place.

(8) Another auxiliary relation, $((\mathscr{A}\,\dot{\cup}\,\mathsf{N})^w)^\theta$ is generated in $T([\mathfrak{A}\,\dot{\cup}\,\emptyset]^w)$. It is not hard to see, directly from our definition of θ, that it is represented by H using the following:

$$(\emptyset)^{\#} \in H,$$

$$x \in (\mathscr{A}^i)^\theta \rightarrow x \in H,$$

$$x \in (\mathsf{N}^j)^\theta \rightarrow x \in H,$$

$$x \in H \rightarrow y \in H \rightarrow z = c(x, y) \rightarrow z \in H,$$

(here we have written $x \in H$ for Hx, in keeping with our other informal notation).

(9) $(=_{(\mathscr{A}\dot{\cup}\mathsf{N})^w})^\theta$ is generated in $T(\lfloor\mathfrak{A}\,\dot{\cup}\,\emptyset\rfloor^w)$. This follows from item (8) by an argument likc that in (5).

(10) $(*)^\theta$ is generated in $T(\lfloor\mathfrak{A}\,\dot{\cup}\,\emptyset\rfloor^w)$. It is simply the c function restricted in $((\mathscr{A}\,\dot{\cup}\,\mathsf{N})^w)^\theta$.

(11) $((y = x^+)^i)^\theta$ is generated in $T\lfloor \mathfrak{A} \dot{\cup} \emptyset \rfloor^w)$. It is, in fact, $\{\langle u, v\rangle \mid$ for some $x \in A$, $u = d(J(x, c^i c^i))$ and $v = d(J(xc^i, c^i c^i))\}$.
This concludes Part II.

Part III: We show the relation $y \in [\theta \restriction (\mathcal{A} \dot{\cup} \emptyset)^w](x)$ is generated in $T([\mathfrak{A} \dot{\cup} \emptyset]^w)$. Well, it is represented by F using the following axioms:

$$x \in \mathcal{A}^i \to t = J(x, c^i) \to y = d(t) \to y \in F(x),$$

$$y_1 \in F(x_1) \to y_2 \in F(x_2) \to x = x_1 x_2 \to y = c(y_1, y_2) \to y \in F(x).$$

This completes the proof.

10. Copies and extensions

We have discussed, in Chapter 5, isomorphisms between productions systems, in particular between recursion and ω-recursion theories. There is a stronger notion we introduce now, which we call *being a copy*.

Let $\mathfrak{A} = \langle \mathcal{A}; \mathcal{R}_1, \dots, \mathcal{R}_k \rangle$ and $\mathfrak{B} = \langle \mathcal{B}; \mathcal{S}_1, \dots, \mathcal{S}_k \rangle$ be two structures. Suppose we have a 1–1, onto coding $\theta : \mathcal{A} \to \mathcal{B}$ such that, for each i, $\mathcal{S}_i = \mathcal{R}_i^\theta$. Then we'll call $\mathfrak{B}$ a *copy* of $\mathfrak{A}$. For example, $\mathfrak{A} \dot{\cup} \emptyset$ is a copy of $\mathfrak{A}$. Now if $\mathfrak{B}$ is a copy of $\mathfrak{A}$, we'll also call rec($\mathfrak{B}$) a copy of rec($\mathfrak{A}$), and likewise for their ω-recursion theories. For all reasonable purposes, we need not distinguish between copies; they are, after all, direct renamings of things. So, from now on, unless there is a strong reason to do otherwise, we'll identify rec($\mathfrak{A} \dot{\cup} \emptyset$) with rec($\mathfrak{A}$). With this in mind, we can restate the result of the last section more simply, as follows.

THEOREM 10.1 (Theorem 9.1 restated). *Let $\mathfrak{A}$ be a structure such that $T(\mathfrak{A})$ is a theory with equality. Then $T([\mathfrak{A} \dot{\cup} \mathfrak{S}(\mathsf{N})]^w)$ is an inessential extension of $T(\mathfrak{A}^w)$ via a* 1–1 *embedding.*

We would like to iterate this result, so that we can throw in many copies of $\mathfrak{S}(\mathsf{N})$, not just one. To do this in the simplest way, we list two independent results.

PROPOSITION 10.2. *If $T(\mathfrak{B})$ is an inessential extension of $T(\mathfrak{A})$ via g, and $T(\mathfrak{C})$ is an inessential extension of $T(\mathfrak{B})$ via f, then $T(\mathfrak{C})$ is an inessential extension of $T(\mathfrak{A})$ via gf.*

PROOF. Both $f : T(\mathfrak{C}) \to T(\mathfrak{B})$ and $g : T(\mathfrak{B}) \to T(\mathfrak{A})$ are embeddings, hence $gf : T(\mathfrak{C}) \to T(\mathfrak{A})$ is an embedding. All that needs to be shown is that the relation $y \in (gf \restriction \mathscr{A})(x)$ is generated in $T(\mathfrak{A})$.

Now, $y \in (f \restriction \mathscr{B})(x)$ is generated in $T(\mathfrak{B})$, and g is an embedding, so g carries this to a relation generated in $T(\mathfrak{A})$; let us call it R. Then

$$R(y, x) \Leftrightarrow \text{ for some } a, b \in \mathscr{B},\ y \in g(b), \\ x \in g(a) \text{ and } b \in (f \restriction \mathscr{B})(a).$$

Recall from Ch. 5, §5 that $f \restriction \mathscr{B} = f\, \mathrm{inj}_{\mathscr{B},\mathscr{C}}$ (we will make use of this and similar things quite a bit). Hence

$$R(y, x) \Leftrightarrow \text{ for some } a, b \in \mathscr{B},\ y \in g(b), \\ x \in g(a) \text{ and } b \in f\, \mathrm{inj}_{\mathscr{B},\mathscr{C}}(a).$$

But then, $R(y, x) \Leftrightarrow$ for some $a \in \mathscr{B}$, $x \in g(a)$ and $y \in gf\, \mathrm{inj}_{\mathscr{B},\mathscr{C}}(a)$.

Also the relation $y \in (g \restriction \mathscr{A})(x)$ is generated in $T(\mathfrak{A})$, hence so is the following:

$$S(y, z) \Leftrightarrow \text{ for some } x \in \mathscr{A},\ R(y, x) \text{ and } x \in (g \restriction \mathscr{A})(z).$$

Writing out the characterization of R, and rewriting $g \restriction \mathscr{A}$,

$$S(y, z) \Leftrightarrow \text{ for some } a \in \mathscr{B} \text{ and some } x \in \mathscr{A},\ x \in g(a), \\ y \in gf\, \mathrm{inj}_{\mathscr{B},\mathscr{C}}(a) \text{ and } x \in g\, \mathrm{inj}_{\mathscr{A},\mathscr{B}}(z).$$

Now, if $x \in g(a)$ and $x \in g(\mathrm{inj}_{\mathscr{A},\mathscr{B}}(z))$, since g is a coding, it follows that $a \in \mathrm{inj}_{\mathscr{A},\mathscr{B}}(z)$. It further follows that

$$S(y, z) \Leftrightarrow y \in gf\, \mathrm{inj}_{\mathscr{B},\mathscr{C}}\, \mathrm{inj}_{\mathscr{A},\mathscr{B}}(z) \\ \Leftrightarrow y \in gf\, \mathrm{inj}_{\mathscr{A},\mathscr{C}}(z) \Leftrightarrow y \in ((gf) \restriction \mathscr{A})(z).$$

Since S is generated in $T(\mathfrak{A})$, we are done.

PROPOSITION 10.3. *If both $T(\mathfrak{A})$ and $T(\mathfrak{B})$ are theories with equality, so is $T(\mathfrak{A}) \dot\cup T(\mathfrak{B}) = T(\mathfrak{A} \dot\cup \mathfrak{B})$.*

PROOF. By hypothesis, $\neq_{\mathscr{A}}$ is generated in $T(\mathfrak{A})$ and $\neq_{\mathscr{B}}$ is generated in $T(\mathfrak{B})$. Now, $\mathfrak{A} \dot\cup \mathfrak{B} = \langle \mathscr{A} \dot\cup \mathscr{B}; \mathscr{A}^i, \mathscr{B}^j, \ldots \rangle$ and both $(\neq_{\mathscr{A}})^i$ and $(\neq_{\mathscr{B}})^j$ are generated in $T(\mathfrak{A} \dot\cup \mathfrak{B})$ since i and j are embeddings. Then $\neq_{\mathscr{A} \dot\cup \mathscr{B}}$ is also generated, since it is the relation $(\neq_{\mathscr{A}})^i \cup (\neq_{\mathscr{B}})^j \cup (\mathscr{A}^i \times \mathscr{B}^j) \cup (\mathscr{B}^j \times \mathscr{A}^i)$.

Now we can give our "iterated" version of Theorem 10.1.

THEOREM 10.4. *Let $\mathfrak{A}$ be a structure such that $T(\mathfrak{A})$ is a theory with* *equality. Then each of the following is an inessential extension of* $T(\mathfrak{A}^w)$ *via a* 1–1 *embedding*:

(1) $T([\mathfrak{A} \dot{\cup} \mathfrak{S}(\mathbb{N})]^w)$,

(2) $T([(\mathfrak{A} \dot{\cup} \mathfrak{S}(\mathbb{N})) \dot{\cup} \mathfrak{S}(\mathbb{N})]^w)$,

(3) $T([((\mathfrak{A} \dot{\cup} \mathfrak{S}(\mathbb{N})) \dot{\cup} \mathfrak{S}(\mathbb{N})) \dot{\cup} \mathfrak{S}(\mathbb{N})]^w)$

(n) etc.

PROOF. Line (1) is a restatement of Theorem 10.1.

$T(\mathfrak{A})$ is a theory with equality, by hypothesis, and we know $T(\mathfrak{S}(\mathbb{N}))$ is a theory with equality too, so by Proposition 10.3, $T(\mathfrak{A} \dot{\cup} \mathfrak{S}(\mathbb{N}))$ is a theory with equality. Applying Theorem 10.1 to it we get that $T([(\mathfrak{A} \dot{\cup} \mathfrak{S}(\mathbb{N})) \dot{\cup} \mathfrak{S}(\mathbb{N})]^w)$ is an inessential extension of $T([\mathfrak{A} \dot{\cup} \mathfrak{S}(\mathbb{N})]^w)$ via a 1–1 embedding. Combining this with the fact of line (1), using Proposition 10.2, gives us line (2). And we may continue in this fashion.

11. Universal machines

For this section $\mathfrak{A}$ is a structure such that $T(\mathfrak{A})$ is a theory with equality. We use $\mathscr{I}(n, m)$ here to denote the collection of all operators in $T(\mathfrak{A}^w)$ of order $\langle n, m\rangle$. What we show in this section is that $\mathfrak{A}$ has an extension $\mathfrak{A}^*$, such that each class $\mathscr{I}(n, m)$ can be "mimicked" by a single operator in $T(\mathfrak{A}^{*w})$. Precisely what this means will become clear as we proceed.

To begin with, we need a context in which $T(\mathfrak{A}^w)$ can *formally* be "talked about" in essentially the same way we do. We "talk about" $T(\mathfrak{A}^w)$ by working with (ω) elementary formal systems. So, we want a setting in which (ω) elementary formal systems, and derivations, are objects which can be formally manipulated. As a start, we need variables, relation symbols, and punctuation, as well as the members of $\mathscr{A}^w$. Well, let $\mathfrak{A}^*$ be the structure $((\mathfrak{A} \dot{\cup} \mathfrak{S}(\mathbb{N})) \dot{\cup} \mathfrak{S}(\mathbb{N})) \dot{\cup} \mathfrak{S}(\mathbb{N})$. In this structure there are three copies of $\mathbb{N}$; we simply identify one copy of $\mathbb{N}$ with the variables [and write var(x) if x is in that copy], we identify a second copy of $\mathbb{N}$ with the relation symbols [and write rel(x) if x is in that copy], and we identify as much of the third copy of $\mathbb{N}$ as is needed with the punctuation symbols [and write punct(x) if x is in that copy]. Also there is a copy of $\mathscr{A}$ "given" in this structure; we simply use $\mathscr{A}$ to represent it.

Thus the domain of $\mathfrak{A}^*$ has all the "letters" we need to construct elementary formal systems over $\mathfrak{A}^w$, and hence $\mathfrak{A}^{*w}$ actually has, in its domain, each elementary formal system axiom over $\mathfrak{A}^w$. We also have that in $T(\mathfrak{A}^{*w})$ each of var, rel, punct and $\mathscr{A}$ is generated (in fact, each is one of the "given" relations). $T(\mathfrak{A}^{*w})$ is an excellent place to try "talking about" $T(\mathfrak{A}^w)$. Incidentally, by Theorem 10.4, $T(\mathfrak{A}^{*w})$ is an inessential extension of $T(\mathfrak{A}^w)$, but this will play no role until Section 14.

We use $\mathscr{A}^*$ as the domain of $\mathfrak{A}^*$; thus $\mathscr{A}^* = ((\mathscr{A} \dot{\cup} \mathbf{N}) \dot{\cup} \mathbf{N}) \dot{\cup} \mathbf{N}$. We take as one of our formal punctuation symbols a *space*, and use # to denote it.

Suppose we have an (ω) elementary formal system over $\mathfrak{A}^w$ with axioms $A_1, A_2, \ldots, A_j$. Then, in accordance with the above discussion, we can consider each A_i to be a word in $\mathscr{A}^{*w}$. Also $\#A_1\#A_2\# \cdots \#A_j\#$ is in $\mathscr{A}^{*w}$. Such a word is called a *base*, with components $A_1, A_2, \ldots, A_j$. In a sense, it codifies the (ω) elementary formal system with axioms $A_1, A_2, \ldots, A_j$.

$\mathscr{I}(n, m)$ is the collection of all (ω) enumeration operators in $T(\mathfrak{A}^w)$ of order $\langle n, m\rangle$. By a *universal machine* for the class $\mathscr{I}(n, m)$ we mean a "black box" (operator) $\mathscr{U}$ of order $\langle n, m+1\rangle$, whose behavior is as follows. Let $\Phi \in \mathscr{I}(n, m)$; let $A_1, A_2, \ldots, A_j$ be axioms for an (ω) elementary formal system for Φ, and let $a = \#A_1\#A_2\# \cdots A_j\#$. Then, if $\mathscr{P}$ is any n-ary relation on $\mathscr{A}^w$, we should have

$$\langle x_1, \ldots, x_m\rangle \in \Phi(\mathscr{P}) \Leftrightarrow \langle a, x_1, \ldots, x_m\rangle \in \mathscr{U}(\mathscr{P}).$$

Thus $\mathscr{U}$ "mimics" each operator in $\mathscr{I}(n, m)$, keeping track of each by "tagging" outputs by, essentially, the elementary formal system axioms used to generate it.

The result promised at the beginning of this section, that in $T(\mathfrak{A}^{*w})$ there was an operator that could "mimic" the entire class $\mathscr{I}(n, m)$, can now be stated properly.

THEOREM 11.1. *Let $\mathscr{U}$ be a universal machine for the class $\mathscr{I}(n, m)$. Then $\mathscr{U}$ is an (ω) enumeration operator in $T(\mathfrak{A}^{*w})$.*

Thus the behavior of all members of $\mathscr{I}(n, m)$ is captured by $\mathscr{U}$. But note, members of $\mathscr{I}(n, m)$ are operators in $T(\mathfrak{A}^w)$, while $\mathscr{U}$ is an operator in $T(\mathfrak{A}^{*w})$. We will simplify this later on.

We devote the rest of this section to a proof of the above theorem, which (simply) amounts to writing an (ω) elementary formal system for $\mathscr{U}$. And to do this, we give formal counterparts of the definitions in Chapter 1, through that of enumeration operator. Before doing so, however, we have some remarks.

First, there is no harm if we use, say, P as both an n-place and a k-place relation symbol. We can tell the two uses apart by looking at the number of terms which follow P. This means we can make do with one list of relation symbols, instead of separate lists for 1-place, 2-place, etc., and this is a technical convenience.

$T(\mathfrak{S}(\mathbb{N}))$ is a theory with equality. It follows that in $T(\mathfrak{A}^{*w})$, the unequals relation for each copy of $\mathbb{N}$ is generated. We write $\neq_{\text{var}}$ and $\neq_{\text{rel}}$ for those on the variables and on the relation symbols respectively.

As punctuation symbols, in addition to $\#$, we need an arrow, a comma, and (if we are talking ω-recursion theory) a universal quantifier. To tell the symbols *of* $\mathcal{A}^{*w}$ from the symbols *we* use, we write

$\Rightarrow$ for the arrow in $\mathcal{A}^{*w}$,

$\circ$ for the comma in $\mathcal{A}^{*w}$,

π for the universal quantifier in $\mathcal{A}^{*w}$.

The symbols we use, from the outside, will be as usual: the above are in punct.

There is a similar problem with relation symbols. Now, $\mathfrak{A} = \langle \mathcal{A}; \mathcal{R}_1, \ldots, \mathcal{R}_k \rangle$. For use in *our* elementary formal systems, certain predicate symbols have been assigned to $\mathcal{R}_1, \ldots, \mathcal{R}_k$; we have been designating them $R_1, \ldots, R_k$. Well, certain members of rel in $\mathcal{A}^{*w}$ will also have to be assigned to $\mathcal{R}_1, \ldots, \mathcal{R}_k$. Let us designate them by $\bar{R}_1, \ldots, \bar{R}_k$. Thus R_i is a symbol we use, while $\bar{R}_i$ is a member of $\mathcal{A}^{*w}$.

Similarly for input and output symbols. We will continue to use I and O for the purpose. We assume two symbols of rel have been designated to play their role "inside" the theory; we denote them $\bar{I}$ and $\bar{O}$.

$\mathcal{U}$ embodies many different enumeration operators. In giving axioms for $\mathcal{U}$ we will need some way of keeping track of these various operators. More precisely, if X is derivable over $T(\mathfrak{A}^{w})$ using axioms $A_1, A_2, \ldots, A_j$, we need some simple way of "remembering" this. Well, $\# A_1 \# A_2 \# \cdots \# A_j \# X$ is a word in $\mathcal{A}^{*w}$; we work with it, thus keeping track of the axioms used, and X at the same time.

Finally, T is either rec(-) or ω-rec(-). In the one case it is an elementary formal system we must construct; in the other, an ω-elementary formal system. The two are very similar, so we list the axioms assuming T(-) is rec(-) and then we indicate the changes necessary if T(-) is ω-rec(-).

Now we list our axioms for $\mathcal{U}$, in rec($\mathfrak{A}^{*w}$). As we introduce each new relation symbol we say what it is intended to represent before giving axioms for it. 0 is the empty word here.

(1) word $x \Leftrightarrow x$ is a word over $\mathcal{A}$ (i.e. x is a member of $\mathcal{A}^{*w}$ made up entirely of letters from the copy of $\mathcal{A}$).

word 0,

$\mathscr{A}x \to \text{word } y \to z = y * x \to \text{word } z.$

(2) term $x \Leftrightarrow x$ is a term.

word $x \to$ term x,

var $x \to$ term x.

(3) pseudoatomic $x \Leftrightarrow x$ is a pseudoatomic formula.

rel $z \to$ term $x \to y = z * x \to$ pseudoatomic y,

pseudoatomic $y \to$ term $x \to w = y * \circ * x \to$ pseudoatomic w.

(4) atomic $x \Leftrightarrow x$ is an atomic formula (no variables).

rel $z \to$ word $x \to y = z * x \to$ atomic y,

atomic $y \to$ word $x \to w = y * \circ * x \to$ atomic w.

(5) pseudoformula $x \Leftrightarrow x$ is an elementary formal system pseudoformula.

pseudoatomic $x \to$ pseudoformula x,

pseudoformula $y \to$ pseudoformula $z \to x = y * \Rightarrow * z$
$\to$ pseudoformula x.

(6) formula $x \Leftrightarrow x$ is a formula (no variables).

atomic $x \to$ formula x,

formula $y \to$ formula $z \to x = y * \Rightarrow * z \to$ formula x.

(7) proper $x \Leftrightarrow x$ is a proper pseudoformula, in that none of $\bar{R}_1, \ldots, \bar{R}_k$ or $\bar{I}$ occurs in the conclusion.

pseudoatomic $x \to x = y * z \to$ rel $y \to y \neq_{\text{rel}} \bar{R}_1 \to \cdots$
$\cdots \to y \neq_{\text{rel}} \bar{R}_k \to y \neq_{\text{rel}} \bar{I} \to$ proper x,

pseudoformula $x \to x = y * z \to$ pseudoatomic $z \to$ proper z
$\to$ proper x.

(8) $Sx, y, z, w \Leftrightarrow x$ and z are words over $\mathscr{A}^*$ (note), y is a variable, and w is the result of substituting z for all occurrences of y in x.

$\mathscr{A}x \to$ var $y \to Sx, y, z, x$,

rel $x \to$ var $y \to Sx, y, z, x$,

$$\text{punct } x \rightarrow \text{var } y \rightarrow Sx, y, z, x,$$

$$\text{var } x \rightarrow Sx, x, z, z,$$

$$\text{var } x \rightarrow \text{var } y \rightarrow x \neq_{\text{var}} y \rightarrow Sx, y, z, x,$$

$$Sx_1, y, z, w_1 \rightarrow Sx_2, y, z, w_2$$
$$\rightarrow x = x_1 * x_2 \rightarrow w = w_1 * w_2 \rightarrow Sx, y, z, w.$$

(9) w is a partial instance of x $\Leftrightarrow$ x is a word over $\mathscr{A}^*$ and w is the result of replacing one or more variables in x by words over $\mathscr{A}$.

$$Sx, y, z, w \rightarrow \text{word } z \rightarrow w \text{ is a partial instance of } x,$$

$$w_1 \text{ is a partial instance of } x \rightarrow Sw_1, y, z, w_2 \rightarrow \text{word } z$$
$$\rightarrow w_2 \text{ is a partial instance of } x.$$

(10) w is an instance of pseudoformula x $\Leftrightarrow$ x is a pseudoformula and w is the result of replacing all variables of x by words over $\mathscr{A}$.

$$\text{pseudoformula } x \rightarrow w \text{ is a partial instance of } x$$
$$\rightarrow \text{formula } w \rightarrow w \text{ is an instance of pseudoformula } x.$$

(11) Bx $\Leftrightarrow$ x is a base.

$$\text{proper } x \rightarrow y = \# * x * \# \rightarrow By,$$

$$\text{proper } x \rightarrow By \rightarrow z = y * x * \# \rightarrow Bz.$$

(12) part x, y $\Leftrightarrow$ word x is part of word y

$$y = u * x * v \rightarrow \text{part } x, y.$$

(13) Cx, y $\Leftrightarrow$ x is a component of base y.

$$\text{pseudoformula } x \rightarrow By \rightarrow z = \# * x * \# \rightarrow \text{part } z, y \rightarrow Cx, y.$$

(14) Fx. The axioms for F, unlike those above, accept input and, when given $\mathscr{P} \subseteq \mathscr{A}^n$ as input, Fx is derivable $\Leftrightarrow$ x is the form $\# A_1 \# A_2 \# \cdots \# A_j \# X$, where X is derivable in $T(\mathfrak{A}^w)$ in the elementary formal system with axioms $A_1, A_2, \ldots, A_j$, using $\mathscr{P}$ as input. (We assume $\mathscr{R}_i$ is n_i-ary.)

$$Ix_1, \ldots, x_n \rightarrow y = \bar{I} * x_1 * \circ * \cdots * \circ * x_n \rightarrow Bz \rightarrow w = z * y \rightarrow Fw,$$

$$R_1x_1, \ldots, x_{n_1} \rightarrow y = \bar{R}_1 * x_1 * \circ * \cdots * \circ * x_{n_1} \rightarrow Bz \rightarrow w = z * y \rightarrow Fw,$$
$$\vdots$$
$$R_kx_1, \ldots, x_{n_k} \rightarrow y = \bar{R}_k * x_1 * \circ * \cdots * \circ * x_{n_k} \rightarrow Bz \rightarrow w = z * y \rightarrow Fw,$$

$$By \to Cx, y \to w \text{ is an instance of}$$

$$\text{pseudoformula } x \to r = y * w \to Fr,$$

$$u = x * y \to v = x * y * \Rightarrow * z \to w = x * z \to Bx \to \text{atomic } y$$

$$\to \text{formula } z \to Fu \to Fv \to Fw.$$

(15) Finally, $\mathcal{U}$ is represented by O using the following.

$$Fw \to w = a * y \to Ba \to y = \bar{O} * x_1 * \circ * x_2 * \circ * \cdots * \circ * x_m$$

$$\to \text{word } x_1 \to \text{word } x_2 \to \cdots \to \text{word } x_m \to Oa, x_1, \ldots, x_m.$$

This completes the proof if $T(\text{-})$ is rec(-). Now we list the modifications necessary if $T(\text{-})$ is ω-rec(-).

In (1) and (2) no changes.

Replace (3) by

(3′) pseudoatomic$_\omega$ x $\Leftrightarrow$ x is a pseudoatomic formula allowing occurrences of π.

$$\text{rel } z \to \text{term } x \to y = z * x \to \text{pseudoatomic}_\omega\, y,$$

$$\text{rel } z \to y = z * \pi \to \text{pseudoatomic}_\omega\, x,$$

$$\text{pseudoatomic}_\omega\, y \to \text{term } x \to w = y * \circ * x \to \text{pseudoatomic}_\omega\, w,$$

$$\text{pseudoatomic}_\omega\, y \to w = y * \circ * \pi \to \text{pseudoatomic}_\omega\, y.$$

Make similar changes in (4) to get (4′) atomic$_\omega$ x. We leave this and other straightforward changes to the reader.

Change (5) and (6) by replacing pseudoatomic by pseudoatomic$_\omega$, and atomic by atomic$_\omega$, to get axioms for:

(5′) pseudoformula$_\omega$ x.

(6′) formula$_\omega$ x.

(7) Requires more attention. In an ω-elementary formal system we don't want π to occur in axiom conclusions. Now, if x is a pseudoformula$_\omega$, and z is the conclusion of x, saying z is just plain pseudoatomic says, in particular, that π does not occur in it. Thus we have the following.

(7′) pseudoformula$_\omega$ $x \to x = y * z \to$ pseudoatomic $z \to$ proper z $\to$ proper$_\omega$ x.

(8) and (9) require no changes.

In (10), replace pseudoformula and formula by their ω-versions to get axioms for:

(10′) w is an instance of pseudoformula$_\omega$ x.

In (11) replace proper by its ω-version to get axioms for:

(11′) $B_\omega x$ $\Leftrightarrow$ x is a base for an ω-elementary formal system.

(12) and (13) require no changes.

Before (14) we insert the following

(13a) $\overline{\text{word}}\ x \Leftrightarrow x$ is not a word over $\mathscr{A}$ (i.e. x is a member of $\mathscr{A}^{*w}$ which includes some letters outside the copy of $\mathscr{A}$).

$$\text{var}\, x \to \overline{\text{word}}\, x,$$

$$\text{rel}\, x \to \overline{\text{word}}\, x,$$

$$\text{punct}\, x \to \overline{\text{word}}\, x,$$

$$\overline{\text{word}}\, x \to w = u * x * v \to \overline{\text{word}}\, w.$$

Now, replace (14) by the following. In these axioms, Gx, y, z is intended to represent: x is $b * r$ where b is a base and r is pseudoatomic$_\omega$, y is a variable, and either

(1) z is a word over $\mathscr{A}$ and we have a derivation of $b * r'$ where r' is the result of replacing y by z in r, or

(2) z is not a word over $\mathscr{A}$.

(14′) $F_\omega x$ (like Fx, but taking the ω-rule into account).

$$Ix_1, \ldots, x_n \to y = \bar{I} * x_1 * \circ * \cdots * \circ * x_n \to B_\omega z \to w = z * y \to F_\omega w,$$

$$R_1 x_1, \ldots, x_{n_1} \to y = \bar{R}_1 * x_1 * \circ * \cdots * \circ * x_{n_1} \to B_\omega z \to w = z * y \to F_\omega w,$$

$$\vdots$$

$$R_k x_1, \ldots, x_{n_k} \to y = \bar{R}_k * x_1 * \circ * \cdots * \circ * x_{n_k} \to B_\omega z \to w = z * y \to F_\omega w,$$

$B_\omega y \to Cx,\ y \to w$ is an instance of

$$\text{pseudoformula}_\omega x \to r = y * w \to F_\omega r,$$

$$u = x * y \to v = x * y * \Rightarrow * z \to w = x * z \to B_\omega x$$

$$\to \text{atomic}_\omega y \to \text{formula}_\omega z \to F_\omega u \to F_\omega v \to F_\omega w,$$

$$\overline{\text{word}}\, z \to \text{var}\, y \to B_\omega b \to \text{pseudoatomic}_\omega r \to x = b * r \to Gx, y, z,$$

$$\text{word}\, z \to \text{var}\, y \to B_\omega b \to \text{pseudoatomic}_\omega r \to x = b * r$$

$$\to Sr, y, z, r' \to x' = b * r' \to F_\omega x' \to Gx, y, z,$$

$$B_\omega b \to \text{pseudoatomic}_\omega r \to x = b * r \to Gx, y, \forall$$

$$\to Sr, y, \pi, r' \to x' = b * r' \to F_\omega r'.$$

And finally

(15′) $\mathscr{U}$ is represented by O using the following.

$$F_\omega w \to w = a * y \to B_\omega a \to y = \bar{O} * x_1 * \circ * \cdots * \circ * x_m$$

$$\to \text{word}\, x_1 \to \text{word}\, x_2 \to \cdots \to \text{word}\, x_m \to Oa, x_1, \ldots, x_m.$$

12. Operator indexing

We introduce some terminology, and restate the result of the previous section.

Let $\mathfrak{A}$ and $\mathfrak{B}$ be structures, with $T(\mathfrak{B})$ an extension of $T(\mathfrak{A})$. By $\mathcal{I}_{T(\mathfrak{A})}(n, m)$ we mean the collection of all operators of $T(\mathfrak{A})$ of order $\langle n, m\rangle$. Let $\mathcal{U}^{\langle n,m\rangle}$ be an operator in $T(\mathfrak{B})$ of order $\langle n, m+1\rangle$. We call $\mathcal{U}^{\langle n,m\rangle}$ *universal for* $\mathcal{I}_{T(\mathfrak{A})}(n, m)$ if: $\Phi \in \mathcal{I}_{T(\mathfrak{A})}(n, m)$ if and only if there is some $b \in \mathcal{B}$ such that

$$\langle x_1, \ldots, x_m\rangle \in \Phi(\mathcal{P}) \iff \langle b, x_1, \ldots, x_m\rangle \in \mathcal{U}(\mathcal{P}), \text{ for all } \mathcal{P} \in [\mathcal{A}]^n.$$

We say $\mathcal{I}_{T(\mathfrak{A})}(n, m)$ has an *indexing* in $T(\mathfrak{B})$ if some operator in $T(\mathfrak{B})$ is universal for $\mathcal{I}_{T(\mathfrak{A})}(n, m)$.

Now suppose, for each n and m, that $\mathcal{U}^{\langle n,m\rangle}$ is an operator in $T(\mathfrak{B})$ which is universal for $\mathcal{I}_{T(\mathfrak{A})}(n, m)$. We call the family $\mathcal{U} = \{\mathcal{U}^{\langle n,m\rangle} \mid n, m \geq 1\}$ an *indexing of* $T(\mathfrak{A})$ *in* $T(\mathfrak{B})$.

Suppose $\mathcal{U}$ is an indexing of $T(\mathfrak{A})$ in $T(\mathfrak{B})$. By $\Phi_b^{\langle n,m\rangle}$ we mean that operator (if one exists) in $T(\mathfrak{A})$ of order $\langle n, m\rangle$ such that, for $\mathcal{P} \in [\mathcal{A}]^n$,

$$\langle x_1, \ldots, x_m\rangle \in \Phi_b^{\langle n,m\rangle}(\mathcal{P}) \iff \langle b, x_1, \ldots, x_m\rangle \in \mathcal{U}^{\langle n,m\rangle}(\mathcal{P}).$$

We call b an *index* of $\Phi_b^{\langle n,m\rangle}$. Note that indexes are relative to the particular indexing chosen.

In our present terms, we may restate Theorem 11.1 (with Theorem 10.4) as follows:

THEOREM 12.1. *Let $\mathfrak{A}$ be a structure for which $T(\mathfrak{A})$ is a theory with equality. Then $T(\mathfrak{A}^w)$ has an indexing in $T(\mathfrak{A}^{*w})$, an inessential extension of $T(\mathfrak{A}^w)$ via a 1–1 embedding.*

By Proposition 3.1, $T(\mathfrak{A}^w)$ is an extension of $T(\mathfrak{A})$, so the operators of $T(\mathfrak{A})$ are among those of $T(\mathfrak{A}^w)$. Thus we immediately have the following.

COROLLARY 12.2. *Under the same hypotheses, $T(\mathfrak{A})$ has an indexing in $T(\mathfrak{A}^{*w})$, an extension, though not necessarily inessential, of $T(\mathfrak{A})$.*

Thus our recursion and ω-recursion theories (with equality) have recursion and ω-recursion theories that can "talk about" them. We have two main questions: (1) what can be "said", and (2) which theories can "talk about" themselves. We look at these questions in the next two sections.

13. Input and output place-fixing theorems

For this section, $T(\mathfrak{A})$ is a theory with equality.

We have constructed an indexing of $T(\mathfrak{A}^w)$ in $T(\mathfrak{A}^{*w})$. Informally, $T(\mathfrak{A}^{*w})$ can "talk about" $T(\mathfrak{A}^w)$. Now we ask what sort of things can it "say"? In this section we establish two such things, from which, in turn, many others will follow in the course of the next two chapters. Thus these play a key role in our development. The two may be thought of as generalizing what is called the s-m-n theorem of ordinary recursion theory, which is due to Kleene. See Kleene [1952], p. 342. Kleene works with functions, which take n-tuples as input, but which give numbers as output. The s-m-n theorem has to do with manipulating input: changing n-tuples to k-tuples ($k < n$) by fixing some of the places. Not much can be done with output: numbers are 1-tuples, and can get no shorter. Here, however, we work with operators, and they can use n-ary relations both as input and as output. Thus both input and output is subject to the kind of "place-fixing" manipulation of the s-m-n theorem. Consequently we have two theorems below, not just one.

NOTE. We call a function f *partial recursive* if its graph is r.e., though its domain may not be total.

THEOREM 13.1 (Output Place Fixing Theorem). *Let Φ be an operator of $T(\mathfrak{A}^w)$ of order $\langle n, q+m\rangle$. There is a q-ary partial recursive function f, in* $\mathrm{rec}(\mathfrak{A}^{*w})$, *such that*

$$\langle a_1, \ldots, a_q, b_1, \ldots, b_m\rangle \in \Phi(\mathscr{P}) \Leftrightarrow \langle b_1, \ldots, b_m\rangle \in \Phi^{\langle n,m\rangle}_{f(a_1,\ldots,a_q)}(\mathscr{P})$$

(for all n-ary $\mathscr{P}$).

REMARKS. The intention in constructing $\mathfrak{A}^{*w}$ was that $\mathrm{rec}(\mathfrak{A}^{*w})$ could not only "discuss" $\mathrm{rec}(\mathfrak{A}^w)$, but could do it in, essentially, the same way we do. The proof below illustrates this.

PROOF. Let $E = \{A_1, A_2, \ldots, A_j\}$ be axioms for an (ω)-elementary formal system for Φ in $T(\mathfrak{A}^w)$, in which I represents input and D output. For each $a_1, \ldots, a_q \in \mathscr{A}^w$, let $E_{a_1,\ldots,a_q}$ be the (ω) elementary formal system with axioms $A_1, A_2, \ldots, A_j$ and $Da_1, \ldots, a_q, x_1, \ldots, x_m \to Ox_1, \ldots, x_m$. It should be clear that

$$\langle a_1, \ldots, a_q, b_1, \ldots, b_m\rangle \in \Phi(\mathscr{P})$$

$$\Leftrightarrow \langle a_1, \ldots, a_q, b_1, \ldots, b_m\rangle \in [E^I_D](\mathscr{P})$$

$$\Leftrightarrow \langle b_1, \ldots, b_m\rangle \in [(E_{a_1,\ldots,a_q})^I_O](\mathscr{P}).$$

Now define a partial function f by: for $a_1, \ldots, a_q \in \mathscr{A}^w$,

$$f(a_1, \ldots, a_q) = \# A_1 \# A_2 \# \cdots \# A_j \# Da_1, \ldots, a_q, x_1, \ldots, x_m$$
$$\rightarrow Ox_1, \ldots, x_m \# .$$

That is, $f(a_1, \ldots, a_q)$ is a base, whose components are the axioms of $E_{a_1,\ldots,a_q}$. Thus $[(E_{a_1,\ldots,a_q})^I_O](\mathscr{P}) = \Phi^{\langle n,m\rangle}_{f(a_1,\ldots,a_q)}(\mathscr{P})$ by the construction of our universal machines in Section 11. We are done once we show f is partial recursive in rec($\mathfrak{A}^{*w}$). This is simple. Let $v_1, \ldots, v_m$ be members of **var**. Now as axioms for f, take those for word (item (1) in Section 11) and the following:

$$\text{word } v_1 \rightarrow \cdots \rightarrow \text{word } v_q$$
$$\rightarrow z = \bar{D} * v_1 * \circ * \cdots * \circ * v_q * \circ * x_1 * \circ * \cdots * \circ * x_m$$
$$* \Rightarrow * \bar{O} * x_1 * \circ * \cdots * \circ * x_m$$
$$\rightarrow y = \# A_1 \# A_2 \# \cdots \# A_j \# z \rightarrow f(v_1, \ldots, v_q) = y.$$

This ends the proof.

THEOREM 13.2 (Input Place Fixing Theorem). *Let Φ be an operator of $T(\mathfrak{A}^w)$ of order $\langle q+n, m\rangle$. There is a q-ary partial recursive function f in* rec($\mathfrak{A}^{*w}$) *such that, for $\mathscr{P} \subseteq (\mathscr{A}^w)^n$,*

$$\Phi(\{\langle a_1, \ldots, a_q\rangle\} \times \mathscr{P}) = \Phi^{\langle n,m\rangle}_{f(a_1,\ldots,a_q)}(\mathscr{P}).$$

PROOF. Let $E = \{A_1, A_2, \ldots, A_j\}$ be axioms for an (ω) elementary formal system for Φ in $T(\mathfrak{A}^w)$ with J as input and O as output. For each $a_1, \ldots, a_q \in \mathscr{A}^w$, let $E_{a_1,\ldots,a_q}$ be the (ω) elementary formal system with axioms:

$$A_1, A_2, \ldots, A_j \quad \text{and} \quad Ix_1, \ldots, x_n \rightarrow Ja_1, \ldots, a_q, x_1, \ldots, x_n.$$

Then

$$\Phi(\{\langle a_1, \ldots, a_q\rangle\} \times \mathscr{P}) = [E^J_O](\{\langle a_1, \ldots, a_q\rangle\} \times \mathscr{P})$$
$$= [(E_{a_1,\ldots,a_q})^I_O](\mathscr{P}).$$

Now proceed as in the above proof.

14. Indexing revisited

This section is the culmination of a long sequence of sections. We have seen that, if $T(\mathfrak{A})$ is a theory with equality, then $T(\mathfrak{A}^{*w})$ can "talk about" $T(\mathfrak{A}^w)$. Here we prove some results that imply the nicer fact that $T(\mathfrak{A}^w)$

can "talk about" itself. Better still, under certain mild conditions, even $T(\mathfrak{A})$ can "talk about" itself.

Let $\mathcal{U}$ be an indexing of $T(\mathfrak{A})$ in $T(\mathfrak{B})$, as in Section 12. Let Φ be an operator in $T(\mathfrak{A})$ of order $\langle n, q+m\rangle$. By an *q-ary output place-fixing function* for Φ we mean a function $f: \mathcal{A}^q \to \mathcal{B}$ such that, for all $\mathcal{P} \subseteq \mathcal{A}^n$,

$$\langle a_1,\dots,a_q,x_1,\dots,x_m\rangle \in \Phi(\mathcal{P}) \Leftrightarrow \langle f(a_1,\dots,a_q),x_1,\dots,x_m\rangle \in \mathcal{U}^{\langle n,m\rangle}(\mathcal{P})$$
$$\Leftrightarrow \langle x_1,\dots,x_m\rangle \in \Phi^{\langle n,m\rangle}_{f(a_1,\dots,a_q)}(\mathcal{P}).$$

We say $\mathcal{U}$ has the *output place-fixing property* if, for each operator Φ of $T(\mathfrak{A})$, and for each q, there is some q-ary output place-fixing function f for Φ which is generated in $T(\mathfrak{B})$. Moreover, if f can always be taken to be r.e. in rec($\mathcal{B}$) we say $\mathcal{U}$ has the *strong* output place-fixing property.

Again, let Φ be an operator in $T(\mathfrak{A})$ of order $\langle q+n, m\rangle$. By a *q-ary input place-fixing function* for Φ we mean a function $f: \mathcal{A}^q \to \mathcal{B}$ such that, for all $\mathcal{P} \subseteq \mathcal{A}^n$,

$$\langle x_1,\dots,x_m\rangle \in \Phi(\{\langle a_1,\dots,a_q\rangle\} \times \mathcal{P})$$
$$\Leftrightarrow \langle f(a_1,\dots,a_q),x_1,\dots,x_m\rangle \in \mathcal{U}^{\langle n,m\rangle}(\mathcal{P})$$
$$\Leftrightarrow \langle x_1,\dots,x_m\rangle \in \Phi^{\langle n,m\rangle}_{f(a_1,\dots,a_q)}(\mathcal{P}).$$

We say $\mathcal{U}$ has the *input place-fixing property* if, for each operator Φ of $T(\mathfrak{A})$, and for each q, there is some q-ary input place-fixing function f for Φ which is generated in $T(\mathfrak{B})$. Moreover, if f can always be taken to be r.e. in rec($\mathfrak{B}$) we say $\mathcal{U}$ has the *strong* input place-fixing property.

Combining work of Sections 11 and 13, we have:

PROPOSITION 14.1. *Let $T(\mathfrak{A})$ be a theory with equality, and let $\mathcal{U}$ be the indexing of Section* 11. *$\mathcal{U}$ is an indexing of $T(\mathfrak{A}^w)$ in $T(\mathfrak{A}^{*w})$ which has both the strong input and the strong output place-fixing properties.*

Now we prove the main result of this section, which will allow us to strengthen the above proposition.

THEOREM 14.2. *Suppose $T(\mathfrak{A})$ has an indexing $\mathcal{U}$ in $T(\mathfrak{B})$, where $T(\mathfrak{B})$ is an inessential extension of $T(\mathfrak{A})$ via a* 1–1 *embedding. Then*

(1) *$T(\mathfrak{A})$ has an indexing $\mathcal{V}$ in itself.*

(2) *If $\mathcal{U}$ has the input (output) place-fixing property, so does $\mathcal{V}$.*

(3) *If* rec($\mathfrak{B}$) *is an inessential extension of* rec($\mathfrak{A}$) *via a* 1–1 *embedding, then if $\mathcal{U}$ has the strong input (output) place-fixing property, so does $\mathcal{V}$.*

REMARK. If $\mathcal{V}$ is an indexing of $T(\mathfrak{A})$ in *itself*, and if f is an input or output place-fixing function for some operator, f is a *total* function.

PROOF. Say $T(\mathfrak{B})$ is an inessential extension of $T(\mathfrak{A})$ via h, where h is a 1–1 embedding. Thus $h : T(\mathfrak{B}) \to T(\mathfrak{A})$ is a 1–1 embedding, and the relation $y \in (h \restriction \mathcal{A})(x)$ is generated in $T(\mathfrak{A})$. By Proposition 4.12.2, $(h \restriction \mathcal{A})_n$ is an operator of $T(\mathfrak{A})$. We note that, for $\mathcal{P} \subseteq \mathcal{A}^n$, $(h \restriction \mathcal{A})_n(\mathcal{P}) = \{\langle y_1, \ldots, y_n\rangle \mid$ for some $\langle x_1, \ldots, x_n\rangle \in \mathcal{P}$, $y_1 \in h(x_1) \wedge \cdots \wedge y_n \in h(x_n)\}$. We will also want an operator like $(h \restriction \mathcal{A})_m^{-1}$, but which leaves one position alone. We get this from the proof, rather than the statement of Proposition 4.12.2. Thus: define a relation $\mathcal{R}$ by

$$\mathcal{R}(y, y_1, \ldots, y_m, x, x_1, \ldots, x_m)$$
$$\Leftrightarrow\ y = x \wedge y_1 \in (h \restriction \mathcal{A})(x_1) \wedge \cdots \wedge y_m \in (h \restriction \mathcal{A})(x_m).$$

Then $\mathcal{R}$ is generated in $T(\mathfrak{A})$ so by Corollary 2.5.2 (1), the following, J_{m+1}, is an operator, of order $\langle m+1, m+1\rangle$, in $T(\mathfrak{A})$: for $\mathcal{P} \subseteq \mathcal{A}^{m+1}$, $J_{m+1}(\mathcal{P}) = \mathcal{R}''(\mathcal{P})$. We note that $J_{m+1}(\mathcal{P}) = \{\langle x, x_1, \ldots, x_m\rangle \mid$ for some $\langle x, y_1, \ldots, y_m\rangle \in \mathcal{P}$, $y_1 \in (h \restriction \mathcal{A})(x_1) \wedge \cdots \wedge y_m \in (h \restriction \mathcal{A})(x_m)\}$.

Now, we turn to the proof of (1). Let $\mathcal{I}_{T(\mathfrak{A})}(n, m)$ be the collection of operators of $T(\mathfrak{A})$ of order $\langle n, m\rangle$. $\mathcal{U}^{\langle n,m\rangle}$ is an indexing for $\mathcal{I}_{T(\mathfrak{A})}(n, m)$ in $T(\mathfrak{B})$. By the preceding, $\mathcal{V}^{\langle n,m\rangle}$ is an operator in $T(\mathfrak{A})$, where:

$$\mathcal{V}^{\langle n,m\rangle} = J_{m+1}(\mathcal{U}^{\langle n,m\rangle})^h (h \restriction \mathcal{A})_n.$$

We claim $\mathcal{V}^{\langle n,m\rangle}$ is an indexing for $\mathcal{I}_{T(\mathfrak{A})}(n, m)$ in $T(\mathfrak{A})$.

Let Φ be an operator in $T(\mathfrak{A})$ of order $\langle n, m\rangle$. Then Φ has indexes in $T(\mathfrak{B})$, using the $\mathcal{U}$ indexing. Let b be one of them, and let $a \in h(b)$. We claim, for $\mathcal{P} \subseteq \mathcal{A}^n$,

$$\langle x_1, \ldots, x_m\rangle \in \Phi(\mathcal{P}) \ \Leftrightarrow\ \langle a, x_1, \ldots, x_m\rangle \in \mathcal{V}^{\langle n,m\rangle}(\mathcal{P}),$$

and thus a is an index, using $\mathcal{V}^{\langle n,m\rangle}$, for Φ in $T(\mathfrak{A})$. Actually, what must be shown is

$$\langle b, x_1, \ldots, x_m\rangle \in \mathcal{U}^{\langle n,m\rangle}(\mathcal{P}) \ \Leftrightarrow\ \langle a, x_1, \ldots, x_m\rangle \in \mathcal{V}^{\langle n,m\rangle}(\mathcal{P}).$$

This is straightforward, and is left to the reader.

Next, (2). Suppose Φ is an operator of $T(\mathfrak{A})$, of order $\langle n, q+m\rangle$, f is a q-ary output place-fixing function for Φ (relative to the $\mathcal{U}$ indexing) and f is generated in $T(\mathfrak{B})$. As we did above, by use of Corollary 2.5.2 (1), we may produce an operator K in $T(\mathfrak{A})$, of order $\langle q+1, q+1\rangle$, such that, for $\mathcal{P} \subseteq \mathcal{A}^{q+1}$, $K(\mathcal{P}) = \{\langle x_1, \ldots, x_q, x\rangle \mid$ for some $\langle y_1, \ldots, y_q, x\rangle \in \mathcal{P}$, $y_1 \in h(x_1) \wedge \cdots \wedge y_q \in h(x_q)\}$.

Now, f is generated in $T(\mathfrak{B})$, hence $f' = K(f^h)$ is generated in $T(\mathfrak{A})$. That f' is a function follows from the fact that h is 1–1. We leave it to the reader to check that f' is a q-ary output place-fixing function for Φ relative to the $\mathcal{V}$ indexing.

Thus $\mathcal{V}$ has the output place-fixing property. The input place-fixing property is treated similarly.

Finally, (3) has a proof similar to (2). We note that if rec($\mathfrak{B}$) is an inessential extension of rec($\mathfrak{A}$) via h, it follows by Corollary 3.5.2 and by Proposition 1.9.1, that $T(\mathfrak{B})$ is an inessential extension of $T(\mathfrak{A})$ via h. We leave the rest of the proof to the reader.

Now for the corollaries we have been after all along.

COROLLARY 14.3. (1) *If $T(\mathfrak{A})$ is a theory with equality, then $T(\mathfrak{A}^w)$ has an indexing in itself having both the input and the output place-fixing properties.*

(2) *If* rec($\mathfrak{A}$) *is a theory with equality, $T(\mathfrak{A}^w)$ has an indexing in itself having both the strong input and the strong output place-fixing properties.*

PROOF. We show (1) only. $T(\mathfrak{A})$ is a theory with equality, so by Theorem 10.4, $T(\mathfrak{A}^{*w})$ is an inessential extension of $T(\mathfrak{A}^w)$ via a 1–1 embedding. By Theorem 11.1, there is an indexing of $T(\mathfrak{A}^w)$ in $T(\mathfrak{A}^{*w})$ which, by Theorems 13.1 and 13.2, has the output and the input place-fixing properties. The result follows from Theorem 14.2.

COROLLARY 14.4. *Every recursion theory has an extension which is a recursion theory having an indexing with the strong input and strong output place-fixing properties. Similarly for ω-recursion theories.*

PROOF. If rec($\mathfrak{A}$) is a theory with equality, we are done, by Corollary 14.3 (and Proposition 3.1). If not, let $\mathfrak{A}' = \langle \mathfrak{A}, \neq_{\mathcal{A}} \rangle$. Then, trivially, rec($\mathfrak{A}'$) is a theory with equality, which is an extension of rec($\mathfrak{A}$). Now finish as before.

COROLLARY 14.5. (1) *If $T(\mathfrak{A})$ is a theory with equality and an effective pairing function (where $\mathcal{A}$ is infinite), then $T(\mathfrak{A})$ has an indexing in itself which has the input and the output place-fixing properties.*

(2) *If* rec($\mathfrak{A}$) *is a theory with equality and an effective pairing function (where $\mathcal{A}$ is infinite), then $T(\mathfrak{A})$ has an indexing in itself which has the strong input and stong output place-fixing properties.*

PROOF. We only prove (1). By Corollary 8.5, since $T(\mathfrak{A})$ has an effective pairing function $T(\mathfrak{A}^w)$ is an inessential extension of $T(\mathfrak{A})$, via a 1–1 embedding. Since $T(\mathfrak{A})$ is a theory with equality, by Corollary 14.3, $T(\mathfrak{A}^w)$ has an indexing in itself, having the input and the output place-fixing properties. It follows easily that $T(\mathfrak{A})$ has an indexing in $T(\mathfrak{A}^w)$ having these properties, and we are done, by Theorem 14.2.

COROLLARY 14.6. (1) *If* $T(\mathfrak{A})$ *is a theory with equality, both* $T(\mathfrak{A}^s)$ *and* $T(\mathfrak{A}^p)$ *have indexings in themselves, having the input and the output place-fixing properties.*

(2) *If* rec$(\mathfrak{A})$ *is a theory with equality, both* $T(\mathfrak{A}^s)$ *and* $T(\mathfrak{A}^p)$ *have indexings in themselves having the strong input and the strong output place-fixing properties.*

PROOF. We only prove (1). By Theorem 6.1, $T(\mathfrak{A}^s)$ has an effective pairing function, and trivially, so does $T(\mathfrak{A}^p)$. Since $T(\mathfrak{A})$ is a theory with equality, so is $T(\mathfrak{A}^s)$, by Lemma 5.1. It is not hard to show $T(\mathfrak{A}^p)$ is also a theory with equality. The result follows by Corollary 14.5.

EXAMPLES. (1) Let $\mathfrak{A}$ be one of our "standard" structures, $\mathfrak{S}(\mathbb{N})$, $\mathfrak{S}(\mathrm{L}_\omega)$ or $\mathfrak{S}(a_1, \ldots, a_n)$. Then rec$(\mathfrak{A})$ is a theory with equality (shown in Ch. 1, §4), and has an effective pairing function (shown in Ch. 4, §10). Hence both the recursion and the ω-recursion theories of these structures have indexings having the strong input and the strong output place-fixing properties.

(2) Let $\mathfrak{S}(\mathbb{R})$ be the structure $\langle \mathbb{R}; +, \times, > \rangle$ where $\mathbb{R}$ is the set of real numbers. In ω-rec$(\mathfrak{S}(\mathbb{R}))$ there is an effective pairing function (Chapter 4, §10) and it is a theory with equality. Hence ω-rec$(\mathfrak{S}(\mathbb{R}))$ has an indexing having the input and the output place-fixing properties.

15. Relational indexing

Suppose $T(\mathfrak{A})$ has an indexing in itself. This is an indexing of *operators*, but it induces various indexings of *relations* that are of interest. Here we say how, then in the next chapter we investigate the structure that results, before returning to operators in Chapter 8.

Let Γ be a collection of sets and relations on $\mathcal{A}$. Γ is said to be *parametrized* if for every $n \geq 1$ there is an $n+1$ place relation $U^n \in \Gamma$ such that

$$\begin{aligned}&\mathcal{P} \text{ is an } n\text{-ary relation in } \Gamma \\ &\quad \Leftrightarrow \text{ for some } i \in \mathcal{A},\ \mathcal{P} = \{\boldsymbol{x} \mid \langle i, \boldsymbol{x} \rangle \in U^n\}.\end{aligned}$$

If Γ is thus parametrized, we write $\mathcal{R}_i^n$ for $\{\mathbf{x} \mid U^n(i, \mathbf{x})\}$ and we call i an *index* of $\mathcal{R}_i^n$.

Suppose, in addition to being parametrized, Γ meets the following conditions: For each n, there are functions $f : \mathcal{A}^n \to \mathcal{A}$ called *basic* such that

(1) the collection of basic functions is closed under composition,

(2) (the graph of) each basic function is a member of Γ,

(3) (iteration property) for each $n+k$-place relation, $\mathcal{Q} \in \Gamma$ there is a basic function $f : \mathcal{A}^n \to \mathcal{A}$ such that

$$\mathcal{Q}(x_1, \ldots, x_n, y_1, \ldots, y_k) \Leftrightarrow \mathcal{R}^k_{f(x_1,\ldots,x_n)}(y_1, \ldots, y_k).$$

If this happens, we will say (the indexing of) Γ has the *iteration property*.

EXAMPLES. (1) Let $T(\mathfrak{A})$ have an indexing $\mathcal{U}$ in itself, which has the output place-fixing property. Let Γ be the collection of relations generated in $T(\mathfrak{A})$ [the r.e. or ω-r.e. relations over $\mathfrak{A}$ depending on whether $T(\text{-})$ is rec(-) or ω-r.e.(-)]. Then Γ is parametrized and has the iteration property. We may show this as follows.

For each $n \geq 1$, define an $n+1$-place relation U^n by:

$$U^n = \mathcal{U}^{\langle 1,n\rangle}(\emptyset).$$

U^n is generated in $T(\mathfrak{A})$, hence is in Γ (Proposition 2.3.1). Further, if $\mathcal{R}$ is any n-ary relation on $\mathcal{A}$ which is generated in $T(\mathfrak{A})$, then for some operator $\Phi \in T(\mathfrak{A})$ of order $\langle 1, n\rangle$, $\mathcal{R} = \Phi(\emptyset)$ (Proposition 2.3.1 again). Now Φ has an operator index, say i. That is,

$$\mathbf{x} \in \Phi(\mathcal{P}) \Leftrightarrow \langle i, \mathbf{x}\rangle \in \mathcal{U}^{\langle 1,n\rangle}(\mathcal{P}).$$

In particular,

$$\mathbf{x} \in \mathcal{R} \Leftrightarrow \mathbf{x} \in \Phi(\emptyset) \Leftrightarrow \langle i, \mathbf{x}\rangle \in \mathcal{U}^{\langle 1,n\rangle}(\emptyset) \Leftrightarrow \langle i, \mathbf{x}\rangle \in U^n.$$

Thus Γ is parametrized. Note that the operator indexing and the relation indexing are connected by

$$\mathcal{R}_i^n = \Phi_i^{\langle 1,n\rangle}(\emptyset).$$

Now, take the basic functions simply to be those functions generated in $T(\mathfrak{A})$. It is easy to see that the iteration property for Γ is an immediate consequence of the output place-fixing property.

(2) Let $T(\text{-})$ be ω-rec, and suppose rec($\mathfrak{A}$) is a theory with equality and an effective pairing function. Then, based on Corollary 14.5, we may construct a parametrization for the set of generated relations of ω-rec($\mathfrak{A}$),

having the iteration property, but in which the basic functions are those which are r.e., rather than ω-r.e. The notion of basic function is thus not always as trivial as it was in Example 1.

(3) It is a standard result of ordinary recursion theory, rec($\mathfrak{S}(\mathbb{N})$), that the generated (r.e.) relations may be parametrized so that the iteration property holds, where the basic functions are the *primitive recursive* ones. This is a rather narrow class of functions that has no exact analog for arbitrary structures. Nonetheless, analogs have essentially been constructed for set structures, Jensen and Karp [1971], and for word structures, Asscr [1960]. (See also Machtey and Young [1978].) We do not continue this topic here, though.

(4) Let $\mathcal{P}$ be an arbitrary relation on $\mathcal{A}$, and let Γ be the collection of relations $\leqslant \mathcal{P}$ (in $T(\mathfrak{A})$). Again, Γ is parametrized and has the iteration property. We may show this exactly as in Example 1, but writing $\mathcal{P}$ wherever $\emptyset$ was written.

Finally, we present a non-example. For all of this chapter, T(-) was either rec(-) or ω-rec(-) while in earlier chapters the theory assignment f.o. was generally given equal billing. The reason for omitting it in the present chapter is that no first order theory can have an operator indexing, since we have the following simple diagonal argument.

If f.o.($\mathfrak{A}$) had an operator indexing then, as in Example 1 above, the generated relations of f.o.($\mathfrak{A}$) would have a parametrization. So there would be a generated (first order definable) relation U^1 so that

$$x \in \mathcal{R}_i^1 \iff \langle i, x \rangle \in U^1.$$

Consider the set

$$\mathcal{P} = \{x \mid U^1(x, x)\}.$$

$\mathcal{P}$ would also be first order definable, hence so would $\bar{\mathcal{P}}$ (since in f.o.($\mathfrak{A}$) negation is available). Let a be an index for $\bar{\mathcal{P}}$, so $\bar{\mathcal{P}} = \mathcal{R}_a^1$. Then

$$a \in \bar{\mathcal{P}} \iff a \in \mathcal{R}_a^1 \iff U^1(a, a) \iff a \in \mathcal{P},$$

which is a contradiction.

16. Finite codes

In recursion theories, every enumeration operator is *compact* (Proposition 1.10.8); loosely speaking this means that, though input may be infinite, it is used in finite chunks. (The same need not be true of ω-enumeration operators, a fact we will elaborate on in Chapter 8.) Consequently, the finite sets can be expected to play a major role in recursion theories. The problem is, as things stand, we have no mechansism for "telling" an enumeration operator that the finite input we may have supplied thus far is all it is ever going to get. It can't tell the difference between finite, and infinite but incomplete yet. Neither, for that matter, can we where outputs are concerned. But there is the feeling that one ought to be able to deal with finite sets as single, complete objects. For this purpose, *finite codes* are introduced. The consequences of their existence are developed in Chapter 8; here we consider only the circumstances.

Let $\mathfrak{B}$ be a structure, with $\mathcal{A} \subseteq \mathcal{B}$. Suppose, to each finite subset $F \subseteq \mathcal{A}$, one or more members of $\mathcal{B}$ have been assigned, so that distinct finite subsets of $\mathcal{A}$ never have the same member of $\mathcal{B}$ assigned to them. Then we say $\mathcal{A}$ has a *finite coding* in $\mathcal{B}$. We call the members of $\mathcal{B}$ assigned to F *codes* for F. We write D_c for the finite subset of $\mathcal{A}$ with code c (if c is not a code, D_c is not defined).

Now suppose $\mathcal{A}$ has a finite coding in $\mathcal{B}$. We say $\mathcal{A}$ has a *positive canonical coding* in rec($\mathfrak{B}$) if

(1) There is an enumeration operator Φ in rec($\mathfrak{B}$) of order $\langle 1, 1\rangle$ such that, for each $\mathcal{P} \subseteq \mathcal{A}$,

$$\Phi(\mathcal{P}) = \{c \mid D_c \subseteq \mathcal{P}\}.$$

(2) The following relation is r.e. in rec($\mathfrak{B}$):

$$M(x, y) \Leftrightarrow y \text{ is a finite code and } x \in D_y.$$

We say, simply, $\mathcal{A}$ has a *canonical coding in* rec($\mathfrak{B}$) if, in addition to the above we also have

(3) The following relation is r.e. in rec($\mathfrak{B}$):

$$N(x, y) \Leftrightarrow y \text{ is a finite code and } x \notin D_y.$$

Let $\mathfrak{A}^*$ be defined from $\mathfrak{A}$ as in Section 11. Our first goal in this section is to show $\mathcal{A}^w$ has a canonical coding in rec($\mathfrak{A}^{*w}$), under the right circumstances. Afterwards, we will improve on the result.

Select a member of punct in $\mathcal{A}^*$ (see Section 11) that we have not used so far, let us denote it by $\diamond$. Now, if $\{a_1, a_2, \dots, a_n\}$ is a finite subset of $\mathcal{A}^w$ we code it by the word $\diamond a_1 \diamond a_2 \diamond \cdots \diamond a_n \diamond$. Note that n element subsets of $\mathcal{A}^w$ thus have $n!$ finite codes in $\mathcal{A}^{*w}$.

LEMMA 16.1. *The finite coding defined above is a positive canonical coding of* $\mathcal{A}^w$ *in* rec($\mathfrak{A}^{*w}$). *If also* rec($\mathfrak{A}$) *is a theory with equality, then it is a canonical coding of* $\mathcal{A}^w$ *in* rec($\mathfrak{A}^{*w}$).

PROOF. *Part* 1. Let Φ be the enumeration operator of order $\langle 1,1\rangle$ in $\mathrm{rec}(\mathfrak{A}^{*w})$ defined by the following axioms (I represents input, O represents output):

$$\text{axioms for word} \qquad \text{(item (1) in Section 11)},$$

$$O\Diamond\Diamond,$$

$$Ix \to \text{word}\ x \to Ow \to y = w * \Diamond * x * \Diamond \to Oy.$$

It should be clear that, for $\mathcal{P} \subseteq \mathcal{A}$, $\Phi(\mathcal{P}) = \{c \mid D_c \subset \mathcal{P}\}$.

Part 2. Let $M(x, y) \Leftrightarrow y$ is a finite code and $x \in D_y$. M is r.e. in $\mathrm{rec}(\mathfrak{A}^{*w})$, having the following axioms:

$$\text{axioms for word},$$

$$y = \Diamond * x * \Diamond \to \text{word}\ x \to M(x, y),$$

$$M(x, v) \to y = v * w * \Diamond \to \text{word}\ w \to M(x, y),$$

$$M(u, v) \to y = u * x * \Diamond \to \text{word}\ x \to M(x, y).$$

(3) Now suppose $\mathrm{rec}(\mathfrak{A})$ is a theory with equality. $\mathrm{rec}(\mathfrak{S}(\mathsf{N}))$ is certainly a theory with equality, so by Proposition 10.3, $\mathrm{rec}(\mathfrak{A}^*)$ is a theory with equality, and then by Lemma 5.1, $\mathrm{rec}(\mathfrak{A}^{*w})$ is a theory with equality. Let $N(x, y) \Leftrightarrow y$ is a finite code and $x \notin D_y$. N is r.e. in $\mathrm{rec}(\mathfrak{A}^{*w})$, having the following axioms.

$$\text{axioms for word},$$

$$\text{axioms for} \neq,$$

$$N(x, \Diamond\Diamond),$$

$$N(x, v) \to y = v * w * \Diamond \to \text{word}\ w \to x \neq w \to N(x, y),$$

Now we set to work on bettering this result.

LEMMA 16.2. *Suppose* $\mathrm{rec}(\mathfrak{B})$ *is an inessential extension of* $\mathrm{rec}(\mathfrak{A})$. *Then*

(1) *If* $\mathcal{A}$ *has a positive canonical coding in* $\mathrm{rec}(\mathfrak{B})$, $\mathcal{A}$ *has a positive canonical coding in* $\mathrm{rec}(\mathfrak{A})$.

(2) *If* $\mathcal{A}$ *has a canonical coding in* $\mathrm{rec}(\mathfrak{B})$, $\mathcal{A}$ *has a canonical coding in* $\mathrm{rec}(\mathfrak{A})$.

PROOF. Say $\mathrm{rec}(\mathfrak{B})$ is an inessential extension of $\mathrm{rec}(\mathfrak{A})$ via h, so that $h : \mathrm{rec}(\mathfrak{B}) \to \mathrm{rec}(\mathfrak{A})$ is an embedding, and the relation $y \in (h \restriction \mathcal{A})(x)$ is r.e. in $\mathrm{rec}(\mathfrak{A})$.

Suppose $\mathcal{A}$ has a (positive) canonical coding in $\mathrm{rec}(\mathfrak{B})$. Using it we define a finite coding of $\mathscr{A}$ in $\mathscr{A}$ as follows. Let F be a finite subset of $\mathscr{A}$, let b be any finite code for F in $\mathscr{B}$, let $c \in h(b)$; we call c a finite code for F in $\mathscr{A}$. We now show this coding is (positive) canonical.

(1) Suppose the original finite coding of $\mathscr{A}$ in $\mathscr{B}$ is positive canonical; we show the finite coding of $\mathscr{A}$ in $\mathscr{A}$ defined above is also positive canonical.

Let Φ_B be the enumeration operator in rec($\mathfrak{B}$) that takes $\mathscr{P} \subseteq \mathscr{A}$ to the collection of $\mathfrak{B}$ codes for finite subsets of $\mathscr{P}$. By Proposition 4.12.2, $(h \upharpoonright \mathscr{A})_1$ is an enumeration operator in rec($\mathfrak{A}$), since the relation $y \in (h \upharpoonright \mathscr{A})(x)$ is r.e. Then the following is also an enumeration operator in rec($\mathfrak{A}$):

$$\Phi = \Phi_B^h(h \upharpoonright \mathscr{A})_1.$$

Now, if $\mathscr{P} \subseteq \mathscr{A}$,

$$\Phi(\mathscr{P}) = h_1\Phi_B h_1^{-1}(h \upharpoonright \mathscr{A})_1(\mathscr{P}) = h_1\Phi_B(\mathscr{P})$$

and this is simply the collection of $\mathscr{A}$ codes for finite subsets of $\mathscr{P}$. Thus we have a suitable finite-subset-operator.

Let $M_B(x, y) \Leftrightarrow y$ is a finite code in $\mathscr{B}$ for a finite subset of $\mathscr{A}$ with x as a member. We are given that M_B is r.e. in rec($\mathfrak{B}$). Let $M_A(x, y) \Leftrightarrow y$ is a finite code in $\mathscr{A}$ for a finite subset of $\mathscr{A}$ with x as a member. We must show that M_A is r.e. in rec($\mathfrak{A}$). We note that

$$M_A(x, y) \Leftrightarrow (\exists y')[y \in h(y') \wedge M_B(x, y')].$$

Now the following is r.e. in rec($\mathfrak{A}$):

$$(\exists z)[\langle z, y\rangle \in M_B^h \wedge z \in (h \upharpoonright \mathscr{A})(x)].$$

This is equivalent to

$$(\exists z)(\exists z')(\exists y')$$
$$[z \in h(z') \wedge y \in h(y') \wedge M_B(z', y') \wedge z \in (h \upharpoonright \mathscr{A})(x)].$$

Now, if $z \in h(z')$ and $z \in (h \upharpoonright \mathscr{A})(x)$, it follows that $x = z'$, so the above is equivalent to

$$(\exists z)(\exists y')[z \in (h \upharpoonright \mathscr{A})(x) \wedge y \in h(y') \wedge M_B(x, y')]$$

and this in turn is equivalent to

$$(\exists y')[y \in h(y') \wedge M_B(x, y')].$$

Thus M_A is r.e. in rec($\mathfrak{A}$). This completes (1).

(2) The N-relation is treated exactly the way the M-relation was treated above. We omit the details.

DEFINITION. If $\mathscr{A}$ has a canonical coding in rec($\mathfrak{A}$) we simply say rec($\mathfrak{A}$) has a canonical coding.

THEOREM 16.3. *If* rec($\mathfrak{A}$) *is a theory with equality,* rec($\mathfrak{A}^w$) *has a canonical coding.*

PROOF. The above lemmas and Theorem 10.4.

THEOREM 16.4. *If* rec($\mathfrak{A}$) *is an infinite theory with equality and an effective pairing function, then* rec($\mathfrak{A}$) *has a canonical coding.*

PROOF. By the previous theorem, rec($\mathfrak{A}^w$) has a canonical coding. It follows easily that $\mathscr{A}$ has a canonical coding in rec($\mathfrak{A}^w$), and the result follows by Lemma 16.2 and Corollary 8.5.

EXAMPLES. By Theorem 16.4, our "standard" recursion theories, rec($\mathfrak{S}(\mathbb{N})$), rec($\mathfrak{S}(L_\omega)$) and rec($\mathfrak{S}(a_1, \ldots, a_n)$) all have canonical codings. In fact, for rec($\mathfrak{S}(L_\omega)$) a simple, direct canonical coding can be defined: code each finite subset of L_ω by itself. For rec($\mathfrak{S}(\mathbb{N})$) there is a well-known canonical coding defined in Rogers [1967], p. 70 (where it is called a canonical indexing). This one is the archetype for all canonical codings. Notice that these are 1–1, onto codings. In general this is too much to hope for, and more than we need.

CHAPTER SEVEN

INDEXED RELATIONAL SYSTEMS

1. Introduction

This is the first of two chapters in which we investigate the consequences of the existence of indexings, as established in the previous chapter. Here we consider just relations; in the next chapter, operators. We proceed axiomatically; Chapter 6 supplies us with many models for our axioms, but it should be kept in mind that other, rather different models will occur later on, most immediately in Section 6. We have chosen the axiom system below because it arises naturally out of our elementary formal system approach, but there are other treatments possible. The most elegant axiomatic development of essentially the material of this chapter is that of Wagner [1969] and Strong [1968]; see also Friedman [1971].

Definitions. Let $\mathscr{A}$ be a non-empty set. By an *indexed relational system* on $\mathscr{A}$; we mean a triple $\mathscr{R}_{\mathscr{A}} = \langle \mathscr{G}, U, \mathscr{F} \rangle$ where:

(1) $\mathscr{G}$ is a collection of relations on $\mathscr{A}$ called *generated* (in $\mathscr{R}_{\mathscr{A}}$), which contains $\emptyset$, singletons, $=_{\mathscr{A}}$, and is closed under transpositions, projections, $\cap$, $\cup$, where defined, and $\times$.

(2) U is a function that assigns, to each integer $n \geqslant 0$, an $n+1$-place *universal relation*, U^n, on $\mathscr{A}$ such that

(a) $U^n \in \mathscr{G}$.

(b) For each n-place $R \in \mathscr{G}$ there is an $i \in \mathscr{A}$, called an *index* of R, such that

$$R(\boldsymbol{x}) \Leftrightarrow U^n(i, \boldsymbol{x}).$$

[We write R_i^n for $\{x \in \mathscr{A}^n \mid U^n(i, \boldsymbol{x})\}$.]

(3) The members of $\mathscr{F}$ are functions from $\mathscr{A}^k$ to $\mathscr{A}$ (for $k = 1, 2, \ldots$), called *basic*, such that

(a) $\mathscr{F}$ is closed under composition.

(b) $\mathscr{F} \subseteq \mathscr{G}$ (that is, the graph of each basic function is a generated relation).

(c) (Iteration property). For each $n+k$-place generated relation Q there is a basic function $f : \mathcal{A}^n \to \mathcal{A}$ such that

$$Q(x_1, \ldots, x_n, y_1, \ldots, y_k) \Leftrightarrow R^k_{f(x_1,\ldots,x_n)}(y_1, \ldots, y_k).$$

EXAMPLES (See Ch. 6, §15). (1) If rec($\mathfrak{A}$) is a theory with equality, and an effective pairing function, then the r.e. relations give us an indexed relational system, taking the basic functions to be the total functions with r.e. graphs (recursive functions).

(2) Again, if rec($\mathfrak{A}$) is a theory with equality and an effective pairing function, and if $\mathcal{P}$ is a relation on $\mathcal{A}$, then the relations $\leqslant \mathcal{P}$ give an indexed relational system, still taking the basic functions to be the recursive functions. This example includes the previous one as the special case $\mathcal{P} = \emptyset$.

(3) Similarly, the relations $\leqslant \mathcal{P}$ in ω-rec($\mathfrak{A}$) give an indexed relational system, where we can still take the basic functions to be the recursive ones, provided rec($\mathfrak{A}$) is a theory with equality and an effective pairing function.

(4) In ordinary recursion theory, rec($\mathfrak{S}(\mathbb{N})$), the r.e. relations give us an indexed relational system, taking the basic functions to be the *primitive* recursive ones. Similarly for hyperarithmetic theory, ω-rec($\mathfrak{S}(\mathbb{N})$).

For the rest of this chapter, $\mathcal{R}_{\mathcal{A}} = \langle \mathcal{G}, U, \mathcal{F} \rangle$ *is some fixed indexed relational system on* $\mathcal{A}$.

2. Immediate consequences

THEOREM 2.1. *There is a set* $\mathcal{P} \subseteq \mathcal{A}$ *such that* $\mathcal{P}$ *is generated in* $\mathcal{R}_{\mathcal{A}}$ *but its complement,* $\bar{\mathcal{P}}$, *is not.*

PROOF. Let U^1 be universal for the 1-place relations, i.e. sets. Let $\mathcal{P} = \{x \mid U^1(x, x)\}$. $\mathcal{P}$ is generated in $\mathcal{R}_{\mathcal{A}}$ since $\mathcal{P} = U^1 \cap (=_{\mathcal{A}})$. Now suppose $\bar{\mathcal{P}}$ were also generated. Then it would have an index, say i. Thus, for all $x \in \mathcal{A}$,

$$\bar{\mathcal{P}}(x) \Leftrightarrow U^1(i, x).$$

But then, setting $x = i$,

$$\bar{\mathcal{P}}(i) \Leftrightarrow U^1(i, i) \Leftrightarrow \mathcal{P}(i)$$

which is a contradiction. Hence $\bar{\mathcal{P}}$ is not generated.

REMARK. This diagonal argument compares nicely with the one at the end of §15 in Ch. 6. In both cases we essentially proved one can't have an indexing and closure under complementation. In Ch. 6, §15 we had closure under complementation, hence there could be no indexing. Above we are given an indexing, hence there is no closure under complementation. The argument is due originally to Cantor, in essentially the following form. Suppose there were a 1–1 correspondence θ between a set S and its power set, $\mathcal{P}(S)$. "Index" each member $\mathcal{B}$ of $\mathcal{P}(S)$ by that member of S which corresponds to it under θ. This gives us an indexing of $\mathcal{P}(S)$ in S. But $\mathcal{P}(S)$ is closed under complementation, so a contradiction results. Hence θ does not exist.

We make two further observations concerning this proof. First, we made no use of the basic functions or the iteration property; the existence of an indexing is sufficient. Second, by comparing this result with Proposition 2.6.2, it follows immediately that an indexed relational system can not be finite.

The following theorem shows that, if we had a second system, V, of universal relations, they could be "mimicked" by the U-system.

THEOREM 2.2. *U is the given system of universal relations of $\mathcal{R}_{\mathcal{A}}$. We write R_i^n for the n-place generated relation having U-index i. Let V be another indexing system for $\mathcal{R}_{\mathcal{A}}$, satisfying condition* (2) *of the definition of indexed relational system. We write S_i^n for the n-place generated relation having V-index i.*

For each n, there is a basic function f such that $S_x^n = R_{f(x)}^n$.

PROOF. V^n is generated in $\mathcal{R}_{\mathcal{A}}$. Also the iteration property holds for the U-indexing, so there is a basic function f such that

$$V^n(x, y_1, \ldots, y_n) \Leftrightarrow R_{f(x)}^n(y_1, \ldots, y_n),$$

and this, rewritten, says

$$S_x^n(y_1, \ldots, y_n) \Leftrightarrow R_{f(x)}^n(y_1, \ldots, y_n).$$

DEFINITION. Keeping the notation of the above theorem we say two indexings U and V are *equivalent* if, for each n, there are basic functions f and g such that $R_x^n = S_{f(x)}^n$ and $S_x^n = R_{g(x)}^n$.

COROLLARY 2.3. *Any two indexings which have the iteration property* (*satisfy condition* (3)) *are equivalent.*

From now on, we use only the "given" indexing, U.

The collection of generated relations has certain closure properties, $\cap$,

$\cup$, etc. In fact, we have an *effective* closure under these operations, in the sense that, for example, given indexes for two relations, we can *calculate* an index for their intersection. We show this in the following theorem; there are similar effective closure results for $\cup$, etc., which we leave to the reader.

THEOREM 2.4. *For each n there is a basic function f such that*

$$R^n_x \cap R^n_y = R^n_{f(x,y)}.$$

PROOF. Define an $n+2$-place relation Q by

$$Q(u, v, \boldsymbol{x}) \Leftrightarrow U^n(u, \boldsymbol{x}) \wedge U^n(v, \boldsymbol{x}).$$

Q is generated, so by the iteration property there is a basic function f such that

$$Q(u, v, \boldsymbol{x}) \Leftrightarrow R^n_{f(u,v)}(\boldsymbol{x}).$$

Now,

$$\begin{aligned} \boldsymbol{x} \in R^n_a \cap R^n_b &\Leftrightarrow R^n_a(\boldsymbol{x}) \wedge R^n_b(\boldsymbol{x}) \\ &\Leftrightarrow U^n(a, \boldsymbol{x}) \wedge U^n(b, \boldsymbol{x}) \\ &\Leftrightarrow Q(a, b, \boldsymbol{x}) \\ &\Leftrightarrow R^n_{f(a,b)}(\boldsymbol{x}). \end{aligned}$$

REMARKS. In the above proof we asserted Q was generated. This will be our practice, omitting the details of verification. Just this once, however, we show Q is, in fact, generated, as an example.

$=_{\mathscr{A}}$ is generated, and we have closure under projection, hence $\mathscr{A}$ is generated.

U^n is generated, and we have closure under $\times$, hence $\mathscr{A} \times U^n$ is generated.

We have closure under transposition, hence $\mathscr{P} = \{\langle x, y, \boldsymbol{x}\rangle \mid \langle y, x, \boldsymbol{z}\rangle \in \mathscr{A} \times U^n\}$ is generated.

We have closure under intersection, hence $\mathscr{P} \cap (\mathscr{A} \times U^n)$ is generated and this, in fact, is Q.

3. Fixed point theorems

We prove, for $\mathscr{R}_{\mathscr{A}}$, two so called fixed point theorems. The first, in ordinary recursion theory, is essentially due to Kleene and is an immediate consequence of his Second Recursion Theorem, see Kleene [1938]. The second, again in ordinary recursion theory, is due to Myhill, see Myhill [1955, 1955A]. In fact, standard proofs from ordinary recursion theory carry over to $\mathcal{R}_{\mathcal{A}}$ with no difficulty. We begin by following the development in Smullyan [1961], in which Myhill's theorem is proved first, and Kleene's theorem is derived from that. We follow this with a short, direct proof of the Kleene result.

We do not present the Kleene Second Recursion Theorem itself here, because we are emphasizing relations, while that concerns itself directly with functions. We note, however, that a functional indexing could be introduced, and then the Second Recursion Theorem is an immediate consequence of Lemma 3.1 below. There is a proof of this in Smullyan [1961], p. 92, for ordinary recursion theory, which works quite well in $\mathcal{R}_{\mathcal{A}}$ too.

LEMMA 3.1 (Smullyan). *Let $M(z, \mathbf{x}, y)$ be a generated $n+2$-place relation. Then there is a basic function f such that*

$$R^n_{f(y)}(\mathbf{x}) \Leftrightarrow M(y, \mathbf{x}, f(y)).$$

PROOF. Consider the generated relation P defined by

$$P(z, \mathbf{x}) \Leftrightarrow R^{n+1}_z(\mathbf{x}, z) \Leftrightarrow U^{n+1}(z, \mathbf{x}, z).$$

Using the iteration property, there is a basic function d such that

$$R^n_{d(i)}(\mathbf{x}) \Leftrightarrow P(i, \mathbf{x}).$$

That is,

$$R^n_{d(i)}(\mathbf{x}) \Leftrightarrow R^{n+1}_i(\mathbf{x}, i) \qquad (*)$$

Next, consider the generated relation

$$(\exists w)[d(y) = w \wedge M(z, \mathbf{x}, w)] \Leftrightarrow M(z, \mathbf{x}, d(y)).$$

Again, by the iteration property, there is a basic function φ such that

$$R^{n+1}_{\varphi(i)}(\mathbf{x}, y) \Leftrightarrow M(i, \mathbf{x}, d(y)). \qquad (**)$$

Finally, set $f = d\varphi$. Since the basic functions are closed under composition, f is basic, and

$$\begin{aligned} R^n_{f(y)}(\mathbf{x}) &\Leftrightarrow R^n_{d\varphi(y)}(\mathbf{x}) \\ &\Leftrightarrow R^{n+1}_{\varphi(y)}(\mathbf{x}, \varphi(y)) \qquad \text{(by (*))} \\ &\Leftrightarrow M(y, \mathbf{x}, d\varphi(y)) \qquad \text{(by (**))} \\ &\Leftrightarrow M(y, \mathbf{x}, f(y)). \end{aligned}$$

COROLLARY 3.2 (Myhill fixed point theorem). *For each n there is a basic f such that*

$$R^n_{f(y)}(\boldsymbol{x}) \Leftrightarrow R^{n+1}_y(\boldsymbol{x}, f(y)).$$

PROOF. Set $M = U^{n+1}$ in the lemma.

COROLLARY 3.3. *For any $n+1$-place generated relation S there is an i such that*

$$R^n_i(\boldsymbol{x}) \Leftrightarrow S(\boldsymbol{x}, i).$$

PROOF. Let a be some index for S, use the previous corollary, and set $i = f(a)$.

COROLLARY 3.4 (Kleene fixed point theorem). *For any function $t : \mathscr{A} \to \mathscr{A}$, whose graph is generated in $\mathscr{R}$, there is an i such that*

$$R^n_i(\boldsymbol{x}) \Leftrightarrow R^n_{t(i)}(\boldsymbol{x}).$$

PROOF. Define S by

$$S(\boldsymbol{x}, y) \Leftrightarrow R^n_{t(y)}(\boldsymbol{x}) \Leftrightarrow (\exists z)[t(y) = z \wedge U^n(z, \boldsymbol{x})].$$

Then S is generated, so by the previous corollary, for some i,

$$R^n_i(x) \Leftrightarrow S(\boldsymbol{x}, i) \Leftrightarrow R^n_{t(i)}(\boldsymbol{x}).$$

REMARKS. The above proof of the Kleene fixed point theorem is rather roundabout, proceeding, as it does, via the Myhill fixed point theorem. So we include a more direct proof of this important result. It will be noted that the basic ideas of the proof are the same.

SECOND PROOF OF COROLLARY 3.4. Consider the $n+1$-place relation

$$P(u, \boldsymbol{x}) \Leftrightarrow U^{n+1}(u, \boldsymbol{x}, u) \Leftrightarrow R^{n+1}_u(\boldsymbol{x}, u).$$

P is generated, so by the iteration property, there is a basic function g such that

$$R^n_{g(u)}(\boldsymbol{x}) \Leftrightarrow P(u, \boldsymbol{x}) \Leftrightarrow R^{n+1}_u(\boldsymbol{x}, u). \tag{$*$}$$

g is generated; so is t, hence so is tg. Let v be an index for the generated relation

$$U^n(tg(y), \boldsymbol{x}) \Leftrightarrow R^n_{tg(y)}(\boldsymbol{x}).$$

That is,

$$R_v^{n+1}(\mathbf{x}, y) \Leftrightarrow R_{tg(y)}^n(\mathbf{x}). \qquad (**)$$

Let $i = g(v)$. Then

$$\begin{aligned} R_i^n(\mathbf{x}) &\Leftrightarrow R_{g(v)}^n(\mathbf{x}) \\ &\Leftrightarrow R_v^{n+1}(\mathbf{x}, v) \qquad \text{(by (*))} \\ &\Leftrightarrow R_{tg(v)}^n(\mathbf{x}) \qquad \text{(by (**))} \\ &\Leftrightarrow R_{t(i)}^n(\mathbf{x}). \end{aligned}$$

We conclude with two simple, and not very important, illustrations of the uses of these theorems. There are others, of greater significance, in the next section.

EXAMPLE 1. There is an i such that $R_i^1 = \{i\}$. (The method of proof below clearly can be modified to produce whole families of related examples.)

Consider the equality relation on $\mathscr{A}$, which for the present we will denote by E. Thus

$$E(x, y) \Leftrightarrow x =_{\mathscr{A}} y.$$

By definition, E is generated in $\mathscr{R}_{\mathscr{A}}$, so by the iteration property, there is a basic function t such that $R_{t(x)}^1(y) \Leftrightarrow E(x, y)$. That is,

$$R_{t(x)}^1 = \{x\}.$$

Now use the Kleene fixed point theorem.

EXAMPLE 2. In Theorem 2.2 we saw that if there were a second system, V, of universal relations, they could be "mimicked" by those of the U system. Here we show the curious fact that, under the same hypotheses and notation of Theorem 2.2, there must be some n-ary generated relation to which U and V assign the *same* index. Indeed, very simply, by Theorem 2.2, there must exist a basic function f, such that $S_x^n = R_{f(x)}^n$. Now use the Kleene fixed point theorem.

4. Rice's Theorem

In ordinary recursion theory there is a result, due to Rice, see Rice [1953] which embodies many separate undecidability results. Rice's Theorem carries over directly to $\mathscr{R}_{\mathscr{A}}$ as does its proof (though the one we present is not Rice's original). After proving it, we illustrate its use, then prove a related result which we will use often, later on.

DEFINITION. Let C be a set of indexes for n-ary generated relations of $\mathscr{A}$. C is called *closed* if

$$i \in C \quad \text{and} \quad R_i^n = R_j^n \ \Rightarrow\ j \in C.$$

C is called *bi-generated* if both C and $\bar{C}$ are generated.

THEOREM 4.1 (Rice's Theorem). *Let C be a set of indexes for n-ary generated relations on $\mathscr{A}$. If C is closed and bi-generated, then either C is empty or C contains every index.*

PROOF. Let a_0 and b_0 be indexes of n-ary generated relations, and suppose
(1) C is closed,
(2) $a_0 \in C$,
(3) $b_0 \notin C$,
(4) C is bi-generated.
We derive a contradiction.

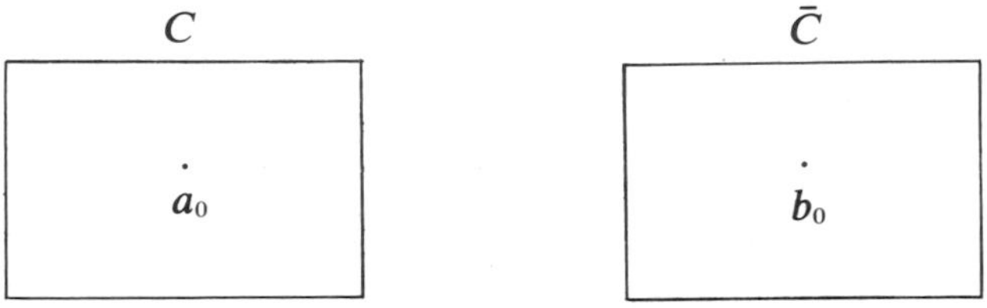

Fig. 1.

Define a function $t : \mathscr{A} \to \mathscr{A}$ by (see Fig. 1)

$$t(x) = \begin{cases} b_0 & \text{if } x \in C, \\ a_0 & \text{if } x \notin C. \end{cases}$$

The graph of t is generated in $\mathscr{R}_{\mathscr{A}}$, since it is

$$(C \times \{b_0\}) \cup (\bar{C} \times \{a_0\}).$$

Now, by Corollary 3.4, for some i, $R_i^n(\boldsymbol{x}) \Leftrightarrow R_{t(i)}^n(\boldsymbol{x})$. But this says i and $t(i)$ index the same n-ary relation. Since C is complete, i and $t(i)$ are both in C, or both in $\bar{C}$, but the definition of t makes this impossible.

EXAMPLES. (1) Let C be the set of all indexes for *infinite* generated subsets of $\mathscr{A}$. C is complete, so C is not bi-generated.

(2) Let C be the set of all indexes for generated subsets of $\mathscr{A}$ containing c (c fixed). C is complete, so C is not bi-generated.

(3) Let R be a fixed, generated relation on $\mathscr{A}$, and let C be the set of all indexes for R. C is complete, so C is not bi-generated.

(4) An example rather intimately connected with generalized recursion theory. It is trivial that, considered as production systems, $\mathrm{rec}(\langle \mathsf{N}; +\rangle) = \mathrm{rec}(\mathfrak{S}(\mathsf{N}))$, and it is equally clear that $\mathrm{rec}(\langle \mathsf{N}; \emptyset\rangle) \neq \mathrm{rec}(\mathfrak{S}(\mathsf{N}))$. But in general, can one tell, for an arbitrary relation $\mathscr{R}$ on the natural numbers, whether or not $\mathrm{rec}(\langle \mathsf{N}; \mathscr{R}\rangle)$ is the same as ordinary recursion theory? Of course, there is no chance of $\mathrm{rec}(\langle \mathsf{N}; \mathscr{R}\rangle)$ being ordinary recursion theory unless $\mathscr{R}$ is r.e. in the ordinary sense. So we may narrow the question to: in ordinary recursion theory, for fixed n, is the set of indexes i such that

$$\mathrm{rec}(\langle \mathsf{N}; R_i^n\rangle) = \mathrm{rec}(\mathfrak{S}(\mathsf{N})) \text{ a recursive set?}$$

Well, the set is complete, by definition, and neither empty nor all indexes, by the examples above. Hence by Rice's Theorem, it is not recursive.

REMARK. The above proof used Corollary 3.4, which itself was proved using Corollary 3.3. It is possible to base a proof of Rice's Theorem directly on Corollary 3.3. We sketch this, because it is similar to, though slightly simpler than the proof of the next theorem.

SECOND PROOF OF THEOREM 4.1. Make assumptions (1)–(4) as in the above proof. Then define an $n+1$-place relation S by

$$S(\boldsymbol{x}, y) \Leftrightarrow [R^n_{a_0}(\boldsymbol{x}) \wedge y \in \bar{C}] \vee [R^n_{b_0}(\boldsymbol{x}) \wedge y \in C].$$

S is generated, so by Corollary 3.3, there is an i such that

$$R^n_i(\boldsymbol{x}) \Leftrightarrow S(\boldsymbol{x}, i).$$

Now, if $i \in C$, then $S(\boldsymbol{x}, i) \Leftrightarrow R^n_{b_0}(\boldsymbol{x})$, so $R^n_i(\boldsymbol{x}) \Leftrightarrow R^n_{b_0}(\boldsymbol{x})$. Thus i and b_0 index the same relation, but $i \in C$ while $b_0 \notin C$.

There is a similar contradiction if $i \in \bar{C}$.

THEOREM 4.2. *Let C be a closed set of indexes for n-ary generated relations on $\mathscr{A}$. If C is generated, then $a_0 \in C$ and $R^n_{a_0} \subseteq R^n_{b_0} \Rightarrow b_0 \in C$.*

PROOF. Suppose

(1) C is closed, (2) C is generated,

(3) $R^n_{a_0} \subseteq R^n_{b_0}$,

(4) $a_0 \in C$, (5) $b_0 \notin C$.

We derive a contradiction.

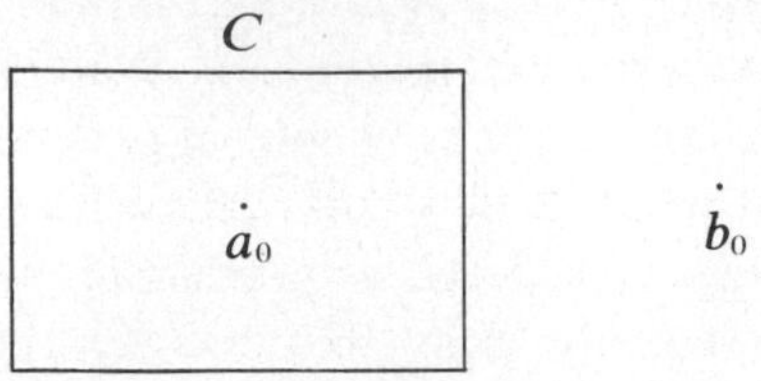

Fig. 2.

Define an $n+1$-place relation S by (see Fig. 2):

$$S(\boldsymbol{x}, y) \Leftrightarrow [R^n_{a_0}(\boldsymbol{x})] \vee [R^n_{b_0}(\boldsymbol{x}) \wedge y \in C].$$

S is generated, so by Corollary 3.3, there is an i such that

$$R^n_i(\boldsymbol{x}) \Leftrightarrow S(\boldsymbol{x}, i).$$

Case 1: $i \notin C$. Then by definition of S,

$$R^n_i(\boldsymbol{x}) \Leftrightarrow S(\boldsymbol{x}, i) \Leftrightarrow R^n_{a_0}(\boldsymbol{x}),$$

so i and a_0 index the same relation. But $a_0 \in C$ while $i \notin C$, which is impossible.

Case 2: $i \in C$. Then, again, using the definition of S together with $R^n_{a_0} \subseteq R^n_{b_0}$,

$$R^n_i(\boldsymbol{x}) \Leftrightarrow S(\boldsymbol{x}, i) \Leftrightarrow R^n_{b_0}(\boldsymbol{x}),$$

so i and b_0 index the same relation, which again is impossible.

5. Creative sets

Myhill [1955A] showed, in ordinary recursion theory, that any two creative sets (definition below) are recursively isomorphic. This is too much to manage in $\mathcal{R}_{\mathcal{A}}$ without assuming more structure, but a portion of Myhill's work does go through in the present generality. In this section we show that, in $\mathcal{R}_{\mathcal{A}}$, any two creative sets are many-one reducible to each other.

DEFINITION. $C \subseteq \mathcal{A}$ is *creative* if C is generated and there is a function $\varphi : \mathcal{A} \to \mathcal{A}$, whose graph is generated, such that if $R^1_i \cap C = \emptyset$ then $\varphi(i) \not\subset R^1_i \cup C$.

REMARKS. In the above definition we assumed φ was a *total* function. In ordinary recursion theory the definition is generally given using a *partial* function, as follows. $C \subseteq \mathcal{A}$ is creative if C is generated and there is a partial function φ whose graph is generated, such that if $R_i^! \cap C = \emptyset$, then i is in the domain of φ, and $\varphi(i) \notin R_i^! \cup C$.

In ordinary recursion theory these two definitions are equivalent. More generally, they are equivalent in every indexed relational system in which every two-place generated relation can be "cut down" to a generated function with the same domain. See Rogers [1967], p. 71. This is probably not the case for all indexed relational systems, however, and the exact status of the two definitions of creative is not known in full generality.

We assume, throughout this section, that φ is total.

We showed in Section 2 that there was a set which was generated but whose complement was not. Creative sets have this feature in an "effective" way. Say C is creative. Then C is generated, by definition. Suppose we try to see if $\bar{C}$ is generated, say we suspect $\bar{C} = R_i^!$. Then $R_i^! \cap C = \emptyset$, so $\varphi(i) \notin R_i^! \cup C$; that is, $\varphi(i) \in \bar{C}$ but $\varphi(i) \notin R_i^!$, so we have "calculated" an instance, $\varphi(i)$, where $R_i^!$ and $\bar{C}$ differ.

PROPOSITION 5.1. *There is a creative set.*

PROOF. The same set which was used in proving Theorem 2.1 works here. Let $\mathcal{P} = \{x \mid U^1(x,x)\} = \{x \mid R_x^!(x)\}$. $\mathcal{P}$ is generated. Let φ be the identity function, $\varphi(x) = x$. The graph of φ is generated since it is simply $=_{\mathcal{A}}$. Now suppose $R_i^! \cap \mathcal{P} = \emptyset$. Then in particular, $i \notin R_i^! \cap \mathcal{P}$. If $i \notin R_i^!$, by definition of $\mathcal{P}$, $i \notin \mathcal{P}$, and conversely. Hence $\varphi(i) = i \notin R_i^! \cup \mathcal{P}$.

DEFINITION. Let $\mathcal{P}, \mathcal{R} \subseteq \mathcal{A}$. $\mathcal{P}$ is *many-one reducible to* $\mathcal{R}$, written $\mathcal{P} \leqslant_m \mathcal{R}$, if there is a function $f: \mathcal{A} \to \mathcal{A}$, with generated graph, such that $x \in \mathcal{P} \Leftrightarrow f(x) \in \mathcal{R}$.

THEOREM 5.2. *Suppose $A \leqslant_m B$, A is creative, and B is generated. Then B is creative.*

PROOF. $A \leqslant_m B$. Let $f: \mathcal{A} \to \mathcal{A}$ be a generated function such that

$$x \in A \Leftrightarrow f(x) \in B. \tag{*}$$

Also, define Q by

$$Q(i,x) \Leftrightarrow (\exists y)[R_i^!(y) \wedge y = f(x)] \Leftrightarrow R_i^!(f(x)).$$

Q is generated, so by the iteration property there is a basic function t such that $R^1_{t(i)}(x) \Leftrightarrow Q(i,x)$, hence

$$x \in R^1_{t(i)} \Leftrightarrow f(x) \in R^1_i. \tag{**}$$

Also A is creative so there is some generated function $g : \mathscr{A} \to \mathscr{A}$ such that

$$R^1_i \cap A = \emptyset \Rightarrow g(i) \notin R^1_i \cup A.$$

We need a function to play a similar role for B. We claim it is $\varphi = fgt$.

First, φ is a function from $\mathscr{A}$ to $\mathscr{A}$ with generated graph, since this is true of f, g and t.

Now, suppose $R^1_i \cap B = \emptyset$. Then by (*) and (**), $R^1_{t(i)} \cap A = \emptyset$. Then $gt(i) \notin R^1_{t(i)} \cup A$, and by (*) and (**) again, $fgt(i) \notin R^1_i \cup B$. This concludes the proof.

LEMMA 5.3. *Let A be a generated set, and let $g : \mathscr{A} \to \mathscr{A}$ be a function with generated graph. There is a basic function f such that, for all i,*

$$i \in A \Rightarrow R^1_{f(i)} = \{gf(i)\}, \qquad i \notin A \Rightarrow R^1_{f(i)} = \emptyset.$$

PROOF. Let $M(z,x,y) \Leftrightarrow z \in A \wedge x = g(y)$. Then M is generated, so by Lemma 3.1 there is a basic function f such that

$$R^1_{f(i)}(x) \Leftrightarrow M(i,x,f(i)) \Leftrightarrow i \in A \wedge x = gf(i).$$

THEOREM 5.4. *Suppose B is creative and A is generated. Then $A \leqslant_m B$.*

PROOF. B is creative, so there is a generated function $g : \mathscr{A} \to \mathscr{A}$ such that, if $R^1_i \cap B = \emptyset$ then $g(i) \notin R^1_i \cup B$. It follows trivially that

$$\text{if } R^1_i = \emptyset \quad \text{then } g(i) \notin B, \tag{*}$$

$$\text{if } R^1_i = \{a\} \text{ and } a \notin B \quad \text{then } g(i) \neq a. \tag{**}$$

Now A is generated, so by Lemma 5.3, there is a basic function f such that

$$i \in A \Rightarrow R^1_{f(i)} = \{gf(i)\}, \qquad i \notin A \Rightarrow R^1_{f(i)} = \emptyset.$$

We claim $x \in A \Leftrightarrow gf(x) \in B$, which says that $A \leqslant_m B$.
Suppose $i \in A$. Then $R^1_{f(i)} = \{gf(i)\}$. It follows by (**) that $gf(i) \in B$.
Suppose $i \notin A$. Then $R^1_{f(i)} = \emptyset$. It follows by (*) that $gf(i) \notin B$.
This concludes the proof.

THEOREM 5.5. *Let B be generated. B is creative if and only if every generated set is many-one reducible to B.*

PROOF. Let B be generated. If B is creative and A is generated, $A \leqslant_m B$ by Theorem 5.4. Next, suppose every generated set is many-one reducible to B. By Proposition 5.1, there is a creative set, call it A. A is generated, so $A \leqslant_m B$. Then B is creative by Theorem 5.2.

COROLLARY 5.6. *Let A and B be creative. Then $A \leqslant_m B$ and $B \leqslant_m A$.*

PROOF. B is creative, A is generated, so $A \leqslant_m B$. Similarly $B \leqslant_m A$.

6. Kleene–Mostowski systems

Kleene [1943] and Mostowski [1947] independently developed, for arithmetic, what is now called the arithmetic hierarchy. See Hinman [1978] for a summary. In this section we show that if $\mathcal{R}_{\mathcal{A}}$ meets certain conditions, a similar construction can be carried out based on $\mathcal{R}_{\mathcal{A}}$, a hierarchy results, and that hierarchy provides us with many more examples of indexed relational systems.

DEFINITION. For each $n = 0, 1, 2, \ldots$, we define two collections, $\Sigma_n(\mathcal{R}_{\mathcal{A}})$ and $\Pi_n(\mathcal{R}_{\mathcal{A}})$ of relations on $\mathcal{A}$ as follows:

$\Sigma_0(\mathcal{R}_{\mathcal{A}}) = \Pi_0(\mathcal{R}_{\mathcal{A}})$ is the collection of relations $\mathcal{P}$ on $\mathcal{A}$ such that both $\mathcal{P}$ and $\bar{\mathcal{P}}$ are generated in $\mathcal{R}_{\mathcal{A}}$.

$\Sigma_{n+1}(\mathcal{R}_{\mathcal{A}})$ is the closure of $\Pi_n(\mathcal{R}_{\mathcal{A}})$ under all projections.

$\Pi_n(\mathcal{R}_{\mathcal{A}})$ is the collection of complements of members of $\Sigma_n(\mathcal{R}_{\mathcal{A}})$.

REMARK. It is our intention to show that if $\mathcal{R}_{\mathcal{A}}$ meets certain conditions, the above is actually a hierarchy (proper definition later), and each $\Sigma_n(\mathcal{R}_{\mathcal{A}})$ is itself an indexed relational system, so that all the preceeding work of this chapter applies to it.

LEMMA 6.1. (1) *For $n > 0$, $\Sigma_n(\mathcal{R}_{\mathcal{A}})$ is closed under projections.*

(2) *For all n, $\Sigma_n(\mathcal{R}_{\mathcal{A}})$ and $\Pi_n(\mathcal{R}_{\mathcal{A}})$ contain $\emptyset$, $\mathcal{A}$, and are closed under $\times$, $\cup$, $\cap$ and transpositions.*

PROOF. (1) is immediate from the definition. (2) is shown by induction on n.

The case $n = 0$. $\emptyset$ and $\mathcal{A}$ are both generated, $\bar{\emptyset} = \mathcal{A}$ and $\bar{\mathcal{A}} = \emptyset$, so $\emptyset$ and $\mathcal{A}$ are in $\Sigma_0(\mathcal{R}_{\mathcal{A}}) = \Pi_0(\mathcal{R}_{\mathcal{A}})$. Suppose $\mathcal{P}$ (j-place) and φ (k-place) are in $\Sigma_0(\mathcal{R}_{\mathcal{A}}) = \Pi_0(\mathcal{R}_{\mathcal{A}})$. Then $\mathcal{P}$, $\mathcal{Q}$, $\bar{\mathcal{P}}$ and $\bar{\mathcal{Q}}$ are generated. Hence so are $\mathcal{P} \times \mathcal{Q}$, and $\overline{\mathcal{P} \times \mathcal{Q}} = (\bar{\mathcal{P}} \times \mathcal{A}^k) \cup (\mathcal{A}^j \times \bar{\mathcal{Q}})$, so $\mathcal{P} \times \mathcal{Q}$ is in $\Sigma_0(\mathcal{R}_{\mathcal{A}}) =$

$\Pi_0(\mathcal{R}_{\mathcal{A}})$. $\cup$ and $\cap$ are similar, using DeMorgan's laws, and transposition is straightforward.

Now, suppose (2) holds for n.

Since $\emptyset \in \Pi_n(\mathcal{R}_{\mathcal{A}})$, projections of $\emptyset$ are in $\Sigma_{n+1}(\mathcal{R}_{\mathcal{A}})$, i.e. $\emptyset \in \Sigma_{n+1}(\mathcal{R}_{\mathcal{A}})$. Then $\mathcal{A} = \bar{\emptyset} \in \Pi_{n+1}(\mathcal{R}_{\mathcal{A}})$. Further, $\mathcal{A} \in \Pi_n(\mathcal{R}_{\mathcal{A}})$, which is closed under $\times$, so $\mathcal{A}^2 \in \Pi_n(\mathcal{R}_{\mathcal{A}})$. Then $\mathcal{A} = P_1^2(\mathcal{A}^2) \in \Sigma_{n+1}(\mathcal{R}_{\mathcal{A}})$, and hence $\emptyset = \bar{\mathcal{A}} \in \Pi_{n+1}(\mathcal{R}_{\mathcal{A}})$. Thus $\Sigma_{n+1}(\mathcal{R}_{\mathcal{A}})$ and $\Pi_{n+1}(\mathcal{R}_{\mathcal{A}})$ contain $\emptyset$ and $\mathcal{A}$.

We show $\Sigma_{n+1}(\mathcal{R}_{\mathcal{A}})$ is closed under $\times$. Say $\mathcal{P}, \mathcal{Q} \in \Sigma_{n+1}(\mathcal{R}_{\mathcal{A}})$. Then $\mathcal{P} = \boldsymbol{M}(\mathcal{P}')$ and $\mathcal{Q} = \boldsymbol{N}(\mathcal{Q}')$ where $\mathcal{P}', \mathcal{Q}' \in \Pi_n(\mathcal{R}_{\mathcal{A}})$ and $\boldsymbol{M}$ and $\boldsymbol{N}$ are strings of projection operators. Say $\mathcal{P}'$ is i-place and $\mathcal{Q}$ is j-place. Define $\boldsymbol{M}'$ is like $\boldsymbol{M}$ but with each P_b^a replaced by P_b^{j+a}, and $\boldsymbol{N}'$ is like $\boldsymbol{N}$ but with each P_b^a replaced by P_{i+b}^{i+a}. Now, $\mathcal{P}' \times \mathcal{Q}' \in \Pi_n(\mathcal{R}_{\mathcal{A}})$ by the induction hypothesis, and it is not hard to see that

$$\mathcal{P} \times \mathcal{Q} = \boldsymbol{M}'\boldsymbol{N}'(\mathcal{P}' \times \mathcal{Q}'),$$

hence $\mathcal{P} \times \mathcal{Q} \in \Sigma_{n+1}(\mathcal{R}_{\mathcal{A}})$. Thus $\Sigma_{n+1}(\mathcal{R}_{\mathcal{A}})$ is closed under $\times$.

We show $\Sigma_{n+1}(\mathcal{R}_{\mathcal{A}})$ is closed under $\cup$. Say $\mathcal{P}, \mathcal{Q} \in \Sigma_{n+1}(\mathcal{R}_{\mathcal{A}})$, where both are k-place. Then

$$\mathcal{P} = \boldsymbol{M}(\mathcal{P}') \quad \text{and} \quad \mathcal{Q} = \boldsymbol{N}(\mathcal{Q}')$$

where $\mathcal{P}', \mathcal{Q}' \in \Pi_n(\mathcal{R}_{\mathcal{A}})$ and $\boldsymbol{M}$ and $\boldsymbol{N}$ are strings of projection operators. Since $\Pi_n(\mathcal{R}_{\mathcal{A}})$ is closed under transpositions, we can choose $\mathcal{P}'$ and $\mathcal{Q}'$ so that we are always projecting off the front, that is, the members of $\boldsymbol{M}$ and $\boldsymbol{N}$ are of the form P_1^a. Now, say $\mathcal{P}'$ is i-place, $\mathcal{Q}'$ is j-place and $i > j$. (The cases $i = j$ and $i < j$ are similar). By the induction hypothesis

$$\mathcal{P}' \cup (\mathcal{A}^{i-j} \times \mathcal{Q}') \in \Pi_n(\mathcal{R}_{\mathcal{A}}),$$

and it is not hard to see that

$$\mathcal{P} \cup \mathcal{Q} = \boldsymbol{M}(\mathcal{P}' \cup (\mathcal{A}^{i-j} \times \mathcal{Q}'))$$

and hence is in $\Sigma_{n+1}(\mathcal{R}_{\mathcal{A}})$. Thus $\Sigma_{n+1}(\mathcal{R}_{\mathcal{A}})$ is closed under $\cup$.

We show $\Sigma_{n+1}(\mathcal{R}_{\mathcal{A}})$ is closed under $\cap$. Say $\mathcal{P}, \mathcal{Q} \in \Sigma_{n+1}(\mathcal{R}_{\mathcal{A}})$, where both are k-place. Then $\mathcal{P} = \boldsymbol{M}(\mathcal{P}')$ and $\mathcal{Q} = \boldsymbol{N}(\mathcal{Q}')$ where $\mathcal{P}', \mathcal{Q}' \in \Pi_n(\mathcal{R}_{\mathcal{A}})$ and $\boldsymbol{M}$ and $\boldsymbol{N}$ are strings of projection operators. Again, since $\Pi_n(\mathcal{R}_{\mathcal{A}})$ is closed under transpositions, we can choose $\mathcal{P}'$ and $\mathcal{Q}'$ so that: in $\boldsymbol{M}$ we are projecting off the front, so its terms are of the form P_1^a; in $\boldsymbol{N}$ we are projecting off the back, so its terms are of the form P_a^a.

Now, say $\mathcal{P}'$ is i-place and $\mathcal{Q}'$ is j-place. By the induction hypothesis,

$$(\mathcal{P}' \times \mathcal{A}^{j-k}) \cap (\mathcal{A}^{i-k} \times \mathcal{Q}') \in \Pi_n(\mathcal{R}_{\mathcal{A}}).$$

Let $\boldsymbol{N}'$ be like $\boldsymbol{N}$ but with each P_a^a replaced by P_{a+i-k}^{a+i-k}. Then

$$\boldsymbol{MN}'((\mathscr{P}' \times \mathscr{A}^{j-k}) \cap (\mathscr{A}^{i-k} \times \mathscr{Q}'))$$

is in $\Sigma_{n+1}(\mathscr{R}_{\mathscr{A}})$, and it is not hard to see that this is $\mathscr{P} \cap \mathscr{Q}$. Thus $\Sigma_{n+1}(\mathscr{R}_{\mathscr{A}})$ is closed under $\cap$.

That $\Sigma_{n+1}(\mathscr{R}_{\mathscr{A}})$ is closed under transpositions follows easily from the corresponding closure of $\Pi_n(\mathscr{R}_{\mathscr{A}})$; we omit details.

Finally, we show the various closure properties of $\Pi_{n+1}(\mathscr{R}_{\mathscr{A}})$.

Suppose $\mathscr{P}, \mathscr{Q} \in \Pi_{n+1}(\mathscr{R}_{\mathscr{A}})$ where $\mathscr{P}$ is i-place and $\mathscr{Q}$ is j-place. Then $\bar{\mathscr{P}}, \bar{\mathscr{Q}} \in \Sigma_{n+1}(\mathscr{R}_{\mathscr{A}})$. By what was just shown,

$$(\bar{\mathscr{P}} \times \mathscr{A}^j) \cup (\mathscr{A}^i \times \bar{\mathscr{Q}}) \in \Sigma_{n+1}(\mathscr{R}_{\mathscr{A}}).$$

This is $\overline{\mathscr{P} \times \mathscr{Q}}$, so the complement, $\mathscr{P} \times \mathscr{Q}$, is in $\Pi_{n+1}(\mathscr{R}_{\mathscr{A}})$. Thus $\Pi_{n+1}(\mathscr{R}_{\mathscr{A}})$ is closed under $\times$.

Closure of $\Pi_{n+1}(\mathscr{R}_{\mathscr{A}})$ under $\cap$ and $\cup$ follows similarly using DeMorgan's laws. And closure under transpositions is simple since complementation and transposition commute.

This ends the proof.

PROPOSITION 6.2. (1) $\Pi_n(\mathscr{R}_{\mathscr{A}}) \subseteq \Sigma_{n+1}(\mathscr{R}_{\mathscr{A}})$,
(2) $\Sigma_n(\mathscr{R}_{\mathscr{A}}) \subseteq \Pi_{n+1}(\mathscr{R}_{\mathscr{A}})$,
(3) $\Sigma_n(\mathscr{R}_{\mathscr{A}}) \subseteq \Sigma_{n+1}(\mathscr{R}_{\mathscr{A}})$,
(4) $\Pi_n(\mathscr{R}_{\mathscr{A}}) \subseteq \Pi_{n+1}(\mathscr{R}_{\mathscr{A}})$.

PROOF. (1) Let $\mathscr{P} \in \Pi_n(\mathscr{R}_{\mathscr{A}})$. By the previous Lemma, $\mathscr{A} \times \mathscr{P} \in \Pi_n(\mathscr{R}_{\mathscr{A}})$, so $\mathscr{P}$, which is a projection of $\mathscr{A} \times \mathscr{P}$, is in $\Sigma_{n+1}(\mathscr{R}_{\mathscr{A}})$.

(2) Let $\mathscr{P} \in \Sigma_n(\mathscr{R}_{\mathscr{A}})$. Then $\bar{\mathscr{P}} \in \Pi_n(\mathscr{R}_{\mathscr{A}})$. By (1), $\bar{\mathscr{P}} \in \Sigma_{n+1}(\mathscr{R}_{\mathscr{A}})$ so $\mathscr{P} \in \Pi_{n+1}(\mathscr{R}_{\mathscr{A}})$.

(3) By induction on n. If $n = 0$, $\Sigma_n(\mathscr{R}_{\mathscr{A}}) = \Pi_n(\mathscr{R}_{\mathscr{A}})$ and the result follows by (1).

Suppose inclusion (3) is true at n. And say $\mathscr{P} \in \Sigma_{n+1}(\mathscr{R}_{\mathscr{A}})$. Then $\mathscr{P} = \boldsymbol{M}(\mathscr{P}')$ where $\mathscr{P}' \in \Pi_n(\mathscr{R}_{\mathscr{A}})$, and $\boldsymbol{M}$ is a string of projections. Then $\overline{\mathscr{P}'} \in \Sigma_n(\mathscr{R}_{\mathscr{A}})$, so by the induction hypothesis, $\overline{\mathscr{P}'} \in \Sigma_{n+1}(\mathscr{R}_{\mathscr{A}})$. Then $\mathscr{P}' \in \Pi_{n+1}(\mathscr{R}_{\mathscr{A}})$ so $\mathscr{P} = \boldsymbol{M}(\mathscr{P}') \in \Sigma_{n+2}(\mathscr{R}_{\mathscr{A}})$. This completes the induction.

(4) Suppose $\mathscr{P} \in \Pi_n(\mathscr{R}_{\mathscr{A}})$. Then $\bar{\mathscr{P}} \in \Sigma_n(\mathscr{R}_{\mathscr{A}})$ so by (3), $\bar{\mathscr{P}} \in \Sigma_{n+1}(\mathscr{R}_{\mathscr{A}})$, hence $\mathscr{P} \in \Pi_{n+1}(\mathscr{R}_{\mathscr{A}})$.

Before proceeding further, we must impose some extra conditions on $\mathscr{R}_{\mathscr{A}}$. We use ordinary recursion theory as a guide, though there are other indexed relational systems which meet our conditions as well. We return briefly to this point in the next section.

In Ch. 4, §10 we gave an example of an effective pairing function for ordinary recursion theory. It has the nice feature that it is onto, i.e. every member codes a pair. Also, since ordinary recursion theory is a theory with equality, and the effective pairing function is total, it is recursive (Lemma 2.6.7).

DEFINITION. We say $\mathcal{R}_{\mathcal{A}}$ has a *strong pairing function* if there is a 1–1 onto function $J : \mathcal{A} \times \mathcal{A} \to \mathcal{A}$ such that both the graph of J, and its complement, are generated in $\mathcal{R}_{\mathcal{A}}$.

PROPOSITION 6.3. *Suppose* $\mathcal{R}_{\mathcal{A}}$ *has a strong pairing function. If* $\mathcal{P} \in \Sigma_{n+1}(\mathcal{R}_{\mathcal{A}})$ *is* k*-place, then there is a* $k+1$*-place relation* $\mathcal{P}' \in \Pi_n(\mathcal{R}_{\mathcal{A}})$ *such that* $\mathcal{P} = P_1^{k+1}(\mathcal{P}')$. [*That is, iterated projections can be collapsed to single ones.*]

PROOF. Suppose $\mathcal{P} \in \Sigma_{n+1}(\mathcal{R}_{\mathcal{A}})$. Then $\mathcal{P} = \boldsymbol{M}(\mathcal{Q})$ for some $\mathcal{Q} \in \Pi_n(\mathcal{R}_{\mathcal{A}})$ and some string $\boldsymbol{M}$ of projections. We suppose $\boldsymbol{M}$ has *two* terms; more than two can be treated by iterating the procedure below. Also we may suppose the projections are off the front, since $\Pi_n(\mathcal{R}_{\mathcal{A}})$ is closed under transpositions. So we have

$$\mathcal{P} = P_1^{k+1} P_1^{k+2}(\mathcal{Q}).$$

Now, $\mathcal{Q} \in \Pi_n(\mathcal{R}_{\mathcal{A}})$, so $\bar{\mathcal{Q}} \in \Sigma_n(\mathcal{R}_{\mathcal{A}})$. Define a relation S by

$$S(w, z) \Leftrightarrow (\exists x)(\exists y)[J(x, y) = w \wedge \bar{\mathcal{Q}}(x, y, z)]$$

where J is a strong pairing function. Since both J and its complement are generated, $J \in \Sigma_0(\mathcal{R}_{\mathcal{A}})$, and hence by Proposition 6.2(3), $J \in \Sigma_n(\mathcal{R}_{\mathcal{A}})$. Then, by the closure properties of Lemma 6.1, $S \in \Sigma_n(\mathcal{R}_{\mathcal{A}})$, and hence $\bar{S} \in \Pi_n(\mathcal{R}_{\mathcal{A}})$. We claim $\mathcal{P} = P_1^{k+1}(\bar{S})$, that is,

$$P_1^{k+1}(\bar{S}) = P_1^{k+1} P_1^{k+2}(\mathcal{Q}).$$

Now

$$\begin{aligned} z \in P_1^{k+1}(\bar{S}) &\Leftrightarrow (\exists w)\bar{S}(w, z) \\ &\Leftrightarrow (\exists w) \sim (\exists x)(\exists y)[J(x, y) = w \wedge \bar{\mathcal{Q}}(x, y, z)] \\ &\Leftrightarrow (\exists w)(\forall x)(\forall y)[J(x, y) = w \supset \mathcal{Q}(x, y, z)] \end{aligned}$$

and since J is a 1–1, onto function, this is equivalent to

$$(\exists x)(\exists y)\mathcal{Q}(x, y, z) \Leftrightarrow z \in P_1^{k+1} P_1^{k+2}(\mathcal{Q}).$$

In ordinary recursion theory there is a result, due to Kleene, that says the r.e. relations are the projections of the recursive relations (indeed, stronger versions are true, but this is all we need).

DEFINITION. We call $\mathcal{R}_{\mathcal{A}}$ a *Kleene–Mostowski system* if
(1) $\mathcal{R}_{\mathcal{A}}$ has a strong pairing function and
(2) the generated relations of $\mathcal{R}_{\mathcal{A}}$ are exactly the members of $\Sigma_1(\mathcal{R}_{\mathcal{A}})$.

THEOREM 6.4. *Let $\mathcal{R}_{\mathcal{A}}$ be a Kleene–Mostowski system. Then for each $n > 0$. $\Sigma_n(\mathcal{R}_{\mathcal{A}})$ is itself an indexed relational system, under a suitable choice of universal relations, and using the same notion of basic function as in $\mathcal{R}_{\mathcal{A}}$.*

PROOF. $\emptyset$, singletons, and $=_{\mathcal{A}}$ are generated in $\mathcal{R}_{\mathcal{A}}$, which is a Kleene–Mostowski system, hence all are in $\Sigma_1(\mathcal{R}_{\mathcal{A}})$. Then by Proposition 6.2(3), all are in $\Sigma_n(\mathcal{R}_{\mathcal{A}})$ for $n > 0$. By Lemma 6.1, $\Sigma_n(\mathcal{R}_{\mathcal{A}})$ is closed under transpositions, projections, $\cap$, $\cup$ and $\times$.

Next there is the matter of universal relations.

Suppose $\Sigma_n(\mathcal{R}_{\mathcal{A}})$ has appropriate universal relations (by hypothesis, $\Sigma_1(\mathcal{R}_{\mathcal{A}})$ does). Say $U^{k+1} \in \Sigma_n(\mathcal{R}_{\mathcal{A}})$ is universal for the $k+1$-place relations of $\Sigma_n(\mathcal{R}_{\mathcal{A}})$. Then $\overline{U^{k+1}} \in \Pi_n(\mathcal{R}_{\mathcal{A}})$, so $P_2^{k+2}(\overline{U^{k+1}}) \in \Sigma_{n+1}(\mathcal{R}_{\mathcal{A}})$. We claim it is universal for the k-place relations of $\Sigma_{n+1}(\mathcal{R}_{\mathcal{A}})$.

Let $\mathcal{P} \in \Sigma_{n+1}(\mathcal{R}_{\mathcal{A}})$ be k-place. By Proposition 6.3, there is a $k+1$-place relation $\mathcal{P}' \in \Pi_n(\mathcal{R}_{\mathcal{A}})$ such that $\mathcal{P} = P_1^{k+1}(\mathcal{P}')$. Further, $\overline{\mathcal{P}'} \in \Sigma_n(\mathcal{R}_{\mathcal{A}})$. Since U^{k+1} is universal in $\Sigma_n(\mathcal{R}_{\mathcal{A}})$, for some i,

$$\boldsymbol{x} \in \overline{\mathcal{P}'} \Leftrightarrow \langle i, \boldsymbol{x} \rangle \in U^{k+1}$$

or, in a handier form,

$$\langle u, \boldsymbol{v} \rangle \in \overline{\mathcal{P}'} \Leftrightarrow \langle i, u, \boldsymbol{v} \rangle \in U^{k+1}.$$

Then,

$$\langle u, \boldsymbol{v} \rangle \in \mathcal{P}' \Leftrightarrow \langle i, u, \boldsymbol{v} \rangle \in \overline{U^{k+1}},$$

so

$$\boldsymbol{v} \in P_1^{k+1}(\mathcal{P}') \Leftrightarrow \langle i, \boldsymbol{v} \rangle \in P_2^{k+2}(\overline{U^{k+1}})$$

or

$$\boldsymbol{v} \in \mathcal{P} \Leftrightarrow \langle i, \boldsymbol{v} \rangle \in P_2^{k+2}(\overline{U^{k+1}}).$$

Thus we have universal relations in $\Sigma_{n+1}(\mathcal{R}_{\mathcal{A}})$.

That the iteration property holds in $\Sigma_n(\mathcal{R}_{\mathcal{A}})$, $n > 0$, is proved similarly. We leave this to the reader.

COROLLARY 6.5. *Let $\mathcal{R}_{\mathcal{A}}$ be a Kleene–Mostowski system. Then the Σ_n, Π_n classification produces a hierarchy, in that, for $n > 0$,*

(1) $\Sigma_n(\mathcal{R}_{\mathcal{A}}) \cup \Pi_n(\mathcal{R}_{\mathcal{A}}) \subseteq \Sigma_{n+1}(\mathcal{R}_{\mathcal{A}}) \cap \Pi_{n+1}(\mathcal{R}_{\mathcal{A}})$,
(2) $\Sigma_n(\mathcal{R}_{\mathcal{A}}) \not\subseteq \Pi_n(\mathcal{R}_{\mathcal{A}})$,
(3) $\Pi_n(\mathcal{R}_{\mathcal{A}}) \not\subseteq \Sigma_n(\mathcal{R}_{\mathcal{A}})$,
(4) $\Sigma_n(\mathcal{R}_{\mathcal{A}}) \neq \Sigma_{n+1}(\mathcal{R}_{\mathcal{A}}) \cap \Pi_{n+1}(\mathcal{R}_{\mathcal{A}})$,
(5) $\Pi_n(\mathcal{R}_{\mathcal{A}}) \neq \Sigma_{n+1}(\mathcal{R}_{\mathcal{A}}) \cap \Pi_{n+1}(\mathcal{R}_{\mathcal{A}})$.

PROOF. (1) Is by Proposition 6.2.

(2) Since $\Sigma_n(\mathcal{R}_{\mathcal{A}})$ is an indexed relational system, by Theorem 2.1 there is a relation $\mathcal{P} \in \Sigma_n(\mathcal{R}_{\mathcal{A}})$ with $\bar{\mathcal{P}} \notin \Sigma_n(\mathcal{R}_{\mathcal{A}})$, i.e. $\mathcal{P} \notin \Pi_n(\mathcal{R}_{\mathcal{A}})$.

(3) Let $\mathcal{P}$ be the relation of (2). $\mathcal{P} \in \Sigma_n(\mathcal{R}_{\mathcal{A}})$, so $\bar{\mathcal{P}} \in \Pi_n(\mathcal{R}_{\mathcal{A}})$, but $\bar{\mathcal{P}} \notin \Sigma_n(\mathcal{R}_{\mathcal{A}})$.

(4) If $\Sigma_n(\mathcal{R}_{\mathcal{A}}) = \Sigma_{n+1}(\mathcal{R}_{\mathcal{A}}) \cap \Pi_{n+1}(\mathcal{R}_{\mathcal{A}})$, then $\Sigma_n(\mathcal{R}_{\mathcal{A}})$ would be closed under complementation, but item (3) says it is not.

(5) Similar to (4).

7. The Rice–Shapiro Theorem

In Section 4 we proved Rice's Theorem, which characterized the closed, *bi-generated* sets of indexes in $\mathcal{R}_{\mathcal{A}}$. In this section, under some additional assumptions about $\mathcal{R}_{\mathcal{A}}$, we characterize the closed *generated* sets of indexes. The result, in ordinary recursion theory, is a conjecture in Rice [1953], which was established by a number of people, Shapiro among others. (See Rice [1956].)

In ordinary recursion theory the Rice–Shapiro characterization is in terms of the *finite* sets. It turns out that, in order to prove the Rice–Shapiro Theorem only certain special properties of the finite sets are needed. What we do here is postulate that there is a collection $\mathcal{D}$ of sets having these properties with respect to $\mathcal{R}_{\mathcal{A}}$, and we state our Rice–Shapiro characterization in terms of $\mathcal{D}$. Whether the members of $\mathcal{D}$ are *actually* finite or not is of no importance. In fact, after establishing our results, we present an important class of examples in which some of the members of $\mathcal{D}$ are infinite. This is the first instance in this book where we consider a generalization of finiteness. We will return to this important concept in the next two chapters.

For the rest of this section we make the following additional assumptions about the indexed relational system $\mathcal{R}_{\mathcal{A}}$:

(1) There is a collection $\mathcal{D}$ of sets and relations on $\mathcal{A}$, called ***$\mathcal{D}$-finite***. We write $\mathcal{D}^n$ for the collection of n-ary members of $\mathcal{D}$.

(2) The $\mathscr{D}$-finite relations are closed under sections. That is, if $D \in \mathscr{D}^{n+1}$, and if $c \in \mathscr{A}$ is fixed, then $\{\boldsymbol{x} \mid \langle \boldsymbol{x}, c \rangle \in D\}$ is $\mathscr{D}$-finite.

(3) To each member of $\mathscr{D}^n$, one or more members of $\mathscr{A}$ have been assigned (called $\mathscr{D}^n$-finite codes) so that different members of $\mathscr{D}^n$ do not share a code. We write D^n_x for the member of $\mathscr{D}^n$ with code x.

(4) The following relations are generated:

$$y \text{ is a } \mathscr{D}^n\text{-finite code and } \boldsymbol{x} \in D^n_y,$$

$$y \text{ is a } \mathscr{D}^n\text{-finite code and } \boldsymbol{x} \notin D^n_y.$$

(5) ($\mathscr{D}$-finite approximation condition). If $\mathscr{P} \subseteq \mathscr{A}^n$ is generated, then there is a chain $\mathscr{C}$ of $\mathscr{D}$-finite n-ary relations such that

(a) $\{x \mid D^n_x \in \mathscr{C}\}$ is generated.

(b) $\bigcup \mathscr{C} = \mathscr{P}$.

(c) If $\mathscr{E}$ is any proper initial segment of $\mathscr{C}$, $\bigcup \mathscr{E}$ is $\mathscr{D}$-finite.

PROPOSITION 7.1. *Under the assumptions of this section, every finite relation on $\mathscr{A}$ is also $\mathscr{D}$-finite.*

PROOF. Since singletons are generated, and the generated relations are closed under $\cup$ and $\times$, each finite relation $\mathscr{P}$ on $\mathscr{A}$ is generated. Then there is a chain $\mathscr{C}$ of $\mathscr{D}$-finite relations with $\bigcup \mathscr{C} = \mathscr{P}$. But since $\mathscr{P}$ is finite, and $\mathscr{C}$ is a chain, it follows that $\mathscr{P} \in \mathscr{C}$, hence $\mathscr{P}$ is $\mathscr{D}$-finite.

Now we start out on the Rice–Shapiro Theorem. Recall, a set C of indexes for n-ary generated relations is called *closed* if $i \in C$ and $R^n_i = R^n_j$ imply $j \in C$.

LEMMA 7.2. *Under the general assumptions of this section. Suppose C is a closed set of indexes for n-ary generated relations, and also that C itself is generated. If $a_0 \in C$ then there is some $b \in C$ such that*

$$R^n_b \text{ is } \mathscr{D}\text{-finite} \quad \text{and} \quad R^n_b \subseteq R^n_{a_0}.$$

PROOF. Let $a_0 \in C$. We produce an appropriate b. Choose two distinct members of $\mathscr{A}$, let us designate them 0 and 1. Now, $R^n_{a_0}$ and C are generated, hence so is the $n+1$-place relation

$$T = (R^n_{a_0} \times \{0\}) \cup (C \times \underbrace{\{0\} \times \cdots \times \{0\}}_{n-1} \times \{1\}).$$

Then there is a chain $\mathscr{C}$ of $\mathscr{D}$-finite $n+1$-ary relations such that $\{x \mid D^{n+1}_x \in$

$\mathscr{C}\}$ is generated, $\bigcup \mathscr{C} = T$, and the union of any proper initial segment of $\mathscr{C}$ is $\mathscr{D}$-finite. Let

$$S(x_1,\ldots,x_n,y)$$
$$\Leftrightarrow (\exists c)[c \in \{x \mid D_x^{n+1} \in \mathscr{C}\} \wedge \langle x_1,\ldots,x_n,0\rangle \in D_c^{n+1} \wedge \langle y,0,\ldots,0,1\rangle \notin D_c^{n+1}].$$

If follows, from our assumptions, that S is generated. Then by Corollary 3.3, there is an index b such that

$$R_b^n(x_1,\ldots,x_n) \Leftrightarrow S(x_1,\ldots,x_n,b).$$

We claim this is the b we were looking for.

First we show $R_b^n \subseteq R_{a_0}^n$. Well, suppose $\boldsymbol{x} \in R_b^n$. Then $S(\boldsymbol{x},b)$, and so for some c, $\langle \boldsymbol{x},0\rangle \in D_c^{n+1}$ where $c \in \{x \mid D_x^{n+1} \in \mathscr{C}\}$. Then $\langle \boldsymbol{x},0\rangle \in \bigcup \mathscr{C} = T$, and so $\boldsymbol{x} \in R_{a_0}^n$.

Next we show $b \in C$. Well, suppose $b \notin C$. Then $\langle b,0,\ldots,0,1\rangle \notin T$, and so $\langle b,0,\ldots,0,1\rangle \notin D_c^{n+1}$ for every $D_c^{n+1} \in \mathscr{C}$. It follows that

$$\begin{aligned} S(x_1,\ldots,x_n,b) &\Leftrightarrow (\exists c)[c \in \{x \mid D_x^{n+1} \in \mathscr{C}\} \wedge \langle x_1,\ldots,x_n,0\rangle \in D_c^{n+1}] \\ &\Leftrightarrow \langle x_1,\ldots,x_n,0\rangle \in \bigcup \mathscr{C} = T \\ &\Leftrightarrow \langle x_1,\ldots,x_n\rangle \in R_{a_0}^n. \end{aligned}$$

Thus $\boldsymbol{x} \in R_b^n \Leftrightarrow \boldsymbol{x} \in R_{a_0}^n$. But $a_0 \in C$ and C is complete, hence $b \in C$ after all.

Finally we show R_b^n is $\mathscr{D}$-finite. Since $b \in C$, then $\langle b,0,\ldots,0,1\rangle \in T$, and so $\mathscr{E} = \{D \in \mathscr{C} \mid \langle b,0,\ldots,0,1\rangle \notin D\}$ is a proper initial segment of $\mathscr{C}$. Thus $\bigcup \mathscr{E}$ is $\mathscr{D}$-finite. But also,

$$S(x_1,\ldots,x_n,b) \Leftrightarrow (\exists c)[c \in \{x \mid D_x^{n+1} \in \mathscr{E}\} \wedge \langle x_1,\ldots,x_n,0\rangle \in D_c^{n+1}]$$

since the other clause of S, $\langle b,0,\ldots,0,1\rangle \notin D_c^{n+1}$ is equivalent to $D_c^{n+1} \in \mathscr{E}$. Thus

$$R_b^n(x_1,\ldots,x_n) \Leftrightarrow S(x_1,\ldots,x_n,b) \Leftrightarrow \langle x_1,\ldots,x_n,0\rangle \in \bigcup \mathscr{E}.$$

Since $\bigcup \mathscr{E}$ is $\mathscr{D}$-finite, and the $\mathscr{D}$-finite relations are closed under sections, R_b^n is $\mathscr{D}$-finite.

LEMMA 7.3. *For each* n,

(1) *There is a basic function* f *such that* $D_a^n = R_{f(a)}^n$.
(2) *The set of* $\mathscr{D}^n$*-finite codes is generated.*

PROOF. Let $\mathscr{Q}(a,x_1,\ldots,x_n) \Leftrightarrow a$ is a $\mathscr{D}^n$-finite code and $\langle x_1,\ldots,x_n\rangle \in D_a^n$. $\mathscr{Q}$ is generated. By the iteration property, there is a basic f such that

$$\mathscr{Q}(a, x_1, \ldots, x_n) \Leftrightarrow R^n_{f(a)}(x_1, \ldots, x_n)$$

and this is (1). The set in (2) is a projection of $\mathscr{Q}$, and hence is generated.

THEOREM 7.4. *Under the general assumptions of this section. Let C be a set of indexes for n-ary generated relations. If C is closed, generated, and non-empty, there is a generated set B whose members are $\mathscr{D}^n$-finite codes, such that*

$$c \in C \Leftrightarrow (\exists a \in B)[D^n_a \subseteq R^n_c].$$

PROOF. By Lemma 7.3 there is a basic function f such that $D^n_a = R^n_{f(a)}$. Define a set B by

$$x \in B \Leftrightarrow x \text{ is a } \mathscr{D}^n\text{-finite code and } f(x) \in C.$$

Then B is generated, and we claim it is the desired set.

Suppose $c \in C$. By Lemma 7.2 there is some $b \in C$ such that R^n_b is $\mathscr{D}$-finite and $R^n_b \subseteq R^n_c$. Since R^n_b is $\mathscr{D}$-finite, there is some $\mathscr{D}^n$-finite code a such that $R^n_b = D^n_a$. But $D^n_a = R^n_{f(a)}$ so $R^n_b = R^n_{f(a)}$. C is closed and $b \in C$, so $f(a) \in C$, and hence $a \in B$.

Conversely, suppose, for some $a \in B$, $D^n_a \subseteq R^n_c$. Well, $D^n_a = R^n_{f(a)}$, so $R^n_{f(a)} \subseteq R^n_c$, and $f(a) \in C$ since $a \in B$. Then $c \in C$ by Theorem 4.2.

APPLICATION. Let C be the set of all indexes for *finite* (in the usual sense) subsets of $\mathscr{A}$. Under the assumptions of this section, C is not generated. For, suppose it were; then there would exist a generated set B such that $c \in C \Leftrightarrow (\exists a \in B)[D^n_a \subseteq R^n_c]$. Now, C is not empty, so B is not empty. But, for any $a \in B$, $D^n_a \subseteq \mathscr{A}$, and $\mathscr{A}$ is generated, hence indexes for $\mathscr{A}$ must be in C, and since $\mathscr{R}_{\mathscr{A}}$ is indexed it follows easily that $\mathscr{A}$ must be infinite.

A little later in this section we present another, rather curious application.

EXAMPLES. We give some indexed relational systems which meet the assumptions of this section. More will be presented in Chapter 9.

I. The r.e. relations of ordinary recursion theory provide an indexed relational system such that, taking $\mathscr{D}$-finite to mean *finite*, all the assumptions of this section are met. This, of course, is the original setting for the Rice–Shapiro Theorem.

We sketch why this is an example. As remarked in Ch. 1, §5, in ordinary recursion theory our non-deterministic elementary formal systems can be replaced by deterministic versions, and these deterministic versions can be "discussed" within ordinary recursion theory, much as in Chapter 6. Then, given a deterministic means of generating a set $\mathscr{P}$, the function f, defined as follows, is itself a recursive function:

$$f(n) = \text{a finite code for } \{x \in \mathscr{P} \mid x \text{ is generated in } < n \text{ steps by our deterministic procedure for generating } \mathscr{P}\}.$$

Now, f has the following properties.

(1) The range of f is a set of finite codes, and is r.e.

(2) If $n < k$ then $D_{f(n)} \subseteq D_{f(k)}$, so the range of f is a chain.

(3) $\bigcup$(range of f) $= \mathscr{P}$.

Thus we have the finite approximation condition for r.e. sets.

In a similar fashion, relations can be handled. A finite coding for finite relations can be introduced by using an effective pairing function to "collapse" them to finite sets. We omit details.

II. Let $\mathfrak{A}$ be a structure such that the following four conditions are met:

(1) ω-rec($\mathfrak{A}$) is a theory with equality

(2) Every "given" relation of $\mathfrak{A}$ is ω-recursive in ω-rec($\mathfrak{A}$).

(3) ω-rec($\mathfrak{A}$) has an effective pairing function. [Then by Proposition 6.8.3, ω-rec($\mathfrak{A}$) contains an ω-r.e. copy of the natural numbers; but we need the following stronger assumption.]

(4) In ω-rec($\mathfrak{A}$) there is an ω-recursive copy of $\langle \mathbb{N}, \leqslant \rangle$.

If these four conditions are met, then the ω-r.e. relations of ω-rec($\mathfrak{A}$) provide an indexed relational system such that, taking $\mathscr{D}$-finite to mean *ω-recursive*, all the assumptions of this section are met.

In all examples of this sort, some sets which are actually infinite are $\mathscr{D}$-finite. In particular, the domain, $\mathscr{A}$, is $\mathscr{D}$-finite. Specific examples of this sort are ω-rec($\mathfrak{S}(\mathbb{N})$), that is, hyperarithemetic theory, and ω-rec($\mathfrak{S}(\mathbb{R})$) where $\mathfrak{S}(\mathbb{R})$ is the structure of the reals: $\langle \mathbb{R}; +, \times, > \rangle$.

Before going through the rather technical verification of our claims in Example II, we pause to present an application.

In ordinary recursion theory, every infinite recursively enumerable set contains an infinite, recursive subset. The analog of this for hyperarithmetic theory is: every infinite Π_1^1 set contains an infinite hyperarithmetic subset. Now, this is true, but the hyperarithmetic sets are the analogs not only of the recursive sets, but also of the finite ones. If we consider them that way, the result becomes a consequence of the Rice–Shapiro Theorem, as we now show.

LEMMA 7.5. *In hyperarithmetic theory, let* $\mathscr{C} = \{a \mid R_a^1 \text{ is infinite}\}$. *Then* $\mathscr{C}$ *is* ω*-r.e. (that is,* Π_1^1*).*

PROOF. $\mathscr{C}$ has ω-elementary formal system axioms:

$$x \in R^1_a \to x > y \to Gy, a,$$

$$G\forall, a \to a \in \mathscr{C}.$$

THEOREM 7.6. *In hyperarithmetic theory, every infinite Π^1_1 set contains an infinite hyperarithmetic subset.*

REMARK. Hyperarithmetic theory $= \omega$-rec($\mathfrak{S}(\mathbb{N})$); $\Pi^1_1 = \omega$-r.e.; hyperarithmetic $= \omega$-recursive.

PROOF. By the remarks above, the finiteness assumptions of the present section apply to hyperarithmetic theory, taking $\mathscr{D}$-finite to mean hyperarithmetic. Let $\mathscr{C} = \{a \mid R^1_a \text{ is infinite}\}$. By the lemma, $\mathscr{C}$ is generated, and it is trivially closed and non-empty. Then by Theorem 7.4, there is a generated set B whose members are $\mathscr{D}^1$-finite codes, such that

$$c \in \mathscr{C} \Leftrightarrow (\exists a \in B)[D^1_a \subseteq R^1_c].$$

Now, let R^1_c be some infinite Π^1_1 set. Then $c \in \mathscr{C}$, so for some $a \in B$, $D^1_a \subseteq R^1_c$. D^1_a is hyperarithmetic. Further, it is easy to see that if D^1_a were finite, then $\mathscr{C}$ would contain some indexes for finite sets (in particular, indexes for D^1_a). Hence D^1_a is infinite and we are done.

We now return to our discussion of the general class of examples under II above. What follows is a sketch of the verification of our claims, that these are, in fact, examples. We rely heavily on results in Moschovakis [1974], and what we will say will probably not be intelligible without an understanding of that book. However, nothing in later chapters here depends on this verification, it is enough to take our word for it that the examples described above really are examples.

We should also mention that it is likely that the four conditions listed above can be weakened somewhat. See the remarks in the last section of Barwise [1975A], which is a review of the Moschovakis book, and in Chapter 6 of Barwise [1975]. We have not followed up on this, however.

We rely, throughout the rest of this section, on Propositions 1.13.6 and 1.13.9, and so we use the terms ω-r.e. and inductive interchangeably; similarly for ω-recursive and hyperelementary. We also remark that, since we are supposing ω-rec($\mathfrak{A}$) is a theory with equality, total functions are ω-recursive iff they are ω-r.e. (Lemma 2.6.7). In particular, this applies to a pairing function. Thus, if ω-rec($\mathfrak{A}$) has an effective (ω-r.e.) pairing function, it has an ω-recursive pairing function.

The definition of an *acceptable structure* is given in Moschovakis [1974], p. 22. If ω-rec($\mathfrak{A}$) meets the four conditions above, it follows as in exercise 1.7 on p. 22 of Moschovakis [1974], that there is an acceptable structure $\mathfrak{A}'$ having the same ω-r.e. relations. But then, using Theorem 3.5.1, the identity map is an embedding of ω-rec($\mathfrak{A}$) to ω-rec($\mathfrak{A}'$), and thus ω-rec($\mathfrak{A}$) = ω-rec($\mathfrak{A}'$) (considered as production systems). So from now on, we simply assume $\mathfrak{A}$ itself is acceptable.

Since ω-rec($\mathfrak{A}$) has an effective pairing function, the ω-r.e. relations provide us with an indexed relational system (Ch. 6, §15). We need to verify that the ω-recursive relations satisfy our $\mathscr{D}$-finiteness assumptions. For this, we simply cite results from Moschovakis [1974].

From now on, by a $\mathscr{D}$-finite relation we mean an ω-recursive relation of ω-rec($\mathfrak{A}$). That these are closed under sections is straightforward.

The notation and results assumed in this paragraph are from Theorem 5D.4, Moschovakis [1974], p. 75. By a $\mathscr{D}^n$-finite code we mean a member of the set I^n, and if $a \in I^n$, by D^n_a we mean H^n_a. Then every n-ary ω-recursive ($=\mathscr{D}$-finite) relation has a $\mathscr{D}^n$-finite code. The relation: a is a $\mathscr{D}^n$-finite code and $\boldsymbol{x} \in D^n_a$, is ω-r.e. since it is: $a \in I^n \wedge \langle a, \boldsymbol{x}\rangle \in H^n$, and both I^n and H^n are inductive. Similarly for the relation: a is a $\mathscr{D}^n$-finite code and $\boldsymbol{x} \notin D^n_a$, since it is: $a \in I^n \wedge \langle a, \boldsymbol{x}\rangle \in \neg \check{H}^n$, and $\check{H}^n$ is coinductive.

Finally we must check that the $\mathscr{D}$-finite approximation condition holds. The terminology and results in this paragraph are from Moschovakis [1974], pp. 38–40. Let $\mathscr{P}$ be an n-ary ω-r.e. relation. By the Prewellordering Theorem (Theorem 3A.3, p. 40), $\mathscr{P}$ admits an inductive norm, σ. Let $\mathscr{C}$ be the collection of resolvents of $\mathscr{P}$ relative to σ. Each member of $\mathscr{C}$ is hyperelementary, that is, $\mathscr{D}$-finite. $\mathscr{C}$ is a chain; $\bigcup \mathscr{C} = \mathscr{P}$; and if $\mathscr{E}$ is a proper initial segement of $\mathscr{C}$, $\bigcup \mathscr{E}$ is $\mathscr{D}$-finite. It remains to be verified that $\{x \mid D^n_x \in \mathscr{C}\}$ is ω-r.e., that is, $\{x \mid D^n_x$ is a resolvent of $\mathscr{P}$ relative to $\sigma\}$. For this we give the following ω-elementary formal system, in which it is represented by O.

axioms for $\mathscr{P}$,

axioms for J_σ and for $\neg \check{J}_\sigma$.

[Let $A\boldsymbol{x}, \boldsymbol{y}, a$ mean a is a $\mathscr{D}^n$-finite code and $\boldsymbol{x} \in \overline{D^n_a} \cup$ resolvent $\boldsymbol{y}$, where resolvent $\boldsymbol{y} = \{\boldsymbol{x} \in \mathscr{P} \mid \sigma(\boldsymbol{x}) \leq \sigma(\boldsymbol{y})\}$.]

$\mathscr{P}\boldsymbol{y} \to J_\sigma(\boldsymbol{x}, \boldsymbol{y}) \to a$ is a $\mathscr{D}^n$-finite code $\to A\boldsymbol{x}, \boldsymbol{y}, a$,

$\mathscr{P}\boldsymbol{y} \to a$ is a $\mathscr{D}^n$-finite code and $\boldsymbol{x} \notin D^n_a \to A\boldsymbol{x}, \boldsymbol{y}, a$,

$A\boldsymbol{\forall}, \boldsymbol{y}, a \to D^n_a \subseteq$ resolvent $\boldsymbol{y}$.

[Let Bx, y, a mean a is a $\mathscr{D}^n$-finite code and $x \in D^n_a \cup \overline{\text{resolvent } y}$.]

$\mathscr{P}y \rightarrow a$ is a $\mathscr{D}^n$-finite code and $x \in D^n_a \rightarrow Bx, y, a$,

$\mathscr{P}y \rightarrow a$ is a finite code $\rightarrow \neg \check{J}(x, y) \rightarrow Bx, y, a$,

$B\forall, y, a \rightarrow$ resolvent $y \subseteq D^n_a$,

$D^n_a \subseteq$ resolvent $y \rightarrow$ resolvent $y \subseteq D^n_a \rightarrow$ resolvent $y = D^n_a$,

$\mathscr{P}y \rightarrow$ resolvent $y = D^n_a \rightarrow Oa$.

Thus all the $\mathscr{D}$-finiteness assumptions are verified.

CHAPTER EIGHT

INDEXED PRODUCTION SYSTEMS

1. Introduction

In Chapter 6 we showed many recursion and ω-recursion theories have operator indexings that have the input and the output place-fixing properties. This is the second of two chapters in which we investigate the consequences. In Chapter 7 we concentrated on relations, now we look at operators themselves. There is a summary presentation of much of this material, from a computer point of view, in Fitting [1979].

Also in Chapter 6 we showed that in many recursion theories there was a "usable" coding of the finite sets. Here we consider the implications of that, one being that for such recursion theories, an analog of the customary definition of enumeration operator holds. (Rogers [1967], p. 147.) It turns out that the compactness of enumeration operators plays a key role in this. We then proceed to find something that can play the role, for ω-enumeration operators that the finite sets do for enumeration operators. We find that, no matter what choice is made, non-compactness is inescapable. It is in trying to deal with this non-compactness phenomena in our axiomatic setting that *partial* production systems are forced on us.

In the following chapter we study some particular production systems of great interest, and which are closely related to those which develop in the present chapter.

2. Indexed production systems

We define the primary objects of study in this chapter, give examples, and establish connections with the previous chapter.

DEFINITION. Let $\boldsymbol{A}$ be a production system. Let $\mathcal{U}^{\langle n,k\rangle}$ be some operator in $\boldsymbol{A}$ of order $\langle n, k+1\rangle$. We say $\mathcal{U}^{\langle n,k\rangle}$ is *universal* for the operators in $\boldsymbol{A}$ of order $\langle n, k\rangle$ if, for each operator $\Phi \in \boldsymbol{A}$ of order $\langle n, k\rangle$, there is at least one $a \in \mathcal{A}$ (called an *index* for Φ) such that, for all $\mathcal{P} \in [\mathcal{A}]^n$,

$$x \in \Phi(\mathcal{P}) \Leftrightarrow \langle a, x\rangle \in \mathcal{U}^{\langle n,k\rangle}(\mathcal{P}).$$

REMARK. The following is an equivalent formulation which some may find "neater". Recall, S^n_a is the a-section operator from Ch. 2, §3. Now, let $\mathcal{U}^{\langle n,k\rangle}$ be an operator of order $\langle n, k+1\rangle$ in $\mathbf{A}$. Then, for each $a \in \mathcal{A}$, $S^{k+1}_a\mathcal{U}^{\langle n,k\rangle}$ is an operator in $\mathbf{A}$ of order $\langle n, k\rangle$. We call $\mathcal{U}^{\langle n,k\rangle}$ *universal* if every operator of order $\langle n, k\rangle$ is of this form.

DEFINITION. By an *operator indexing* for $\mathbf{A}$ we mean a family of operators in $\mathbf{A}$, $\mathcal{U} = \{\mathcal{U}^{\langle n,k\rangle} \mid n, k \geqslant 1\}$ where, for each n, k, $\mathcal{U}^{\langle n,k\rangle}$ is universal for the operators in $\mathbf{A}$ of order $\langle n, k\rangle$. We say $\mathbf{A}$ is an *indexed production system* if there is some operator indexing, $\mathcal{U}$, for $\mathbf{A}$. We often write $\Phi^{\langle n,k\rangle}_a$ for the operator in $\mathbf{A}$ of order $\langle n, k\rangle$ and index a. That is, $\Phi^{\langle n,k\rangle}_a = S^{k+1}_a\mathcal{U}^{\langle n,k\rangle}$.

EXAMPLES. By Corollary 6.14.5 every infinite recursion or ω-recursion theory with equality and an effective pairing function is an indexed production system. Also see Corollary 6.14.4. Additional examples will be given in the next chapter.

We begin discussing connections with the notion of indexed relational systems, as defined in Ch. 7, §1. This discussion concludes in the next section.

Let $\mathbf{A}$ be a production system, and let $\mathcal{P} \in [\mathcal{A}]^n$. Consider the collection of relations $\leqslant \mathcal{P}$. It is easy to verify that this collection satisfies part 1 of the definition of indexed relational systems. We note that if $\mathcal{P} = \emptyset$, then by Proposition 2.3.1, the relations involved are exactly the generated relations of $\mathbf{A}$.

Now suppose further that $\mathbf{A}$ has an operator indexing $\mathcal{U}$. For each k, define a $k+1$-place relation U^k, by

$$U^k = \mathcal{U}^{\langle n,k\rangle}(\mathcal{P}).$$

It is quite easy to see that these relations satisfy (2) of the definition of indexed relational systems. In fact, the resulting relational indexes are connected with the operator indexes by $R^k_a = \Phi^{\langle n,k\rangle}_a(\mathcal{P})$.

3. The output place-fixing property

In Chapter 6 we showed that many recursion and ω-recursion theories had indexings which had the output and the input place-fixing properties. In this section we investigate the consequences of the output place-fixing property, then we go on to the input place-fixing property in the next section.

DEFINITION. Let $\boldsymbol{A}$ be an indexed production system with operator indexing $\mathcal{U}$. Let $\mathcal{B}$ be a collection of (total) functions $f:\mathcal{A}^n \to \mathcal{A}$ for $n = 1, 2, 3, \ldots$, which we call *basic*, meeting the conditions

(1) $\mathcal{B}$ is closed under composition,

(2) (the graph of) each member of $\mathcal{B}$ is generated in $\mathbf{A}$.

We say the triple $\langle \mathbf{A}, \mathcal{U}, \mathcal{B}\rangle$ has the *output place-fixing property* if, for each operator $\Phi \in \mathbf{A}$, of order $\langle n, q+m\rangle$ say, there is a basic function $f:\mathcal{A}^q \to \mathcal{A}$ such that, for all $\mathcal{P} \in [\mathcal{A}]^n$,

$$\langle a_1, \ldots, a_q, x_1, \ldots, x_m\rangle \in \Phi(\mathcal{P}) \iff \langle x_1, \ldots, x_m\rangle \in \Phi^{\langle n,m\rangle}_{f(a_1,\ldots,a_q)}.$$

Generally, we will say $\mathbf{A}$ has the output place-fixing property, rather than $\langle \mathbf{A}, \mathcal{U}, \mathcal{B}\rangle$, leaving $\mathcal{U}$ and $\mathcal{B}$ to be understood by context.

EXAMPLES. By Corollary 6.14.5, every infinite recursion or ω-recursion theory with equality and an effective pairing function has an indexing such that the output place-fixing property holds, taking basic to mean recursive, or ω-recursive respectively. But in these examples, the notion of basic function is trivial in that it is as broad as possible. It is interesting generally, to take a smaller collection for the basic functions. In fact, Corollary 6.14.5 shows that if rec($\mathfrak{A}$) is a theory with equality and an effective pairing function, then ω-rec($\mathfrak{A}$) is an indexed production system having the output place-fixing property, even if the basic functions are restricted to the recursive functions, rather than including all ω-recursive ones. Further, in both ordinary recursion theory and in hyperarithmetic theory one can narrow things down to the *primitive recursive* functions as basic. See Kleene [1952] for a definition. Also, the class of *rudimentary* functions, from Smullyan [1961] will do. This example can be generalized, using the word structures we introduced in Chapter 6, but we do not do so here. At any rate, there are many indexed production systems in which the output place-fixing property holds, and in which the basic functions are a rather narrow, interesting subclass of the collection of generated functions. More examples will occur in the next chapter.

REMARKS. Let $\langle \mathbf{A}, \mathcal{U}, \mathcal{B}\rangle$ be an indexed production system having the output place-fixing property. And let $\mathcal{P} \in [\mathcal{A}]^n$. Then the collection of relations $\preccurlyeq \mathcal{P}$ is an indexed relational system, taking basic in the relational sense, to mean member of $\mathcal{B}$. We checked part of this in the last section, and now merely note that the iteration property is an immediate consequence of the output place-fixing property.

Now we turn to consequences. We have assumed that the operators of **A** are closed under $\cap$, $\cup$ and $\times$. We begin by showing that under the circumstances of this section, the closure is *effective*. We postpone a treatment of composition till the next section.

THEOREM 3.1. *Let* $\mathbf{A} = \langle \mathbf{A}, \mathcal{U}, \mathcal{B}\rangle$ *be an indexed production system having the output place-fixing property. For each* $\langle n, k\rangle$ *and* $\langle n, j\rangle$ *there are basic functions f, g and h such that*

(1) $\Phi_a^{\langle n,k\rangle} \cap \Phi_b^{\langle n,k\rangle} = \Phi_{f(a,b)}^{\langle n,k\rangle}$,

(2) $\Phi_a^{\langle n,k\rangle} \cup \Phi_b^{\langle n,k\rangle} = \Phi_{g(a,b)}^{\langle n,k\rangle}$,

(3) $\Phi_a^{\langle n,k\rangle} \times \Phi_b^{\langle n,j\rangle} = \Phi_{h(a,b)}^{\langle n,k+j\rangle}$.

PROOF. (1) Let S be the operator of order $\langle n, k+2\rangle$ given by

$$S = T_{1,2}^{k+2} A^{k+1}\mathcal{U}^{\langle n,k\rangle} \cap A^{k+1}\mathcal{U}^{\langle n,k\rangle}.$$

(Here A^{k+1} and $T_{1,2}^{k+2}$ are place-adding and transposition operators as defined in Chapter 1. $\mathcal{U}^{\langle n,k\rangle}$ is, of course, universal.) Tracing through the definitions, it is not hard to see that, for any $\mathcal{P} \in [\mathcal{A}]^n$,

$$\langle a, b, \boldsymbol{x}\rangle \in S(\mathcal{P}) \Leftrightarrow \boldsymbol{x} \in (\Phi_a^{\langle n,k\rangle} \cap \Phi_b^{\langle n,k\rangle})(\mathcal{P}).$$

Now, by the output place-fixing property, there is a basic function f such that

$$\langle a, b, \boldsymbol{x}\rangle \in S(\mathcal{P}) \Leftrightarrow \boldsymbol{x} \in \Phi_{f(a,b)}^{\langle n,k\rangle}(\mathcal{P})$$

and thus

$$\Phi_{f(a,b)}^{\langle n,k\rangle} = \Phi_a^{\langle n,k\rangle} \cap \Phi_b^{\langle n,k\rangle}.$$

(2) Now let S be the operator of order $\langle n, k+2\rangle$ defined by

$$S = T_{1,2}^{k+2} A^{k+1}\mathcal{U}^{\langle n,k\rangle} \cup A^{k+1}\mathcal{U}^{\langle n,k\rangle}.$$

Then, for $\mathcal{P} \in [\mathcal{A}]^n$,

$$\langle a, b, \boldsymbol{x}\rangle \in S(\mathcal{P}) \Leftrightarrow \boldsymbol{x} \in (\Phi_a^{\langle n,k\rangle} \cup \Phi_b^{\langle n,k\rangle})(\mathcal{P})$$

and the result follows as in (1).

(3) Let L be the logical operator of order $\langle k+j+2, k+j+2\rangle$ such that

$$L : \langle q, r_1, \ldots, r_k, s, t_1, \ldots, t_j\rangle \rightarrow \langle q, s, r_1, \ldots, r_k, t_1, \ldots, t_j\rangle.$$

Let S be the operator of order $\langle n, k+j+2\rangle$ defined by

$$S = L(\mathcal{U}^{\langle n,k\rangle} \times \mathcal{U}^{\langle n,j\rangle}).$$

It follows that, for $\mathcal{P} \in [\mathcal{A}]^n$, and any k-tuple $\boldsymbol{x}$ and j-tuple $\boldsymbol{y}$,

$$\begin{aligned}\langle a, b, \boldsymbol{x}, \boldsymbol{y}\rangle \in S(\mathcal{P}) &\Leftrightarrow \boldsymbol{x} \in \Phi_a^{\langle n,k\rangle}(\mathcal{P}) \text{ and } \boldsymbol{y} \in \Phi_b^{\langle n,j\rangle}(\mathcal{P})\\ &\Leftrightarrow \langle \boldsymbol{x}, \boldsymbol{y}\rangle \in (\Phi_a^{\langle n,k\rangle} \times \Phi_b^{\langle n,j\rangle})(\mathcal{P}).\end{aligned}$$

Now finish as before.

Next we establish an analog, for operators, of the Kleene fixed point theorem, Theorem 7.3.4. Its proof is essentially the second proof of Theorem 7.3.4, transferred to operators.

THEOREM 3.2. *Let* **A** *be an indexed production system having the output place-fixing property. Let* $t : \mathcal{A} \to \mathcal{A}$ *be a (total) function whose graph is generated in* **A**. *Then for each order,* $\langle n, k\rangle$, *there is an* i *such that* $\Phi_i^{\langle n,k\rangle} = \Phi_{t(i)}^{\langle n,k\rangle}$.

PROOF. Using the output place-fixing property, there is a basic function d such that

$$\langle y, \boldsymbol{x}\rangle \in P_1^{k+2} E_{1,2}^{k+2} \mathcal{U}^{\langle n,k+1\rangle}(\mathcal{P}) \Leftrightarrow \boldsymbol{x} \in \Phi_{d(y)}^{\langle n,k\rangle}(\mathcal{P}).$$

Tracing things out, this means

$$\langle y, \boldsymbol{x}\rangle \in \Phi_y^{\langle n,k+1\rangle}(\mathcal{P}) \Leftrightarrow \boldsymbol{x} \in \Phi_{d(y)}^{\langle n,k\rangle}(\mathcal{P})$$

or, more briefly, using the section operator, for all y,

$$S_y^{k+1} \Phi_y^{\langle n,k+1\rangle} = \Phi_{d(y)}^{\langle n,k\rangle}.$$

Next, since d is a generated function, as is t, then the composition, $td : \mathcal{A} \to \mathcal{A}$ is generated. Let H be a constant operator, of order $\langle n, 2\rangle$ and value (the graph of) td. Then there is an index e such that

$$\Phi_e^{\langle n,k+1\rangle} = P_2^{k+2} P_2^{k+3} E_{2,3}^{k+3}(H \times \mathcal{U}^{\langle n,k\rangle}).$$

Again, tracing things out, this says that for all y,

$$S_y^{k+1} \Phi_e^{\langle n,k+1\rangle} = \Phi_{td(y)}^{\langle n,k\rangle}.$$

Finally, let $i = d(e)$. Then

$$\Phi_i^{\langle n,k\rangle} = \Phi_{d(e)}^{\langle n,k\rangle} = S_e^{k+1} \Phi_e^{\langle n,k+1\rangle} = \Phi_{td(e)}^{\langle n,k\rangle} = \Phi_{t(i)}^{\langle n,k\rangle}.$$

COROLLARY 3.3 (Rice's Theorem for operators). *Let* **A** *be an indexed production system having the output place-fixing property. Let* C *be a closed* *set of indexes for operators of* **A** *of order* $\langle n, k\rangle$. *If both* C *and* $\bar{C}$ *are generated in* **A** *then either* C *is empty or* $C = \mathscr{A}$.

PROOF. Similar to the proof of Theorem 7.4.1, but using the above instead of Theorem 7.3.4.

4. The input place-fixing property

We continue the work of the previous section, adding the input place-fixing property as a new assumption.

DEFINITION. Let $\boldsymbol{A}$ be an indexed production system with operator indexing $\mathscr{U}$, and let $\mathscr{B}$ be a collection of *basic* functions (each is total, generated in **A**, and $\mathscr{B}$ is closed under composition). We say the triple $\langle \mathbf{A}, \mathscr{U}, \mathscr{B}\rangle$ has the *input place-fixing property* if, for each operator $\Phi \in \mathbf{A}$, of order $\langle q+n, m\rangle$ say, there is a basic function $f : \mathscr{A}^q \to \mathscr{A}$ such that, for all $\mathscr{P} \in [\mathscr{A}]^n$,

$$\boldsymbol{y} \in \Phi(\{\langle a_1, \ldots, a_q\rangle\} \times \mathscr{P}) \Leftrightarrow \boldsymbol{y} \in \Phi^{\langle n,m\rangle}_{f(a_1,\ldots,a_q)}(\mathscr{P}).$$

EXAMPLES. Every example given in the previous section for the output place-fixing property is also an example in which the present assumption holds.

Now we discuss consequences. We begin with some more results on "effective" combinations of operators.

LEMMA 4.1. *Let* $\mathbf{A} = \langle \mathbf{A}, \mathscr{U}, \mathscr{B}\rangle$ *be an indexed production system having the output place-fixing property. Let* Ψ *be an operator in* **A** *of order* $\langle n, k\rangle$. *For each* q *there is a basic function* $h : \mathscr{A} \to \mathscr{A}$ *such that* $\Phi^{\langle k,q\rangle}_x \Psi = \Phi^{\langle n,q\rangle}_{h(x)}$.

PROOF. For $\mathscr{P} \in [\mathscr{A}]^n$,

$$\boldsymbol{z} \in \Phi^{\langle k,q\rangle}_x \Psi \Leftrightarrow \langle x, \boldsymbol{z}\rangle \in \mathscr{U}^{\langle k,q\rangle}\Psi(\mathscr{P}).$$

Now use the output place-fixing property on $\mathscr{U}^{\langle k,q\rangle}\Psi$.

REMARK. This lemma, and the input place-fixing property, imply the output place-fixing property.

THEOREM 4.2. *Let* $\mathbf{A} = \langle \mathbf{A}, \mathscr{U}, \mathscr{B}\rangle$ *be an indexed production system having* *both the input and the output place-fixing properties. Then for each n there is a basic function $f : \mathscr{A}^n \to \mathscr{A}$ such that $f(a_1, \ldots, a_n)$ is an index for a constant operator of order $\langle 1, n\rangle$ and value $\{\langle a_1, \ldots, a_n\rangle\}$. [Thus "constants" are effective.]*

PROOF. Using the input place-fixing property on the operator P^{n+1}_{n+1}, there is a basic function $g : \mathscr{A}^n \to \mathscr{A}$ such that, for $\mathscr{P} \in [\mathscr{A}]^1$,

$$P^{n+1}_{n+1}(\{\langle a_1, \ldots, a_n\rangle\} \times \mathscr{P}) = \Phi^{\langle 1,n\rangle}_{g(a_1,\ldots,a_n)}(\mathscr{P}).$$

Further, if $\mathscr{P} \neq \emptyset$,

$$P^{n+1}_{n+1}(\{\langle a_1, \ldots, a_n\rangle\} \times \mathscr{P}) = \{\langle a_1, \ldots, a_n\rangle\}.$$

Now, let c be some member of $\mathscr{A}$. Let Ψ be a constant operator of order $\langle 1, 1\rangle$ and value $\{c\}$. Then, for all $\mathscr{P} \in [\mathscr{A}]^1$, $\Psi(\mathscr{P})$ is never empty. So by the above, for all $\mathscr{P} \in [\mathscr{A}]^1$,

$$\Phi^{\langle 1,n\rangle}_{g(a_1,\ldots,a_n)}\Psi(\mathscr{P}) = \{\langle a_1, \ldots, a_n\rangle\}.$$

Now by Lemma 4.1, there is a basic function h such that

$$\Phi^{\langle 1,n\rangle}_{g(a_1,\ldots,a_n)}\Psi = \Phi^{\langle 1,n\rangle}_{hg(a_1,\ldots,a_n)}.$$

Basic functions are closed under composition, so $f = hg$ is the desired funtion.

The universal operators we have been discussing sort things out by "tagging" the output. We now produce universal operators that "tag" input instead. At first thought this should mean we want an operator $\mathscr{V}^{\langle n,k\rangle}$ such that, for each a, and each $\mathscr{P} \in [\mathscr{A}]^n$,

$$\mathscr{V}^{\langle n,k\rangle}(\{a\} \times \mathscr{P}) = \Phi^{\langle n,k\rangle}_a(\mathscr{P}).$$

But further thought shows that this is not going to work when $\mathscr{P}$ is empty, since then $\{a\} \times \mathscr{P} = \{b\} \times \mathscr{P} = \emptyset$, and the "tag" is lost. So what we do instead is replace $\mathscr{P}$ by $(\{0\} \times \mathscr{P}) \cup \{\langle 1, 1, \ldots, 1\rangle\}$. This is never empty, and the original members of $\mathscr{P}$ can be recovered from it by using S^{n+1}_0.

PROPOSITION 4.3. *Let **A** be an indexed production system with operator indexing $\mathscr{U}$ (the place-fixing properties are not assumed). Choose two distinct members, $0, 1 \in \mathscr{A}$. For $\mathscr{P} \in [\mathscr{A}]^n$, by $\mathscr{P}'$ we mean*

$$(\{0\} \times \mathscr{P}) \cup \{\langle \underbrace{1, 1, \ldots, 1}_{n+1}\rangle\}$$

Then for each order $\langle n, k\rangle$ there is an operator $\mathcal{V}^{\langle n,k\rangle}$ of order $\langle n+2, k\rangle$ such that, for $\mathcal{P} \in [\mathcal{A}]^n$,

$$\mathcal{V}^{\langle n,k\rangle}(\{a\}\times\mathcal{P}') = \Phi_a^{\langle n,k\rangle}(\mathcal{P}).$$

PROOF. Let L be the logical operator of order $\langle n+2, 1\rangle$ such that $L:\langle x_1, x_2, \ldots, x_{n+2}\rangle \rightarrow \langle x_1\rangle$. Then set

$$\mathcal{V}^{\langle n,k\rangle} = P_1^{k+1}P_1^{k+2}E_{1,2}^{k+2}[L\times\mathcal{U}^{\langle n,k\rangle}S_0^{n+1}P_1^{n+2}].$$

We leave to the reader the task of checking that this has the right properties.

Finally, we are in a position to treat the "effectiveness" of composition of operators, the only method of combining them that we have not dealt with so far.

THEOREM 4.4. *Let* **A** *be an indexed production system having both the input and the output place-fixing properties. For each n, k and q, there is a basic function* $f:\mathcal{A}^2\rightarrow\mathcal{A}$ *such that* $\Phi_b^{\langle k,q\rangle}\Phi_a^{\langle n,k\rangle} = \Phi_{f(a,b)}^{\langle n,q\rangle}$.

PROOF. First, let Φ be defined by

$$\Phi = \mathcal{U}^{\langle k,q\rangle}\mathcal{V}^{\langle n,k\rangle}$$

where $\mathcal{V}^{\langle n,k\rangle}$ is as in the above proposition, and $\mathcal{U}^{\langle k,q\rangle}$ is the standard universal operator. Φ is of order $\langle n+2, q+1\rangle$. For $\mathcal{P}\in[\mathcal{A}]^n$, let $\mathcal{P}'$ be defined as above, namely

$$\mathcal{P}' = (\{0\}\times\mathcal{P})\cup\{\langle\underbrace{1, 1, \ldots, 1}_{n+1}\rangle\}.$$

Then, simple checking shows

$$\langle b, \boldsymbol{x}\rangle \in \Phi(\{a\}\times\mathcal{P}') \Leftrightarrow \boldsymbol{x}\in\Phi_b^{\langle k,q\rangle}\Phi_a^{\langle n,k\rangle}(\mathcal{P}). \tag{1}$$

By the input place-fixing property, there is a basic function $g:\mathcal{A}\rightarrow\mathcal{A}$ such that, for $\mathcal{Q}\in[\mathcal{A}]^{n+1}$,

$$\begin{aligned} \boldsymbol{z}\in\Phi(\{a\}\times\mathcal{Q}) &\Leftrightarrow \boldsymbol{z}\in\Phi_{g(a)}^{\langle n+1,q+1\rangle}(\mathcal{Q}) \\ &\Leftrightarrow \langle g(a), \boldsymbol{z}\rangle\in\mathcal{U}^{\langle n+1,q+1\rangle}(\mathcal{Q}). \end{aligned} \tag{2}$$

Also, by the output place-fixing property, there is a basic function $h:\mathcal{A}^2\rightarrow\mathcal{A}$ such that, for $\mathcal{Q}\in[\mathcal{A}]^{n+1}$,

$$\langle u, v, \boldsymbol{w}\rangle \in \mathscr{U}^{\langle n+1,q+1\rangle}(\mathscr{Q}) \Leftrightarrow \boldsymbol{w} \in \Phi^{\langle n+1,q\rangle}_{h(u,v)}(\mathscr{Q}). \tag{3}$$

Thus,

$$\begin{aligned} \boldsymbol{x} \in \Phi^{\langle k,q\rangle}_b \Phi^{\langle n,k\rangle}_a(\mathscr{P}) &\Leftrightarrow \langle b, \boldsymbol{x}\rangle \in \Phi(\{a\}\times \mathscr{P}') && \text{(by (1))} \\ &\Leftrightarrow \langle g(a), b, \boldsymbol{x}\rangle \in \mathscr{U}^{\langle n+1,q+1\rangle}(\mathscr{P}') && \text{(by (2))} \\ &\Leftrightarrow \boldsymbol{x} \in \Phi^{\langle n+1,q\rangle}_{h(g(a),b)}(\mathscr{P}'). && \text{(by (3))} \end{aligned}$$

In short,

$$\Phi^{\langle k,q\rangle}_b \Phi^{\langle n,k\rangle}_a(\mathscr{P}) = \Phi^{\langle n+1,q\rangle}_{h(g(a),b)}(\mathscr{P}').$$

Now, let A be the constant operator of order $\langle n, 1\rangle$ and value $\{0\}$; let B be the constant operator of order $\langle n, n+1\rangle$ and value

$$\{\langle \underbrace{1, 1, \ldots, 1}_{n+1} \rangle\};$$

I^n is the identity operator, as usual. Set

$$\Psi = (A \times I^n) \cup B.$$

Then, for $\mathscr{P} \in [\mathscr{A}]^n$, $\Psi(\mathscr{P}) = \mathscr{P}'$ and thus

$$\Phi^{\langle k,q\rangle}_b \Phi^{\langle n,k\rangle}_a = \Phi^{\langle n+1,q\rangle}_{h(g(a),b)} \Psi.$$

The result now follows, using the closure of basic functions under composition, and Lemma 4.1.

We remarked in Sections 2 and 3 that, if $\boldsymbol{A}$ is a production system having the output place-fixing property, then, for each $\mathscr{P} \in [\mathscr{A}]^n$, the collection of relations $\leqslant \mathscr{P}$ is an indexed relational system, using, for $k = 1, 2, \ldots$

$$U^k = \mathscr{U}^{\langle n,k\rangle}(\mathscr{P})$$

as universal relations. We will write $R^{k\mathscr{P}}_a$ for the k-ary relation $\leqslant \mathscr{P}$ having index a under this indexing. Briefly this means $R^{k\mathscr{P}}_a = \Phi^{\langle n,k\rangle}_a(\mathscr{P})$. If $\mathscr{P} = \emptyset$, the relations involved are just the generated relations of $\boldsymbol{A}$, and we will generally write R^k_a for $R^{k\emptyset}_a$. With this understood, we have the following.

COROLLARY 4.5. *Let $\boldsymbol{A}$ be an indexed production system having the input and the output place-fixing properties. Let $\mathscr{P} \in [\mathscr{A}]^n$. For each $\langle k, q\rangle$ there is a basic function f such that $\Phi^{\langle k,q\rangle}_a(R^{k\mathscr{P}}_b) = R^{q\mathscr{P}}_{f(b,a)}$.*

PROOF. This simply asserts $\Phi^{\langle k,q\rangle}_a \Phi^{\langle n,k\rangle}_b(\mathscr{P}) = \Phi^{\langle n,q\rangle}_{f(b,a)}(\mathscr{P})$ for some basic f, and thus is a special case of the theorem above.

COROLLARY 4.6. *Under the assumptions of Corollary 4.5. For each operator Φ of order $\langle k, q\rangle$ there is a basic function g such that $\Phi(R_b^{k\mathcal{P}}) = R_{g(b)}^{q\mathcal{P}}$.*

5. Pointwise generated functions, again

In Ch. 2, §5 we introduced two plausible notions of a function f being "computable" in a production system $\mathbf{A}$. We repeat the definitions for convenience.

DEFINITION. Let f be a partial function from $\mathcal{A}^n$ to $\mathcal{A}^k$.

(1) f is *generated in* $\mathbf{A}$ if the graph of f is generated in $\mathbf{A}$.

(2) f is *generated pointwise* in $\mathbf{A}$ if there is an operator $\Phi \in \mathbf{A}$ of order $\langle n, k\rangle$ such that

$$\Phi(\{\boldsymbol{v}\}) = \begin{cases} \{f(\boldsymbol{v})\} & \boldsymbol{v} \in \operatorname{dom} f, \\ \emptyset & \boldsymbol{v} \notin \operatorname{dom} f. \end{cases}$$

Corollary 2.5.5 says that every generated function is also pointwise generated. Now we show that, under the indexing assumptions of the present chapter, the converse also holds.

PROPOSITION 5.1. *Let* $\mathbf{A}$ *be an indexed production system having the input place-fixing property, and let Φ be an operator in* $\mathbf{A}$ *of order $\langle q+k, n\rangle$. For a fixed $\mathcal{P} \in [\mathcal{A}]^k$, define a $q+n$-place relation $\mathcal{R}$ by*

$$\mathcal{R}(x_1, \ldots, x_q, y_1, \ldots, y_n) \Leftrightarrow \langle y_1, \ldots, y_n\rangle \in \Phi(\{\langle x_1, \ldots, x_q\rangle\} \times \mathcal{P}).$$

Then $\mathcal{R} \leqslant \mathcal{P}$.

PROOF. By the input place-fixing property, there is a basic function f such that

$$\begin{aligned} &\langle y_1, \ldots, y_n\rangle \in \Phi(\{x_1, \ldots, x_q\} \times \mathcal{P}) \\ &\Leftrightarrow \langle y_1, \ldots, y_n\rangle \in \Phi_{f(x_1, \ldots, x_q)}^{\langle k,n\rangle}(\mathcal{P}) \\ &\Leftrightarrow \langle f(x_1, \ldots, x_q), y_1, \ldots, y_n\rangle \in \mathcal{U}^{\langle k,n\rangle}(\mathcal{P}). \end{aligned}$$

Since f is basic, it is generated. Let F be a constant operator of order $\langle k, q+1\rangle$ and value the graph of f. Then it is easy to see that the above is further equivalent to

$$\langle x_1, \ldots, x_q, y_1, \ldots, y_n \rangle \in P^{q+n+1}_{q+1} P^{q+n+2}_{q+1} E^{q+n+2}_{q+1,q+2}(F \times \mathcal{U}^{\langle k,n \rangle})(\mathcal{P})$$

and this is a relation $\leqslant \mathcal{P}$ by Proposition 2.4.7 (1).

COROLLARY 5.2. *Let* **A** *be an indexed production system having the input place-fixing property, and let* Φ *be an operator in* **A** *of order* $\langle q, n \rangle$. *Define a* $q+n$*-place relation* $\mathcal{R}$ *by*

$$\mathcal{R}(x_1, \ldots, x_q, y_1, \ldots, y_n) \Leftrightarrow \langle y_1, \ldots, y_n \rangle \in \Phi(\{x_1, \ldots, x_q\}).$$

Then $\mathcal{R}$ *is a generated relation in* **A**.

PROOF. Define an operator Φ' of order $\langle q+1, n \rangle$ by $\Phi' = \Phi P^{q+1}_{q+1}$. Let $c \in \mathcal{A}$ be fixed. Then certainly

$$\boldsymbol{y} \in \Phi(\{\boldsymbol{x}\}) \Leftrightarrow \boldsymbol{y} \in \Phi'(\{\boldsymbol{x}\} \times \{c\}).$$

Then by the above proposition, $\mathcal{R} \leqslant \{c\}$. But $\{c\}$ is generated (Axiom 3 of production systems) so $\{c\} \leqslant \emptyset$ (*Proposition* 2.3.1). *Thus* $\mathcal{R} \leqslant \emptyset$, so $\mathcal{R}$ is generated.

COROLLARY 5.3. *Again, suppose* **A** *is an indexed production system having the input place-fixing property.*

(1) $\mathcal{R}$ *is generated in* **A** *if and only if there is an operator* $\Phi \in$ **A** *such that*

$$\boldsymbol{x} \in \mathcal{R} \Rightarrow \Phi(\{\boldsymbol{x}\}) = \{0\}, \qquad \boldsymbol{x} \notin \mathcal{R} \Rightarrow \Phi(\{\boldsymbol{x}\}) = \emptyset.$$

(2) *For* $\mathcal{Q} \neq \emptyset$, *allowable in* **A**, $\mathcal{R} \leqslant \mathcal{Q}$ *if and only if there is an operator* $\Phi \in$ **A** *such that*

$$\boldsymbol{x} \in \mathcal{R} \Rightarrow \Phi(\mathcal{Q} \times \{\boldsymbol{x}\}) = \{0\},$$

$$\boldsymbol{x} \notin \mathcal{R} \Rightarrow \Phi(\mathcal{Q} \times \{\boldsymbol{x}\}) = \emptyset.$$

PROOF. We show (1); (2) is similar.

If $\mathcal{R}$ is generated, the existence of a suitable operator Φ is by Corollary 2.5.3.

Now suppose we have an operator Φ in **A** whose output is either $\{0\}$ or $\emptyset$. Define a relation $\mathcal{R}'$ by $\mathcal{R}'(\boldsymbol{x}, y) \Leftrightarrow y \in \Phi(\{\boldsymbol{x}\})$. [$y$ can be only 0, of course.] By the above corollary, $\mathcal{R}'$ is generated in **A** hence, using the section operator and transpositions, so is $\mathcal{R} = \{\boldsymbol{x} \mid \mathcal{R}'(\boldsymbol{x}, 0)\}$. It is easy to see $\mathcal{R}$ has the desired properties.

COROLLARY 5.4. ***A** is an indexed production system having the input place-fixing property. Let f be a partial function from $\mathscr{A}^n$ to $\mathscr{A}^k$. f is generated in **A** if and only if f is generated pointwise in **A**.*

PROOF. Half is by Corollary 2.5.5.

Conversely, suppose f is generated pointwise in $\boldsymbol{A}$. Let $\Phi \in \boldsymbol{A}$ be an operator of order $\langle n, k\rangle$ such that

$$\Phi(\{\boldsymbol{v}\}) = \begin{cases} \{f(\boldsymbol{v})\} & \boldsymbol{v} \in \operatorname{dom} f, \\ \emptyset & \boldsymbol{v} \notin \operatorname{dom} f. \end{cases}$$

By Corollary 5.2, the following relation is generated in $\boldsymbol{A}$.

$$\mathscr{R}(\boldsymbol{x}, \boldsymbol{y}) \Leftrightarrow \boldsymbol{y} \in \Phi(\{\boldsymbol{x}\}).$$

But $\boldsymbol{y} \in \Phi(\{\boldsymbol{x}\}) \Leftrightarrow \boldsymbol{y} = f(\boldsymbol{x})$ so $\mathscr{R}$ is the graph of f, hence f is generated in $\boldsymbol{A}$.

6. $\mathscr{D}$-finiteness

Finite sets are important for recursion theories since enumeration operators are compact. And we saw, in Ch. 7, §7, that for certain ω-recursion theories, there were sets that played a role somewhat analogous to that which the finite sets play for recursion theories. In this section we introduce a kind of generalized finiteness notion into the machinery of production systems. The consequences will occupy us for the rest of this chapter. We note that our present assumptions are not the same as in Ch. 7, §7.

DEFINITION. Let $\boldsymbol{A}$ be a production system, and let $\mathscr{D}$ be a collection of generated subsets of $\mathscr{A}$, which we will call *$\mathscr{D}$-finite.*

Suppose, to each member D of $\mathscr{D}$, one or more members of $\mathscr{A}$ have been assigned, called *$\mathscr{D}$-codes* for D, so that distinct members of $\mathscr{D}$ never have the same member of $\mathscr{A}$ assigned to them. Then we say we have a *$\mathscr{D}$-finite coding* in $\mathscr{A}$. We write D_c for the member of $\mathscr{D}$ with $\mathscr{D}$-code c. (If c is not a $\mathscr{D}$-code, D_c is not defined.)

Suppose now that a $\mathscr{D}$-finite coding exists. We say it is a *positive canonical coding* in $\boldsymbol{A}$ if

(1) There is an operator $F \in A$ of order $\langle 1,1\rangle$ such that, for $\mathcal{P} \in [\mathcal{A}]^1$,

$$F(\mathcal{P}) = \{c \mid D_c \subseteq \mathcal{P}\}.$$

(2) The following relation is generated in $\boldsymbol{A}$:

$$y \text{ is a } \mathcal{D}\text{-finite code and } x \in D_y.$$

We say we have a *canonical coding* if, in addition to above items, the following relation is generated in $\boldsymbol{A}$:

$$y \text{ is a } \mathcal{D}\text{-finite code and } x \notin D_y.$$

EXAMPLES. I. By Theorem 6.16.4, in any recursion theory with equality and an effective pairing function, there is a canonical coding for the $\mathcal{D}$-finite sets, where the $\mathcal{D}$-finite sets are simply the finite subsets of the domain. In particular, this is the case for ordinary recursion theory. In fact, Rogers [1967], p. 70, gives a finite coding under which each finite set has exactly one code, and every number is a code. In general, this is too much to hope for, however.

IIA. In Ch. 7, §7, we gave some conditions on $\mathfrak{A}$ that would ensure ω-rec($\mathfrak{A}$) met the special assumptions of that section, taking $\mathcal{D}$-finite to mean ω-recursive. In fact, the same four conditions also ensure that we have a canonical coding, taking $\mathcal{D}$-finite $= \omega$-recursive. Recall, the four conditions were:

(1) ω-rec($\mathfrak{A}$) is a theory with equality.

(2) Every "given" relation of $\mathfrak{A}$ is ω-recursive in ω-rec($\mathfrak{A}$).

(3) ω-rec($\mathfrak{A}$) has an effective pairing function.

(4) In ω-rec($\mathfrak{A}$) there is a ω-recursive copy of $\langle \mathbb{N}, \leq \rangle$.

In fact, in Ch. 7, §7, it was shown that if the four conditions hold then both

$$y \text{ is a } \mathcal{D}\text{-finite code and } x \in D_y,$$

$$y \text{ is a } \mathcal{D}\text{-finite code and } x \notin D_y,$$

are ω-r.e. All that we need now is the existence of an ω-enumeration operator F such that $F(\mathcal{P}) = \{x \mid D_x \subseteq \mathcal{P}\}$. But, in fact, $y \in F(\mathcal{P}) \Leftrightarrow D_y \subseteq \mathcal{P} \Leftrightarrow \bar{D}_y \cup \mathcal{P}$ is the entire domain. So we are interested in the relation $Nx, y \Leftrightarrow x \in \bar{D}_y \cup \mathcal{P}$. With this said, it should be clear that the desired operator F is $[E^I_O]$ where E consists of:

axioms for the relations

y is a $\mathscr{D}$-finite code and $x \in D_y$,

y is a $\mathscr{D}$-finite code and $x \notin D_y$,

and

y is a $\mathscr{D}$-finite code and $x \in D_y \rightarrow y$ is a $\mathscr{D}$-finite code,

y is a $\mathscr{D}$-finite code and $x \notin D_y \rightarrow y$ is a $\mathscr{D}$-finite code,

y is a $\mathscr{D}$-finite code $\rightarrow Ix \rightarrow Nx, y$,

y is a $\mathscr{D}$-finite code and $x \notin D_y \rightarrow Nx, y$,

$N\forall, y \rightarrow Oy$.

This family of examples is extremely important, and in one way or another, will occupy us for much of this chapter and the next. It includes hyperarithmetic theory, ω-rec($\mathfrak{S}(\mathbb{N})$), as well as ω-rec($\mathfrak{S}(\mathbb{R})$), where $\mathfrak{S}(\mathbb{R})$ is the structure of the reals; $\langle \mathbb{R}; +, \times, > \rangle$.

IIB. If we want a canonical coding, but do not care about the special assumptions of Ch. 7, §7, we can manage with just the first three of the four conditions above for ω-rec($\mathfrak{A}$). We argue for this using material from Barwise [1975], in particular, the notion of $\mathrm{HYP}_{\mathfrak{A}}$, an extremely fundamental topic which must, otherwise, be ignored in this book.

Since ω-rec($\mathfrak{A}$) is assumed to be a theory with equality, and in it, each "given" relation of $\mathfrak{A}$ is ω-recursive, then ω-r.e. $=$ inductive over $\mathfrak{A}$ and ω-recursive $=$ hyperelementary over $\mathfrak{A}$.

By Barwise [1975], Theorem 5.1, p. 230,

(1) ω-r.e. on $\mathfrak{A} = \Sigma$ on $\mathrm{HYP}_{\mathfrak{A}}$.

(2) ω-recursive on $\mathfrak{A} =$ member of $\mathrm{HYP}_{\mathfrak{A}}$.

(3) $\mathrm{HYP}_{\mathfrak{A}}$ is projectible into $\mathfrak{A}$. The definition of *projectible* is on p. 168, Definition 5.1, of Barwise [1975]. We use the notation from there.

Let Π be a notation system for $\mathrm{HYP}_{\mathfrak{A}}$, projecting into $\mathfrak{A}$. Recall, $\mathscr{D}$-finite is to mean ω-recursive. Now suppose $\mathscr{R}$ is $\mathscr{D}$-finite. Then by (1) we have $\mathscr{R} \in \mathrm{HYP}_{\mathfrak{A}}$. And then $\Pi(\mathscr{R}) \subseteq \mathscr{A}$. Let the members of $\Pi(\mathscr{R})$ be *$\mathscr{D}$-finite codes for $\mathscr{R}$.*

By Lemma 5.2, Barwise [1975], p. 169, D_Π is $\mathrm{HYP}_{\mathfrak{A}}$-r.e., that is, Σ on $\mathrm{HYP}_{\mathfrak{A}}$. Then by (2) D_Π is ω-r.e on $\mathfrak{A}$. Then

x is a $\mathscr{D}$-finite code

is ω-r.e. over $\mathfrak{A}$. Consider the relation

x is a $\mathscr{D}$-finite code and $y \in D_x$.

We claim this is ω-r.e. over $\mathfrak{A}$. It is equivalent to

$$x \in D_\Pi \wedge (\exists \mathscr{R})[y \in \mathscr{R} \wedge |x|_\Pi = \mathscr{R}].$$

This is Σ on $\mathrm{HYP}_\mathfrak{A}$, using Lemma 5.2, p. 169, of Barwise [1975], hence it is ω-r.e. over $\mathfrak{A}$ by (1). Similarly, the relation

x is a $\mathscr{D}$-finite code $\wedge\ y \notin D_x$

is equivalent to

$$x \in D_\Pi \wedge (\exists \mathscr{R})[y \notin \mathscr{R} \wedge |x|_\Pi = \mathscr{R}]$$

which is Σ on $\mathrm{HYP}_\mathfrak{A}$, hence is ω-r.e. over $\mathfrak{A}$.

Finally, the existence of an operator F, such that $F(\mathscr{P}) = \{x \mid D_x \subseteq \mathscr{P}\}$, is established as in example IIA.

7. Rogers' form for operators

In ordinary recursion theory, enumeration operators generally are defined quite differently than we did in Chapter 1. The definition from Rogers [1967], p. 147 (for the special case of operators of order $\langle 1, 1\rangle$) amounts to this:

> Φ is an enumeration operator iff there is an r.e. relation $\mathscr{R}(y, x)$ such that $x \in \Phi(\mathscr{P}) \Leftrightarrow$ for some $y, D_y \subseteq \mathscr{P}$ and $\mathscr{R}(y, x)$.

We will show, in this section, as a consequence of our axiomatic assumptions, that the operators of order $\langle 1, 1\rangle$ that satisfy Roger's condition are precisely the operators that are monotone and compact relative to our generalized notion of finiteness. In the next section we introduce some assumptions about effective pairing functions, and extend the Rogers' style characterization to operators of arbitrary order.

We note that, since in recursion theories all operators are compact, it follows from the results below that our definition of enumeration operators agrees with the customary one for ordinary recursion theory.

We begin with a few elementary results about $\mathscr{D}$-finiteness. A *positive* canonical coding is enough for the work of this section.

PROPOSITION 7.1. *Let* **A** *be a production system with a notion of* $\mathcal{D}$*-finiteness for which there is a positive canonical coding. Then*

(1) *There is an operator* $G \in \mathbf{A}$ *of order* $\langle 1,1\rangle$ *such that*

$$G(\mathcal{P}) = \bigcup\{D_y \mid y \text{ is a } \mathcal{D}\text{-finite code and } y \in \mathcal{P}\}.$$

In particular

$$G(\{y\}) = \begin{cases} D_y & \text{if } y \text{ is a } \mathcal{D}\text{-finite code,} \\ \emptyset & \text{otherwise.} \end{cases}$$

If, further, **A** is an indexed production system for which the input place-fixing property holds, then

(2) The relation $\mathcal{Q}(x,y) \Leftrightarrow D_x \subseteq D_y$, is generated in **A**.

(3) The relation $\mathcal{S}(x,y) \Leftrightarrow D_x = D_y$, is generated in **A**.

PROOF. (1) Let $\mathcal{R}(y,x) \Leftrightarrow y$ is a $\mathcal{D}$-finite code and $x \in D_y$. $\mathcal{R}$ is generated in **A**. Now use Theorem 2.5.1.

(2) Let F be the operator of order $\langle 1,1\rangle$ such that $F(\mathcal{P}) = \{x \mid D_x \subseteq \mathcal{P}\}$, and let G be the operator from (1). Then, $D_x \subseteq D_y \Leftrightarrow x \in (FG)(\{y\})$. Now use Corollary 5.2.

(3) Is immediate from (2).

DEFINITION. Let **A** be a production system with a notion of $\mathcal{D}$-finiteness. We call an operator $\Phi \in \mathbf{A}$ of order $\langle 1,1\rangle$ *$\mathcal{D}$-compact* if $x \in \Phi(\mathcal{P})$ implies $x \in \Phi(\mathcal{F})$ for some $\mathcal{D}$-finite set $\mathcal{F} \subseteq \mathcal{P}$.

Recall, Φ is *monotone* if $\mathcal{P} \subseteq \mathcal{Q}$ implies $\Phi(\mathcal{P}) \subseteq \Phi(\mathcal{Q})$.

Now let us return to the Rogers' definition, given above. The defining clause was

$$x \in \Phi(\mathcal{P}) \Leftrightarrow \text{for some } y, D_y \subseteq \mathcal{P} \text{ and } \mathcal{R}(y,x).$$

Now, using the operator F, having the property $F(\mathcal{P}) = \{x \mid D_x \subseteq \mathcal{P}\}$, this can be rewritten:

$$x \in \Phi(\mathcal{P}) \Leftrightarrow \text{for some } y \in F(\mathcal{P}),\ \mathcal{R}(y,x).$$

Further, using the notation of Ch. 2, §5, namely

$$\mathcal{R}''(\mathcal{P}) = \{v \mid \text{for some } w \in \mathcal{P},\ \mathcal{R}(w,v)\},$$

this may be rewritten

$$x \in \Phi(\mathcal{P}) \Leftrightarrow x \in \mathcal{R}''F(\mathcal{P})$$

or, simply $\Phi = \mathcal{R}''F$.

PROPOSITION 7.2. *Let* **A** *be a production system with a notion of* $\mathcal{D}$*-finiteness for which there is a positive canonical coding. Let* $\mathcal{R}$ *be a generated, two place relation on* $\mathcal{A}$*. Define* Φ *of order* $\langle 1,1 \rangle$ *by* $\Phi = \mathcal{R}''F$.

Then Φ *is an operator in* **A**, *which is monotone, and* $\mathcal{D}$*-compact.*

PROOF. $\mathcal{R}''$ is an operator in **A** by Corollary 2.5.2, hence so is Φ. That Φ is monotone and $\mathcal{D}$-compact is straightforward.

DEFINITION. We say an operator Φ of order $\langle 1,1 \rangle$ can be put in *Rogers' form* if $\Phi = \mathcal{R}''F$ for some generated binary relation $\mathcal{R}$.

REMARKS. In the next section we will extend this definition to handle operators of arbitrary orders.

The above proposition says that Rogers' form always gives monotone, $\mathcal{D}$-compact operators. The following says when it gives all of them.

THEOREM 7.3. *Let* **A** *be an indexed production system having the input place-fixing property, and with a notion of* $\mathcal{D}$*-finiteness for which there is a positive canonical coding. The operators of order* $\langle 1,1 \rangle$ *in* **A** *which can be put in Rogers' form are precisely the monotone,* $\mathcal{D}$*-compact operators.*

PROOF. In one direction the result is Proposition 7.2.

Now suppose $\Phi \in$ **A** is of order $\langle 1,1 \rangle$, and is monotone and $\mathcal{D}$-compact. Define a relation $\mathcal{R}$ by

$$\mathcal{R}(y,x) \Leftrightarrow x \in (\Phi G)(\{y\})$$

where G is from Proposition 7.1. By Corollary 5.2, $\mathcal{R}$ is generated in **A**. We claim $\Phi = \mathcal{R}''F$. Let $\mathcal{P} \in [\mathcal{A}]^1$.

(1) Suppose $x \in \mathcal{R}''F(\mathcal{P})$. Then, for some y,

$$y \in F(\mathcal{P}) \quad \text{and} \quad \mathcal{R}(y,x)$$

and hence

$$D_y \subseteq \mathcal{P} \quad \text{and} \quad \mathcal{R}(y, x),$$

$$D_y \subseteq \mathcal{P} \quad \text{and} \quad x \in \Phi(G(\{y\})),$$

$$D_y \subseteq \mathcal{P} \quad \text{and} \quad x \in \Phi(D_y),$$

but Φ is monotone, so $x \in \Phi(\mathcal{P})$. Thus $R''F(\mathcal{P}) \subseteq \Phi(\mathcal{P})$.

(2) Suppose $x \in \Phi(\mathcal{P})$. Since Φ is $\mathcal{D}$-compact, for some y,

$$D_y \subseteq \mathcal{P} \quad \text{and} \quad x \in \Phi(D_y),$$

$$D_y \subseteq \mathcal{P} \quad \text{and} \quad x \in (\Phi G)(\{y\}),$$

$$y \in F(\mathcal{P}) \quad \text{and} \quad \mathcal{R}(y, x),$$

so $x \in \mathcal{R}''F(\mathcal{P})$. Thus $\Phi(\mathcal{P}) \subseteq \mathcal{R}''F(\mathcal{P})$.

This concludes the proof.

8. Effective pairing

We have been discussing the consequences of $\mathcal{D}$-finiteness for *sets*. The simplest way to extend our work to relations is to introduce a pairing function and use it to collapse relations to sets. After all, we do have effective pairing functions in all the recursion and ω-recursion theories that are models for our assumptions up to now. Also, introducing a pairing function provides an easy way to extend Rogers' form to arbitrary operator orders, not just to those of order $\langle 1, 1 \rangle$.

DEFINITION. Let **A** be a production system. We say **A** has an *effective pairing function* J if J is a 1–1 function, $J : \mathcal{A} \times \mathcal{A} \to \mathcal{A}$ which is generated in **A**.

If J is an effective pairing function, we define

$$J^1(x_1) = x_1,$$

$$J^2(x_1, x_2) = J(x_1, x_2),$$

$$J^3(x_1, x_2, x_3) = J(J^2(x_1, x_2), x_3),$$

$$\vdots$$

$$J^{n+1}(x_1, \ldots, x_n, x_{n+1}) = J(J^n(x_1, \ldots, x_n), x_{n+1}),$$

$$\vdots$$

PROPOSITION 8.1. *If J is an effective pairing function in* **A**, *then $J^n(x_1, \ldots, x_n) = J^n(y_1, \ldots, y_n)$ implies $x_1 = y_1$ and $\cdots$ and $x_n = y_n$. Further, each J^n is a generated function in* **A**.

PROOF. J is 1–1; and generated functions are closed under composition.

DEFINITION. Let $\boldsymbol{A}$ be a production system with effective pairing function J. For each n we define operators $[J^n]$ and $[J^n]^{-1}$ of orders $\langle n, 1\rangle$ and $\langle 1, n\rangle$ respectively by

$$[J^n](\mathscr{P}) = \{J^n(x_1,\ldots,x_n) \mid \langle x_1,\ldots,x_n\rangle \in \mathscr{P}\} \quad \text{for } \mathscr{P} \in [\mathscr{A}]^n,$$

$$[J^n]^{-1}(\mathscr{P}) = \{\langle x_1,\ldots,x_n\rangle \mid J^n(x_1,\ldots,x_n) \in \mathscr{P}\} \quad \text{for } \mathscr{P} \in [\mathscr{A}]^1.$$

PROPOSITION 8.2. *Let $\boldsymbol{A}$ be a production system with effective pairing function J.*

(1) *$[J^n]$ and $[J^n]^{-1}$ are operators in $\boldsymbol{A}$.*
(2) *$[J^n]$ and $[J^n]^{-1}$ are monotone.*
(3) $[J^n](\mathscr{P} \cup \mathscr{Q}) = [J^n](\mathscr{P}) \cup [J^n](\mathscr{Q})$,
$[J^n]^{-1}(\mathscr{P} \cup \mathscr{Q}) = [J^n]^{-1}(\mathscr{P}) \cup [J^n]^{-1}(\mathscr{Q})$.
(4) $[J^n]^{-1}[J^n] = I^n$.

PROOF. $[J^n]$ and $[J^n]^{-1}$ are operators in $\boldsymbol{A}$ using Corollary 2.5.2. The other assertions are straightforward.

The introduction of a pairing function gives us a simple means of extending the notion of $\mathscr{D}$-finiteness to relations. Let us consider some examples.

Suppose rec($\mathfrak{A}$) is a recursion theory with a pairing function J. For $\mathscr{P} \subseteq \mathscr{A}^n$, $\mathscr{P}$ and $[J^n](\mathscr{P})$ are of the same cardinality, hence if either is finite, both are; and for recursion theories we have been taking $\mathscr{D}$-finite to mean finite.

Suppose ω-rec($\mathfrak{A}$) is an ω-recursion theory with equality and an effective pairing function J. We claim that for $\mathscr{P} \in [\mathscr{A}]^n$, $\mathscr{P}$ is ω-recursive iff $[J^n](\mathscr{P})$ is ω-recursive. This is significant since, in such situations we have been taking the $\mathscr{D}$-finite sets to be the ω-recursive ones.

We now turn to verifying our claim. We first observe that *range J^n is ω-recursive*. This follows from the equivalences

$$x \in \text{range } J^n \iff (\exists y_1,\ldots,y_n)[J^n(y_1,\ldots,y_n) = x],$$

$$x \notin \text{range } J^n \iff (\forall y_1,\ldots,y_n)(\exists z)[J^n(y_1,\ldots,y_n) = z \wedge z \neq x]$$

and both of these use only machinery available in ω-rec($\mathfrak{A}$). We omit details.

Next, suppose $\mathscr{P}$ is ω-recursive in ω-rec($\mathfrak{A}$). Then $\mathscr{P}$ is ω-r.e., hence so is $[J^n](\mathscr{P})$ since $[J^n]$ is an operator in ω-rec($\mathfrak{A}$), which must take generated relations to generated relations. Likewise, $\bar{\mathcal{P}}$ is ω-r.e., hence so is $[J^n](\bar{\mathcal{P}})$. But,

$$\overline{[J^n](\mathcal{P})} = [J^n](\bar{\mathcal{P}}) \cup \overline{\text{range } J^n}$$

and hence is ω-r.e. Thus $[J^n](\mathcal{P})$ is ω-recursive.

Conversely, suppose $[J^n](\mathcal{P})$ is ω-recursive. Then it is ω-r.e., hence so is $\mathcal{P} = [J^n]^{-1}[J^n](\mathcal{P})$, since $[J^n]^{-1}$ is an operator in ω-rec($\mathfrak{A}$). But also $\overline{[J^n](\mathcal{P})}$ is ω-r.e., hence so is

$$\begin{aligned}[J^n]^{-1}(\overline{[J^n](\mathcal{P})}) &= [J^n]^{-1}([J^n](\bar{\mathcal{P}}) \cup \overline{\text{range } J^n}) \\ &= [J^n]^{-1}[J^n](\bar{\mathcal{P}}) \cup [J^n]^{-1}(\overline{\text{range } J^n}) = \bar{\mathcal{P}}.\end{aligned}$$

Thus $\mathcal{P}$ is ω-recursive.

All this suggests the following as a reasonable addition to our machinery.

DEFINITION. Let $\boldsymbol{A}$ be a production system with an effective pairing function and a notion of $\mathcal{D}$-finiteness for which there is a $\mathcal{D}$-finite coding. Let $\mathcal{P}$ be an allowable n-ary relation. We call $\mathcal{P}$ a *$\mathcal{D}$-finite relation* if $[J^n](\mathcal{P})$ is a $\mathcal{D}$-finite set. By a *$\mathcal{D}$-finite code* for $\mathcal{P}$ we mean any $\mathcal{D}$-finite code for the set $[J^n](\mathcal{P})$. We write D_y^n for the $\mathcal{D}$-finite n-ary relation with $\mathcal{D}$-finite code y. Then $D_y^1 = D_y$.

This also allows an extension of $\mathcal{D}$-compactness. We say an operator $\Phi \in \boldsymbol{A}$ of order $\langle n, m\rangle$ is *$\mathcal{D}$-compact* if $\boldsymbol{x} \in \Phi(\mathcal{P})$ implies $\boldsymbol{x} \in \Phi(D_y^n)$ for some $D_y^n \subseteq \mathcal{P}$.

PROPOSITION 8.3. *Let* $\boldsymbol{A}$ *be a production system with an effective pairing function J and a notion of* $\mathcal{D}$*-finiteness for which there is a* $\mathcal{D}$*-finite coding.*

(1) $[J^n](D_y^n) = D_y$,
(2) $[J^n]^{-1}(D_y) = D_y^n$,
(3) *for* $\mathcal{P} \in [\mathcal{A}]^n$, $D_y \subseteq [J^n](\mathcal{P}) \Leftrightarrow D_y^n \subseteq \mathcal{P}$,
(4) $D_x^n \subseteq D_y^n \Leftrightarrow D_x \subseteq D_y$.

PROOF. (1) is by definition. (2) follows from (1) by Proposition 8.2 (2). (3) follows from (1) and (2) using monotonicity of $[J^n]$ and $[J^n]^{-1}$, as does (4).

Now it is easy to generalize the work of the previous section.

DEFINITION. Let Φ be an operator of order $\langle n, m\rangle$. We say Φ has been written in *Rogers' form* if there is a generated $m+1$-place relation $\mathcal{R}$ such that $\Phi = \mathcal{R}''F[J^n]$.

THEOREM 8.4. *Let **A** be an indexed production system having the input place-fixing property, an effective pairing function, and a notion of $\mathscr{D}$-finiteness, for which there is a positive canonical coding. The operators of **A** that can be written in Rogers' form are precisely the monotone, $\mathscr{D}$-compact operators.*

PROOF. If Φ can be written in Rogers' form, it is simple to show it is a $\mathscr{D}$-compact, monotone operator.

Conversely, suppose Φ is of order $\langle n, m\rangle$, is $\mathscr{D}$-compact and monotone. Let

$$\mathscr{R}(y, x_1, \ldots, x_m) \Leftrightarrow \langle x_1, \ldots, x_m\rangle \in (\Phi[J^n]^{-1}G)(\{y\}).$$

By Corollary 5.2, $\mathscr{R}$ is a generated relation. We claim $\Phi = \mathscr{R}''F[J^n]$.

Suppose $\boldsymbol{x} \in \mathscr{R}''F[J^n](\mathscr{P})$, for some $\mathscr{P} \in [\mathscr{A}]^n$. Then for some y, $y \in F[J^n](\mathscr{P})$ and $\mathscr{R}(y, \boldsymbol{x})$. Since $y \in F[J^n](\mathscr{P})$, $D_y \subseteq [J^n](\mathscr{P})$ so by Proposition 8.3, $D_y^n \subseteq \mathscr{P}$. Also since $\mathscr{R}(y, \boldsymbol{x})$,

$$\boldsymbol{x} \in \Phi[J^n]^{-1}G(\{y\}) \quad \text{or} \quad \boldsymbol{x} \in \Phi[J^n]^{-1}(D_y),$$

so by Proposition 8.3 again, $\boldsymbol{x} \in \Phi(D_y^n)$. Since Φ is monotone, $\boldsymbol{x} \in \Phi(\mathscr{P})$.

Conversely, suppose $\boldsymbol{x} \in \Phi(\mathscr{P})$. Since Φ is $\mathscr{D}$-compact, $\boldsymbol{x} \in \Phi(D_y^n)$ for some $D_y^n \subseteq \mathscr{P}$. Now, $D_y^n \subseteq \mathscr{P}$, so $D_y \subseteq [J^n](\mathscr{P})$ and hence $y \in F[J^n](\mathscr{P})$. Also, $\boldsymbol{x} \in \Phi(D_y^n)$ so $\boldsymbol{x} \in \Phi[J^n]^{-1}(D_y)$, so $\boldsymbol{x} \in \Phi[J^n]^{-1}G(\{y\})$, and thus $\mathscr{R}(y, \boldsymbol{x})$. It follows that $\boldsymbol{x} \in \mathscr{R}''F[J^n](\mathscr{P})$.

9. Non-compact operators

We have seen, in the two preceding sections, that monotone, $\mathscr{D}$-compact operators play a special role. As the chapter goes on, we will see that such operators play a deep and important role indeed. It is not being facetious to ask: are these notions trivial in the sense that, in the production systems that most concern us, all operators are monotone and $\mathscr{D}$-compact. In every recursion or ω-recursion theory, all operators are monotone. In every recursion theory, all operators are $\mathscr{D}$-compact, taking $\mathscr{D}$-finite to mean just plain finite. What about $\mathscr{D}$-compactness in ω-recursion theories? In this section we show that, in the most important of all ω-recursion theories, namely hyperarithmetic theory or ω-rec($\mathfrak{S}(\mathbb{N})$), there must be operators that are *not* $\mathscr{D}$-compact, under any reasonable meaning of $\mathscr{D}$-finite. Much of the rest of this chapter will be devoted to dealing with the existence of these non-compact operators.

All the following takes place in ω-rec($\mathfrak{S}(\mathbb{N})$).

Let $\Phi = [E^{I}_{O}]$ where E consists of the following axioms:

$$\text{axioms for } < \qquad \text{(see Ch. 1, §4)},$$

$$Iy \to x < y \to Gx,$$

$$G\forall \to O5.$$

It should be clear that, if $\mathscr{P}$ is an infinite set of numbers, $\Phi(\mathscr{P}) = \{5\}$, but if $\mathscr{P}$ is finite, $\Phi(\mathscr{P}) = \emptyset$. Thus Φ is not compact, in the ordinary sense. We show Φ is not $\mathscr{D}$-compact for *any* reasonable notion of $\mathscr{D}$-finiteness.

PROPOSITION 9.1. *Let $S_1, S_2, S_3, \ldots$ be a sequence of infinite sets of numbers. There is a set T which is infinite, but for no n do we have $S_n \subseteq T$.*

PROOF. Construct T as follows. Let n_1 be the smallest member of S_1, put $n_1 + 1$ in T. Next, let n_2 be the smallest member of S_2 bigger than $n_1 + 1$ (which exists since S_2 in infinite), put $n_2 + 1$ in T. And so on. Clearly T is infinite. But we can not have $S_i \subseteq T$ since $n_i \in S$ by construction, $n_i \notin T$.

Now, suppose a notion of $\mathscr{D}$-finiteness has been specified for ω-rec($\mathfrak{S}(\mathbb{N})$), for which there is a $\mathscr{D}$-finite coding. We show that the ω-enumeration operator Φ is not $\mathscr{D}$-compact.

Since there is a $\mathscr{D}$-finite coding, there can be only countably many $\mathscr{D}$-finite sets. The example above shows that if $\mathscr{D}$-finite means finite, Φ is a non-compact operator, so now suppose $\mathscr{D}$-finite includes some sets which are actually infinite. Let $S_1, S_2, S_3, \ldots$ be a listing of all the infinite, $\mathscr{D}$-finite sets (with repetitions if necessary to make the sequence infinite). By the proposition, there is an infinite set T which has none of the S_i as subset. Now, using the ω-enumeration operator Φ constructed above,

$$\Phi(T) = \{5\} \quad \text{since } T \text{ is infinite.}$$

But, if D is $\mathscr{D}$-finite, and $D \subseteq T$, D must be actually finite, since no S_i is a subset of T. But then $\Phi(D) = \emptyset$. It follows that Φ is not $\mathscr{D}$-compact.

10. All the $\mathscr{D}$-finiteness assumptions together

We have been investigating the consequences of certain $\mathscr{D}$-finiteness assumptions for indexed production systems **A**. But (assuming the output place-fixing property) the generated relations of **A** constitute an indexed relational system, and our $\mathscr{D}$-finiteness assumptions for relations in the previous chapter were quite different than the assumptions we have been considering here. Now we put all of this together, and find that, under the combined $\mathscr{D}$-finiteness assumptions, there is a nice connection between non-compact and compact operators. We will make use of this connection in the next section. We note that, by our remarks in Ch. 7, §7, and in Section 6 in this chapter, our present work applies to many ω-recursion theories, to hyperarithmetic theory in particular.

We begin with a few results about indexed relational systems. Much of the work in this section is based on Myhill and Shepherdson [1955].

DEFINITION. In an indexed relational system $\mathscr{R}_{\mathscr{A}}$, we call a function g *n-m-extensional* if

$$R_a^n = R_b^n \quad \text{implies} \quad R_{g(a)}^m = R_{g(b)}^m.$$

LEMMA 10.1. *Let* $\mathscr{R}_{\mathscr{A}}$ *be an indexed relational system. If* g *is an n-m-extensional generated function, then* g *is* monotone *in the sense that*

$$R_a^n \subseteq R_b^n \quad \textit{implies} \quad R_{g(a)}^m \subseteq R_{g(b)}^m.$$

PROOF. Suppose $R_a^n \subseteq R_b^n$, and $\boldsymbol{c} \in R_{g(a)}^m$; we show $\boldsymbol{c} \in R_{g(b)}^m$. Define a set C by $C = \{x \mid \boldsymbol{c} \in R_{g(x)}^m\}$.

(1) C is generated, since $x \in C \Leftrightarrow (\exists y)[g(x) = y \wedge U^m(y, \boldsymbol{c})]$.

(2) Thought of as a set of n-ary indexes, C is closed. For, suppose $x \in C$ and $R_x^n = R_y^n$; we show $y \in C$. Well, since $R_x^n = R_y^n$ and g is extensional, $R_{g(x)}^m = R_{g(y)}^m$. Since $x \in C$, $\boldsymbol{c} \in R_{g(x)}^m$, hence $\boldsymbol{c} \in R_{g(y)}^m$ and so $y \in C$.

Now, $a \in C$ since $\boldsymbol{c} \in R_{g(a)}^m$, and $R_a^n \subseteq R_b^n$, so by Theorem 7.4.2, $b \in C$ which means $\boldsymbol{c} \in R_{g(b)}^m$.

LEMMA 10.2. *Let* $\mathscr{R}_{\mathscr{A}}$ *be an indexed relational system meeting the* $\mathscr{D}$-*finiteness assumptions of Ch.* 7, §7. *If* g *is an n-m-extensional generated function then* g *is* compact *in the sense that, if* $\boldsymbol{c} \in R_{g(a)}^m$ *then there is some* $\mathscr{D}$-*finite relation* $R_b^n \subseteq R_a^n$ *with* $\boldsymbol{c} \in R_{g(b)}^m$.

PROOF. Say $\boldsymbol{c} \in R_{g(a)}^m$. Again let $C = \{x \mid \boldsymbol{c} \in R_{g(x)}^m\}$. As in Lemma 10.1, C is generated and closed (as a set of n-ary indexes). And $a \in C$. The result follows by Lemma 7.7.2.

In Ch. 7, §7, when we discussed $\mathscr{D}$-finiteness, we took n-ary $\mathscr{D}$-finite relations as primitive, not just $\mathscr{D}$-finite sets. It seemed simplest to avoid dealing with pairing functions at that time. But now we have introduced them, so relations can be collapsed to sets. The following deals with this issue.

LEMMA 10.3. *Let* **A** *be a production system with an effective pairing function J and a notion of $\mathscr{D}$-finiteness for which there is a $\mathscr{D}$-coding (not necessarily positive or canonical). If any of the following holds for $n = 1$, it also holds for arbitrary n.*

(1) *The relation*

$$y \text{ is a } \mathscr{D}\text{-finite code and } \mathbf{x} \in D_y^n$$

is generated.

(2) *The relation*

$$y \text{ is a } \mathscr{D}\text{-finite code and } \mathbf{x} \notin D_y^n$$

is generated.

(3) *If $\mathscr{P} \subseteq \mathscr{A}^n$ is generated, then there is a chain $\mathscr{C}$ of $\mathscr{D}$-finite n-ary relations such that*

(a) $\{x \mid D_x^n \in \mathscr{C}\}$ *is generated,*

(b) $\bigcup \mathscr{C} = \mathscr{P}$,

(c) *if $\mathscr{E}$ is any proper initial segment of $\mathscr{C}$, $\bigcup \mathscr{E}$ is $\mathscr{D}$-finite.*

PROOF. Left to the reader.

Now, finally, our main result.

THEOREM 10.4. *Let* **A** *be an indexed production system having both the input and the output place-fixing properties, and an effective pairing function. Suppose also that* **A** *has a notion of $\mathscr{D}$-finiteness meeting the following conditions.*

(1) *There is a canonical coding of the $\mathscr{D}$-finite sets.*

(2) *The collection of $\mathscr{D}$-finite relations is closed under sections.*

(3) *If $\mathscr{P}$ is a generated set then there is a chain $\mathscr{C}$ of $\mathscr{D}$-finite sets such that*

(a) $\{x \mid D_x \in \mathscr{C}\}$ *is generated,*

(b) $\bigcup \mathscr{C} = \mathscr{P}$,

(c) $\bigcup \mathscr{E}$ *is $\mathscr{D}$-finite, for any proper initial segment $\mathscr{E}$ of $\mathscr{C}$.*

If these hypotheses are met, then if g is an n-m-extensional generated function, there is a monotone, $\mathscr{D}$-compact operator Φ in **A** *of order $\langle n, m\rangle$ such that $\Phi(R_a^n) = R_{g(a)}^m$ for every generated relation R_a^n.*

PROOF. The generated sets and relations of **A** constitute an indexed relational system, and by Lemma 10.3, all the finiteness assumptions of Ch. 7, §7 hold for it; and consequently Lemmas 10.2 and 10.1 may be applied.

Let F be the operator such that $F(\mathcal{P}) = \{x \mid D_x \subseteq \mathcal{P}\}$. Let f be a generated function such that $D^n_x = R^n_{f(x)}$ (see Lemma 7.7.3). Define a relation P by

$$P(x, y_1, \ldots, y_m) \Leftrightarrow \langle y_1, \ldots, y_m \rangle \in R^m_{gf(x)}.$$

P is generated, so the following is an operator in $\mathbf{A}$:

$$\Phi = P''F[J]^n.$$

We leave it to the reader to verify that Φ is monotone and $\mathcal{D}$-compact. We show $\Phi(R^n_a) = R^n_{g(a)}$.

(1) Suppose $\mathbf{v} \in \Phi(R^n_a)$. That is, $\mathbf{v} \in P''F[J]^n(R^n_a)$. Then for some $w \in F[J^n](R^n_a)$, $P(w, \mathbf{v})$. Now $w \in F[J^n](R^n_a)$ implies $D_w \subseteq [J^n](R^n_a)$ implies $D^n_w \subseteq R^n_a$. Also $P(w, \mathbf{v})$ so $\mathbf{v} \in R^m_{gf(w)}$. Let $z = f(w)$. By definition of f, $D^n_w = R^n_z$, hence we have $\mathbf{v} \in R^m_{g(z)}$ and $R^n_z \subseteq R^n_a$. Then by Lemma 10.1, $\mathbf{v} \in R^m_{g(a)}$.

(2) Suppose $\mathbf{v} \in R^m_{g(a)}$. Then by Lemma 10.2, there is some $\mathcal{D}$-finite $R^n_b \subseteq R^n_a$ with $\mathbf{v} \in R^m_{g(b)}$. Since R^n_b is $\mathcal{D}$-finite it has a $\mathcal{D}$-finite code, say c. Thus $D^n_c = R^n_b$. Then $D^n_c \subseteq R^n_a$, so $D_c \subseteq [J^n](R^n_a)$ and hence $c \in F[J^n](R^n_a)$. Now by definition of f we have $D^n_c = R^n_{f(c)}$, hence $R^n_b = R^n_{f(c)}$. Since g is n-m-extensional, $R^m_{g(b)} = R^m_{gf(c)}$. Also $\mathbf{v} \in R^m_{g(b)}$ so $\mathbf{v} \in R^m_{gf(c)}$, so $P(c, \mathbf{v})$. Since also $c \in F[J^n](R^n_a)$ we have $\mathbf{v} \in P''F[J^n](R^n_a)$ or $\mathbf{v} \in \Phi(R^n_a)$.

This concludes the proof.

COROLLARY 10.5. *Under the assumptions of the above theorem, for every operator $\Phi \in \mathbf{A}$ of order $\langle n, m \rangle$, there is an operator $\Phi^* \in \mathbf{A}$ of the same order, which is monotone and $\mathcal{D}$-compact, and which agrees with Φ on the generated relations.*

PROOF. By Corollary 4.6 there is a basic function g such that $\Phi(R^n_a) = R^m_{g(a)}$. It is trivial that g is n-m-extensional, hence by the above there is a monotone, $\mathcal{D}$-compact Φ^* such that $\Phi^*(R^n_a) = R^m_{g(a)}$.

Φ^* and Φ agree on generated relations.

11. The least fixed point theorem

In Chapter 1, we showed that a least fixed point result held for every recursion and ω-recursion theory (Theorem 1.12.1). Now we ask, can such a result be derived in the axiomatic setting we are presently considering. We have not been able to do so in a version strong enough to have Theorem 1.12.1, as a consequence. However, in this section we present a least fixed point result for the montone, $\mathscr{D}$-compact operators of production systems meeting certain conditions; and in fact those conditions hold in every recursion and ω-recursion theory with an effective pairing function. Also, under some additional hypotheses, the $\mathscr{D}$-compactness restriction may be dropped; and these additional hypotheses hold in every ω-recursion theory meeting the four conditions given under Example IIA in Section 6.

In Section 15 we present a second axiomatic least fixed point result, with rather different hypotheses, and an elegant proof due to Scott and Park. But its range of applicability seems to be more restricted.

Yet another axiomatic least fixed point result may be found in Moschovakis [1971]. It is the first of its kind. The axiomatization followed there abstracts from the notion of *derivation length*, while we abstract from the notion of *finite*.

We begin by recalling the pertinent definitions from Ch. 1, for convenience.

Let Φ be an operator in $\mathbf{A}$ of order $\langle n, n\rangle$. $\mathscr{P}$ is a *fixed point* for Φ if $\Phi(\mathscr{P}) = \mathscr{P}$. If $\mathscr{P}$ is a fixed point, and $\mathscr{P} \subseteq \mathscr{R}$ for every fixed point $\mathscr{R}$, then $\mathscr{P}$ is the *least fixed point*.We are after a result that very loosely says, under the right circumstances, operators have *generated* least fixed points.

The plan is as follows. We first show that, under certain mild conditions on $\mathbf{A}$, every operator Φ of order $\langle n, n\rangle$ has at least one generated fixed point. Next we show, under more restrictive hypotheses, that a monotone, $\mathscr{D}$-compact operator has a generated least fixed point. We do this by associating with Φ a certain operator Ψ which, by construction, can't have more than one fixed point. By the result just mentioned, Ψ must have exactly one generated fixed point. And we will construct Ψ in such a way that its single fixed point provides us with the least fixed point of Φ.

THEOREM 11.1. *Let $\mathbf{A}$ be an indexed production system having both the input and the output place-fixing properties. Every operator $\Phi \in \mathbf{A}$ of order $\langle n, n\rangle$ has at least one generated fixed point.*

PROOF. Since $\mathbf{A}$ is an indexed production system having the output place-fixing property, the collection of generated relations constitutes an indexed relational system, so the work in Chapter 7, may be applied. Then, simply, by Corollary 4.6, there is a basic function g such that $\Phi(R^n_x) = R^n_{g(x)}$

and by Corollary 7.3.4 there is an i such that $R^n_{g(i)} = R^n_i$. Then R^n_i is a generated fixed point for Φ.

DEFINITION. Let $\mathbf{A}$ be a production system with a notion of $\mathscr{D}$-finiteness, and an effective pairing function J. And let R be an $n+m$-place relation on $\mathscr{A}$. We say R *satisfies weak replacement* if, for each $\mathscr{D}$-finite $F \subseteq \mathscr{A}^n$ such that

$$(\forall \mathbf{x} \in F)(\exists \mathbf{y}) R(\mathbf{x}, \mathbf{y})$$

there is a $\mathscr{D}$-finite $G \subseteq \mathscr{A}^m$ such that

$$(\forall \mathbf{x} \in F)(\exists \mathbf{y} \in G) R(\mathbf{x}, \mathbf{y})$$

and

$$(\forall \mathbf{y} \in G)(\exists \mathbf{x} \in F) R(\mathbf{x}, \mathbf{y}).$$

REMARK. This notion will play an important role in the next few sections.

THEOREM 11.2. *Let $\mathbf{A}$ be an indexed production system having the input and the output place-fixing properties, a notion of $\mathscr{D}$-finiteness (we do not require a coding of the $\mathscr{D}$-finite sets, however) and an effective pairing function. Suppose $\mathbf{A}$ meets the following additional conditions:*

(1) *There is a set $S \subseteq \mathscr{A}$ and a transitive, well-founded relation $<$ such that both $<$ and S are generated.*

(2) *For every $\mathscr{D}$-finite subset F of S there is a member of S bigger than anything in F (in the $<$ sense).*

(3) *Every generated relation satisfies weak replacement.*

Then, if Φ is any monotone, $\mathscr{D}$-compact operator in $\mathbf{A}$ of order $\langle n, n\rangle$, Φ has a generated least fixed point. In fact, for each $\alpha \in S$, let $\Phi_\alpha = \bigcup_{\beta<\alpha} \Phi(\Phi_\beta)$. (Recall, $<$ is well founded.) Then each Φ_α is generated, and the least fixed point of Φ is $\bigcup_{\alpha \in S} \Phi_\alpha$.

PROOF. Let Φ be an operator in $\mathbf{A}$ of order $\langle n, n\rangle$ which is monotone and $\mathscr{D}$-compact.

Now, $\Phi P^{n+1}_1 P^{n+2}_1 E^{n+2}_{1,2}$ is an operator in $\mathbf{A}$ of order $\langle n+2, n\rangle$ having the property that, for $\mathscr{P} \in [\mathscr{A}]^{n+1}$,

$$\Phi P^{n+1}_1 P^{n+2}_1 E^{n+2}_{1,2}(\{a\} \times \mathscr{P}) = \Phi(\{z \mid \langle a, z\rangle \in \mathscr{P}\}) = \Phi S^{n+1}_a(\mathscr{P}).$$

By the input place-fixing property, there is a basic function f such that

$$\mathbf{y} \in \Phi P^{n+1}_1 P^{n+2}_1 E^{n+2}_{1,2}(\{a\} \times \mathscr{P}) \iff \mathbf{y} \in \Phi^{\langle n+1, n\rangle}_{f(a)}(\mathscr{P}).$$

And thus, for each a, $\Phi S^{n+1}_a = \Phi^{\langle n+1, n\rangle}_{f(a)}$.

Since f is generated, there is a constant operator I_f of order $\langle n+1,2\rangle$ and value the graph of f. Also the well-founded relation $<$ is generated: let $I_<$ be the constant operator of order $\langle n+1,2\rangle$ and value $<$. Finally, S is generated, hence so is $S\times\mathcal{A}^n$. Let I_S be the constant operator of order $\langle n+1,n+1\rangle$ and value $S\times\mathcal{A}^n$.

Now, let Ψ be the operator in $\boldsymbol{A}$ of order $\langle n+1,n+1\rangle$ given by

$$\Psi = I_S \cap P_2^{n+2}P_2^{n+3}P_2^{n+4}P_1^{n+5}E_{4,5}^{n+5}E_{1,3}^{n+5}(I_< \times I_f \times \mathcal{U}^{\langle n+1,n\rangle})$$

[$\mathcal{U}^{\langle n+1,n\rangle}$ is the universal operator]. Checking this definition through, one sees that, for $\mathcal{P}\in[\mathcal{A}]^{n+1}$,

$$\Psi(\mathcal{P}) = \{\langle\alpha,\mathbf{y}\rangle \mid \alpha\in S \wedge (\exists\beta)(\exists w)[\beta<\alpha \wedge f(\beta)=w \wedge \langle w,\mathbf{y}\rangle\in\mathcal{U}^{\langle n+1,n\rangle}(\mathcal{P})]\}.$$

From this, two important facts follow.

(1) If $\langle\alpha,\mathbf{y}\rangle\in\Psi(\mathcal{P})$ then $\alpha\in S$.

Also, if $\alpha\in S$, we have

$$\begin{aligned}\langle\alpha,\mathbf{y}\rangle\in\Psi(\mathcal{P}) &\Leftrightarrow (\exists\beta<\alpha)[\langle f(\beta),\mathbf{y}\rangle\in\mathcal{U}^{\langle n+1,n\rangle}(\mathcal{P})]\\ &\Leftrightarrow (\exists\beta<\alpha)[\mathbf{y}\in\Phi_{f(\beta)}^{\langle n+1,n\rangle}(\mathcal{P})]\\ &\Leftrightarrow (\exists\beta<\alpha)[\mathbf{y}\in\Phi S_\beta^{n+1}(\mathcal{P})],\end{aligned}$$

and thus:

(2) For $\alpha\in S$, $S_\alpha^{n+1}\Psi(\mathcal{P}) = \bigcup_{\beta<\alpha}\Phi S_\beta^{n+1}(\mathcal{P})$.

From these follow several major items; we argue each separately.

(3) Ψ can't have two fixed points.

PROOF OF (3). Let $\mathcal{P},\mathcal{Q}\in[\mathcal{A}]^{n+1}$ be two fixed points for Ψ. If $\langle\alpha,\mathbf{y}\rangle\in\mathcal{P}=\Psi(\mathcal{P})$ then by (1), $\alpha\in S$. Similarly if $\langle\alpha,\mathbf{y}\rangle\in\mathcal{Q}$ then $\alpha\in S$. So, if $\mathcal{P}$ and $\mathcal{Q}$ are to differ, there must be some $\alpha\in S$ such that $S_\alpha^{n+1}(\mathcal{P})\neq S_\alpha^{n+1}(\mathcal{Q})$. Let α_0 be a minimal member of S such that

$$S_{\alpha_0}^{n+1}(\mathcal{P})\neq S_{\alpha_0}^{n+1}(\mathcal{Q}).$$

Then, if $\beta<\alpha_0$, $S_\beta^{n+1}(\mathcal{P})=S_\beta^{n+1}(\mathcal{Q})$. But,

$$\begin{aligned}S_{\alpha_0}^{n+1}(\mathcal{P}) &= S_{\alpha_0}^{n+1}\Psi(\mathcal{P}) && \text{(since } \mathcal{P} \text{ is a fixed point)}\\ &= \bigcup_{\beta<\alpha_0}\Phi S_\beta^{n+1}(\mathcal{P}) && \text{(by (2))}\\ &= \bigcup_{\beta<\alpha_0}\Phi S_\beta^{n+1}(\mathcal{Q})\\ &= S_{\alpha_0}^{n+1}\Psi(\mathcal{Q}) && \text{(by (2))}\\ &= S_{\alpha_0}^{n+1}(\mathcal{Q}) && \text{(since } \mathcal{Q} \text{ is a fixed point).}\end{aligned}$$

From this contradiction it follows that $\mathcal{P} = \mathcal{Q}$.

Next, by Theorem 11.1, Ψ must have at least one generated fixed point, hence Ψ has exactly one such. Let $\mathcal{R}$ be the unique generated fixed point of Ψ.

(4) The section operator is monotone on $\mathcal{R}$ in the sense that, if $\delta, \gamma \in S$, then $\delta < \gamma \Rightarrow S_\delta^{n+1}(\mathcal{R}) \subseteq S_\gamma^{n+1}(\mathcal{R})$.

PROOF OF (4). Let $\delta, \gamma \in S$ with $\delta < \gamma$. Then

$$\begin{aligned} S_\delta^{n+1}(\mathcal{R}) &= S_\delta^{n+1}\Psi(\mathcal{R}) \qquad (\mathcal{R} \text{ is a fixed point}) \\ &= \bigcup_{\beta<\delta} \Phi S_\beta^{n+1}(\mathcal{R}) \qquad (\text{by } (2)) \\ &\subseteq \bigcup_{\beta<\gamma} \Phi S_\beta^{n+1}(\mathcal{R}) \qquad (\text{since } < \text{ is transitive}) \\ &= S_\gamma^{n+1}\Psi(\mathcal{R}) \qquad (\text{by } (2)) \\ &= S_\gamma^{n+1}(\mathcal{R}) \qquad (\mathcal{R} \text{ is a fixed point}). \end{aligned}$$

Now, let $\mathcal{L} = P_1^{n+1}(\mathcal{R})$. Then $\mathcal{L}$ is generated. We show it is the least fixed point of Φ.

If $\langle \alpha, \mathbf{y} \rangle \in \mathcal{R}$ then $\langle \alpha, \mathbf{y} \rangle \in \Psi(\mathcal{R})$ so by (1), $\alpha \in S$. It follows that $\mathcal{L} = \bigcup_{\alpha \in S} S_\alpha^{n+1}(\mathcal{R})$, a fact we will use frequently.

(5) $\mathcal{L} \subseteq \Phi(\mathcal{L})$.

PROOF OF (5). For each $\beta \in S$, $S_\beta^{n+1}(\mathcal{R}) \subseteq \mathcal{L}$. Φ is monotone, hence

$$\Phi S_\beta^{n+1}(\mathcal{R}) \subseteq \Phi(\mathcal{L}).$$

Hence for each $\alpha \in S$,

$$\bigcup_{\beta<\alpha} \Phi S_\beta^{n+1}(\mathcal{R}) \subseteq \Phi(\mathcal{L})$$

or, by (2),

$$S_\alpha^{n+1}\Psi(\mathcal{R}) \subseteq \Phi(\mathcal{L}), \qquad S_\alpha^{n+1}(\mathcal{R}) \subseteq \Phi(\mathcal{L}).$$

Since α was arbitrary, $\bigcup_{\alpha \in S} S_\alpha^{n+1}(\mathcal{R}) \subseteq \Phi(\mathcal{L})$; so $\mathcal{L} \subseteq \Phi(\mathcal{L})$.

(6) $\Phi(\mathcal{L}) \subseteq \mathcal{L}$.

PROOF OF (6). Since Φ is $\mathcal{D}$-compact, its behavior is determined by what it does with $\mathscr{D}$-finite relations. We will show $F \subseteq \mathscr{L}$ and F is $\mathscr{D}$-finite $\Rightarrow$ $\Phi(F) \subseteq \mathscr{L}$ from which it follows that $\Phi(\mathscr{L}) \subseteq \mathscr{L}$.

So, let $F \subseteq \mathscr{L}$ be $\mathscr{D}$-finite. $F \subseteq P_1^{n+1}(\mathscr{R})$ so

$$(\forall \mathbf{y} \in F)(\exists \alpha)\mathscr{R}(\alpha, \mathbf{y}).$$

Then by hypothesis (3) there is a $\mathscr{D}$-finite set G such that

(a) $(\forall \mathbf{y} \in F)(\exists \alpha \in G)\mathscr{R}(\alpha, \mathbf{y})$,

(b) $(\forall \alpha \in G)(\exists \mathbf{y} \in F)\mathscr{R}(\alpha, \mathbf{y})$.

By (b), if $\alpha \in G$ then $\langle \alpha, \mathbf{y}\rangle \in \mathscr{R}$ for some $\mathbf{y}$, hence $\langle \alpha, \mathbf{y}\rangle \in \Psi(\mathscr{R})$ so $\alpha \in S$ by item (1). Thus $G \subseteq S$. Then by hypothesis (2) there is an upper bound, say α_0, for G; $\alpha \in G \Rightarrow \alpha < \alpha_0$. Then by (a)

$$(\forall \mathbf{y} \in F)(\exists \alpha < \alpha_0)\mathscr{R}(\alpha, \mathbf{y})$$

or

$$F \subseteq \bigcup_{\alpha < \alpha_0} S_\alpha^{n+1}(\mathscr{R}).$$

Then by (4), $F \subseteq S_{\alpha_0}^{n+1}(\mathscr{R})$. Since Φ is monotone,

$$\Phi(F) \subseteq \Phi S_{\alpha_0}^{n+1}(\mathscr{R}).$$

As a trivial application of hypothesis (2), $\{\alpha_0\}$ has a strict upper bound, say β. Then $\alpha_0 < \beta$ so

$$\Phi S_{\alpha_0}^{n+1}(\mathscr{R}) \subseteq \bigcup_{\alpha < \beta} \Phi S_\alpha^{n+1}(\mathscr{R}).$$

Thus

$$\Phi(F) \subseteq \bigcup_{\alpha < \beta} \Phi S_\alpha^{n+1}(\mathscr{R})$$

$$= S_\beta^{n+1}\Psi(\mathscr{R}) = S_\beta^{n+1}(\mathscr{R}) \subseteq \mathscr{L}.$$

It follows from (5) and (6) that $\mathscr{L}$ is a fixed point of Φ. We now show it is the *least* fixed point.

(7) Let $\mathscr{Q} \subseteq [\mathscr{A}]^n$ be some fixed point for Φ. Then $\mathscr{L} \subseteq \mathscr{Q}$.

PROOF OF (7). Suppose we didn't have $\mathscr{L} \subseteq \mathscr{Q}$, that is, we didn't have

$$\bigcup_{\alpha \in S} S_\alpha^{n+1}(\mathscr{R}) \subseteq \mathscr{Q}.$$

Take a minimal member $\alpha_0 \in S$ such that

$$\text{not-}S_{\alpha_0}^{n+1}(\mathscr{R}) \subseteq \mathscr{Q}.$$

Then if $\beta < \alpha_0$, $S_\beta^{n+1}(\mathscr{R}) \subseteq \mathscr{Q}$ and since Φ is monotone,

$$\Phi S^{n+1}_{\beta}(\mathcal{R}) \subseteq \Phi(\mathcal{Q}) = \mathcal{Q}.$$

Thus $\bigcup_{\beta<\alpha_0} \Phi S^{n+1}_{\beta}(\mathcal{R}) \subseteq \mathcal{Q}$; so by (2)

$$S^{n+1}_{\alpha_0}\Psi(\mathcal{R}) \subseteq \mathcal{Q}, \qquad S^{n+1}_{\alpha_0}(\mathcal{R}) \subseteq \mathcal{Q},$$

a contradiction which establishes (7).

(8) For each $\alpha \in S$, the relation Φ_α given by

$$\Phi_\alpha = \bigcup_{\beta<\alpha} \Phi(\Phi_\beta)$$

is generated, and $\mathcal{L} = \bigcup_{\alpha\in S} \Phi_\alpha$.

PROOF OF (8). Let us write, for each $\alpha \in S$, $\Phi_\alpha = S^{n+1}_{\alpha}(\mathcal{R})$. Then, of course, each Φ_α is generated. But also

$$\Phi_\alpha = S^{n+1}_{\alpha}(\mathcal{R}) = S^{n+1}_{\alpha}\Psi(\mathcal{R})$$

$$= \bigcup_{\beta<\alpha} \Phi S^{n+1}_{\beta}(\mathcal{R}) = \bigcup_{\beta<\alpha} \Phi(\Phi_\beta).$$

And finally, $\mathcal{L} = \bigcup_{\alpha\in S} S^{n+1}_{\alpha}(\mathcal{R}) = \bigcup_{\alpha\in S} \Phi_\alpha$.

This completes the proof.

EXAMPLES. We give examples of recursion and ω-recursion theories meeting the hypotheses of the above theorem. Production systems of a rather different sort will be given in the next chapter, and will also provide us with examples.

I. Any recursion theory with an effective pairing function meets the hypotheses, taking $\mathcal{D}$-finite to mean finite.

Conditions (1) and (2) are obviously satisfied by any such recursion theory in which ordinary recursion theory has an embedding. Simply take S to be the image of $\mathbb{N}$, the natural numbers, under the embedding, and $<$ can be the image of the ordering of $\mathbb{N}$ under the embedding. But by Proposition 6.8.3, this can be done if there is an effective pairing function.

If $\mathcal{D}$-finite means finite, condition (3) is trivial since every relation will satisfy weak replacement. Thus, if $(\forall \boldsymbol{x} \in F)(\exists \boldsymbol{y})R(\boldsymbol{x}, \boldsymbol{y})$ where F is finite, simply choose one $\boldsymbol{y}$, such that $R(\boldsymbol{x}, \boldsymbol{y})$, for each $\boldsymbol{x} \in F$, and let G be the set of such $\boldsymbol{y}$. G is finite, having the size of F or smaller, and certainly G meets the necessary conditions.

II. Any ω-recursion theory with an effective pairing function meets the conditions, taking $\mathcal{D}$-finite to mean ω-recursive. This time, we rely on work in Moschovakis [1974]. We merely cite results from that book and leave it to the reader to verify that they do the job.

Let S be an ω-r.e. but not ω-recursive set (our Theorem 7.2.1). Let σ be an inductive norm on S, which exists by Moschovakis [1974], Theorem 3A.3. Let $<$ be the relation defined by

$$x < y \;\Leftrightarrow\; x \in S \wedge [y \notin S \vee \sigma(x) < \sigma(y)].$$

Then $<$ is ω-r.e. by Moschovakis [1974], Theorem 3A.1. Now, if $F \subseteq S$ is $\mathscr{D}$-finite, F is ω-recursive = hyperelementary, so F is coinductive, hence contained in some resolvent S^{δ}_{σ}, by Moschovakis [1974], Theorem 3C.3. Since S is not hyperelementary, $S^{\delta}_{\sigma} \neq S$, so take a member of S outside S^{δ}_{σ}; it is an upper bound for F.

Thus we have conditions (1) and (2) of the above theorem. Condition (3) is a direct consequence of Moschovakis [1974], Corollary 3B.2.

Finally we add some conditions which enable us to drop the $\mathscr{D}$-compactness restriction. Note Corollary 10.5 in this context.

COROLLARY 11.3. *Suppose that, in addition to the hypotheses of Theorem* 11.2, *every monotone operator of* **A** *agrees with a monotone,* $\mathscr{D}$*-compact operator on the generated relations. Then every monotone operator in* **A** *of order* $\langle n, n\rangle$ *has a generated least fixed point.*

PROOF. Let $\Phi^* \in \mathbf{A}$ be monotone, of order $\langle n, n\rangle$. Then there is a monotone $\mathscr{D}$-compact operator Φ agreeing with Φ^* on generated relations. By Theorem 11.2, Φ has a generated least fixed point $\mathscr{L}$. Since $\mathscr{L}$ is generated, $\Phi^*(\mathscr{L}) = \Phi(\mathscr{L}) = \mathscr{L}$, so $\mathscr{L}$ is a fixed point for Φ^* too. We show it is least for Φ^* as well as for Φ.

Let $\mathscr{P}$ be any fixed point for Φ^*; we show $\mathscr{L} \subseteq \mathscr{P}$, that is, using the notation of Theorem 11.2,

$$\bigcup_{\alpha \in S} \Phi_\alpha \subseteq \mathscr{P}.$$

If this is not true, there is a minimal $\alpha_0 \in S$ such that not $\Phi_{\alpha_0} \subseteq \mathscr{P}$. Then, if $\beta < \alpha_0$, $\Phi_\beta \subseteq \mathscr{P}$. Since Φ^* is monotone,

$$\Phi^*(\Phi_\beta) \subseteq \Phi^*(\mathscr{P}) = \mathscr{P}$$

and since (by Theorem 11.2), Φ_β is generated, $\Phi(\Phi_\beta) \subseteq \mathscr{P}$. Thus

$$\bigcup_{\beta < \alpha_0} \Phi(\Phi_\beta) \subseteq \mathscr{P},$$

so $\Phi_{\alpha_0} \subseteq \mathscr{P}$. This contradiction ends the proof.

12. Metacompactness

We have seen that compactness is important. It has come up in connection with Rogers' forms, and in connection with the Least Fixed Point Theorem. Yet we have also seen, in Section 9, that even in the best known of ω-recursion theories there are non-compact operators. In this section and the next two we look into the possiblity of taking a production system with non-compact operators and "cutting it down" to one in which all operators are compact. We will find *partial* production systems starting to play a major role.

A reasonable first attempt at getting rid of the non-compact operators of a production system $\boldsymbol{A}$ is to simply toss them out. But this does not work in general because what is left may not be closed under composition and so may not be a production system. The following example illustrates this.

EXAMPLE. Consider hyperarithmetic theory, ω-rec($\mathfrak{S}(\mathbb{N})$), where we take $\mathcal{D}$-finite to mean ω-recursive = hyperarithmetic. What we do is break the example of a non-compact operator in Section 9, into two parts.

Let $[A^I_G]$ be the ω-enumeration operator with axioms A consisting of

axioms for $<$,

$Iy \to x < y \to Gx.$

Then $[A^I_G](\mathcal{P}) = \{x \mid x < y \text{ for some } y \in \mathcal{P}\}$. It is trivial that $[A^I_G]$ is monotone. Also, if $x \in [A^I_G](\mathcal{P})$ then $x \in [A^I_G](\{y\})$ for some $y \in \mathcal{P}$, and singletons are hyperarithmetic. Hence $[A^I_G]$ is $\mathcal{D}$-compact.

Let $[B^G_O]$ be the ω-enumeration operator in which B consists of the single axiom

$G\forall \to O5.$

Then

$$[B^G_O](\mathcal{P}) = \begin{cases} \emptyset & \text{if } \mathcal{P} \neq \mathbb{N}, \\ \{5\} & \text{if } \mathcal{P} = \mathbb{N}. \end{cases}$$

$[B^G_O]$ is trivially monotone. It is also $\mathcal{D}$-compact, for, if $x \in [B^G_O](\mathcal{P})$ then x must be 5 and $\mathcal{P}$ must be $\mathbb{N}$. But $\mathbb{N}$ itself is hyperarithmetic, hence $\mathcal{D}$-finite.

But, the composition, $[B^G_O][A^I_G]$ is easily seen to be the operator $[E^I_O]$ of Section 9, which is not $\mathcal{D}$-compact for this (or any other) notion of $\mathcal{D}$-finite.

What we do, then, is introduce a stronger version of compactness, the one mentioned in the title of this section, in order to get around the difficulty illustrated by the above example.

DEFINITION. Let $\boldsymbol{A}$ be a production system with a notion of $\mathcal{D}$-finiteness and an effective pairing function J (hence there is a notion of $\mathcal{D}$-finiteness for relations as well as for sets). Let $\Phi \in \boldsymbol{A}$ be of order $\langle n, m\rangle$. We call Φ *$\mathcal{D}$-metacompact* if, for each $\mathcal{P} \in [\mathcal{A}]^n$, $F \subseteq \Phi(\mathcal{P})$ where F is $\mathcal{D}$-finite $\Rightarrow$ $F \subseteq \Phi(G)$ for some $\mathcal{D}$-finite $G \subseteq \mathcal{P}$.

It is easy to check that the compostion of two $\mathcal{D}$-metacompact operators is $\mathcal{D}$-metacompact. But still we can't just take a production system $\boldsymbol{A}$, toss out all the operators that aren't $\mathcal{D}$-metacompact, and expect to wind up with another production system. The following example shows this.

EXAMPLE. Again consider hyperarithmetic theory, taking $\mathcal{D}$-finite to mean hyperarithmetic. The projection operator (of order $\langle 2, 1\rangle$) is not $\mathcal{D}$-metacompact. For, let $\mathcal{P}$ be any infinite, non-hyperarithmetic set. $\mathcal{P}$ is countable; let

$$f : \mathbb{N} \xrightarrow[\text{onto}]{1-1} \mathcal{P}.$$

Then f is certainly not hyperarithmetic either. Now, $P_2^2(f) = \mathbb{N}$ which is hyperarithmetic hence $\mathcal{D}$-finite. If P_2^2 were $\mathcal{D}$-metacompact, since $\mathbb{N} \subseteq P_2^2(f)$, we would have $\mathbb{N} \subseteq P_2^2(G)$ for some hyperarithmetic $G \subseteq f$. But by construction of f, if $\mathbb{N} \subseteq P_2^2(G)$ where $G \subseteq f$, we must have $G = f$, which is not hyperarithmetic.

Incidentally, if singletons are $\mathcal{D}$-finite, it is trivial that every $\mathcal{D}$-metacompact operator is $\mathcal{D}$-compact. The above example shows the converse need not be true, since it is easy to see P_2^2 is $\mathcal{D}$-compact in ω-rec$(\mathfrak{S}(\mathbb{N}))$.

Next we show that there is a variation of what we called Rogers' form available for metacompact operators.

THEOREM 12.1. *Let $\boldsymbol{A}$ be an indexed production system having the input place-fixing property, and a notion of $\mathcal{D}$-finiteness for which there is a positive canonical coding, and an effective pairing function. If $\Phi \in \boldsymbol{A}$ is of order $\langle n, m\rangle$, is monotone and $\mathcal{D}$-metacompact, then there is a generated two-place relation R such that, for allowable $\mathcal{P}$,*

$$D_a^m \subseteq \Phi(\mathcal{P}) \Leftrightarrow (\exists b)[R(b, a) \wedge D_b^n \subseteq \mathcal{P}].$$

PROOF. Using the operator G of Proposition 7.1, define a relation R by

$$R(b, a) \Leftrightarrow a \in F[J^m]\Phi[J^n]^{-1}G(\{b\}).$$

By Corollary 5.2, R is generated.

(1) Suppose $R(b, a) \wedge D_b^n \subseteq \mathscr{P}$. Then, using various items from Section 8, we have

$$R(b, a),$$

$$a \in F[J^m]\Phi[J^n]^{-1}G(\{b\}),$$

$$D_a \subseteq [J^m]\Phi[J^n]^{-1}G(\{b\}),$$

$$D_a^m \subseteq \Phi[J^n]^{-1}G(\{b\}),$$

$$D_a^m \subseteq \Phi[J^n]^{-1}(D_b),$$

$$D_a^m \subseteq \Phi(D_b^n),$$

but Φ is monotone, and $D_b^n \subseteq \mathscr{P}$, hence $D_a^m \subseteq \Phi(\mathscr{P})$.

(2) Suppose $D_a^m \subseteq \Phi(\mathscr{P})$. Since Φ is $\mathscr{D}$-metacompact, for some $\mathscr{D}$-finite $D_b^n \subseteq \mathscr{P}$ we have $D_a^m \subseteq \Phi(D_b^n)$. But then every step above reverses and we have $R(b, a)$. Thus $(\exists b)[R(b, a) \wedge D_b^n \subseteq \mathscr{P}]$.

Thus, every $\mathscr{D}$-metacompact, monotone operator "corresponds" to a generated relation (under appropriate hypotheses). The converse is not true, however. Not every generated binary relation need determine an operator in the above way.

EXAMPLE. In ordinary recursion theory, rec($\mathfrak{S}$(N)). Let D_a and D_b be two finite sets, with $D_a \neq \emptyset$. Also, let c be a finite code for $\emptyset$, so $D_c = \emptyset$. Let R be the r.e. relation $R = \{\langle b, a\rangle\}$. Suppose there were an enumeration operator Φ such that

$$D_x \subseteq \Phi(\mathscr{P}) \Leftrightarrow (\exists y)[R(y, x) \wedge D_y \subseteq \mathscr{P}].$$

Then what is $\Phi(D_b)$? Certainly $\emptyset \subseteq \Phi(D_b)$, that is $D_c \subseteq \Phi(D_b)$. So we should have $(\exists y)[R(y, c) \wedge D_y \subseteq \mathscr{P}]$, but this is not possible since R holds only in the case $R(b, a)$.

DEFINITION. Let $\boldsymbol{A}$ be a production system with a notion of $\mathscr{D}$-finiteness for which there is a coding, and with an effective pairing function. Let $\mathscr{P} \in [\mathscr{A}]^n$ and $\mathscr{Q} \in [\mathscr{A}]^m$. We say $\mathscr{Q}$ is *strongly reducible to* $\mathscr{P}$ if there is some generated binary relation R such that

$$D_x^m \subseteq \mathscr{Q} \Leftrightarrow (\exists y)[R(y, x) \wedge D_y^n \subseteq \mathscr{P}].$$

REMARK. Then Theorem 12.1, implies that, in **A**, for monotone, $\mathcal{D}$-metacompact Φ, if $\Phi(\mathcal{P}) = \mathcal{Q}$ then $\mathcal{Q}$ is strongly reducible to $\mathcal{P}$. The following is a weak converse.

THEOREM 12.2. *Let* **A** *be an indexed production system having the input place-fixing property, and a notion of $\mathcal{D}$-finiteness for which there is a positive canonical coding, and with an effective pairing function. Suppose also that singletons are $\mathcal{D}$-finite. Let $\mathcal{P} \in [\mathcal{A}]^n$ and $\mathcal{Q} \in [\mathcal{A}]^m$. If $\mathcal{Q}$ is strongly reducible to $\mathcal{P}$ then there is a monotone, $\mathcal{D}$-compact (sic) operator $\Phi \in$* **A** *such that $\Phi(\mathcal{P}) = \mathcal{Q}$.*

PROOF. Suppose $\mathcal{Q}$ is strongly reducible to $\mathcal{P}$. Then there is a generated relation R such that

$$D_x^m \subseteq \mathcal{Q} \Leftrightarrow (\exists y)[R(y,x) \wedge D_y^n \subseteq \mathcal{P}].$$

Define Φ of order $\langle n, m\rangle$ by

$$\Phi = [J^m]^{-1} GR''F[J^n].$$

Then $\Phi \in \mathbf{A}$ and it is straightforward to check that Φ is $\mathcal{D}$-compact. We show $\Phi(\mathcal{P}) = \mathcal{Q}$.

(1) Suppose $\mathbf{w} \in \mathcal{Q}$. Then $\mathbf{w} \in \{\mathbf{w}\} \subseteq \mathcal{Q}$. Singletons are $\mathcal{D}$-finite; let $\{\mathbf{w}\} = D_a^m$. Then $\mathbf{w} \in D_a^m$ where $D_a^m \subseteq \mathcal{Q}$. Since $\mathcal{Q}$ is strongly reducible to $\mathcal{P}$, then for some b, $R(b,a) \wedge D_b^n \subseteq \mathcal{P}$. Now

$$D_b^n \subseteq \mathcal{P} \Leftrightarrow D_b \subseteq [J^n](\mathcal{P}) \Leftrightarrow b \in F[J^n](\mathcal{P}).$$

Further, since $R(b,a)$, then $a \in R''F[J^n](\mathcal{P})$. But then $D_a \subseteq GR''F[J^n](\mathcal{P})$ and it follows that $D_a^m \subseteq [J^m]^{-1}GR''F[J^n](\mathcal{P})$ or $D_a^m \subseteq \Phi(\mathcal{P})$. Since $\mathbf{w} \in D_a^m$, $\mathbf{w} \in \Phi(\mathcal{P})$. Thus $\mathcal{Q} \subseteq \Phi(\mathcal{P})$.

(2) Suppose $\mathbf{w} \in \Phi(\mathcal{P})$. Then $\mathbf{w} \in [J^m]^{-1}GR''F[J^n](\mathcal{P})$, so for some $z \in GR''F[J^n](\mathcal{P})$, $z = J^m(\mathbf{w})$. But then, $z \in D_q$ for some $q \in R''F[J^n](\mathcal{P})$. Since $z \in D_q$ then $\mathbf{w} \in D_q^m$. And since $q \in R''F[J^n](\mathcal{P})$, for some t, $R(t,q)$ where $t \in F[J^n](\mathcal{P})$. But then, $D_t \subseteq [J^n](\mathcal{P})$ so $D_t^n \subseteq \mathcal{P}$. Thus, for this t, $R(t,q)$ and $D_t^n \subseteq \mathcal{P}$. It follows that $D_q^m \subseteq \mathcal{Q}$. Since $\mathbf{w} \in D_q^m$, $\mathbf{w} \in \mathcal{Q}$. Thus $\Phi(\mathcal{P}) \subseteq \mathcal{Q}$. This concludes the proof.

13. Semi-hyperregularity

We continue our investigation of metacompactness by asking, what happens if we require that every operator be metacompact. It is here that restricting the inputs of production systems begins to play a role.

We begin with a few definitions, the first repeated from Section 11 for convenience.

DEFINITION. Let **A** be a production system with a notion of $\mathscr{D}$-finiteness and an effective pairing function, and let R be an $n+m$ place relation on $\mathscr{A}$. We say R *satisfies weak replacement* if, for each $\mathscr{D}$-finite $F \subseteq \mathscr{A}^n$ such that

$$(\forall \boldsymbol{x} \in F)(\exists \boldsymbol{y}) R(\boldsymbol{x}, \boldsymbol{y})$$

there is a $\mathscr{D}$-finite $G \subseteq \mathscr{A}^m$ such that

$$(\forall \boldsymbol{x} \in F)(\forall \boldsymbol{y} \in G) R(\boldsymbol{x}, \boldsymbol{y})$$

and

$$(\forall \boldsymbol{y} \in G)(\exists \boldsymbol{x} \in F) R(\boldsymbol{x}, \boldsymbol{y}).$$

DEFINITION. Again let **A** be a production system with a notion of $\mathscr{D}$-finiteness and an effective pairing function. Let $\mathscr{P} \in [\mathscr{A}]^k$. We call $\mathscr{P}$ *semi-hyperregular* if, for each $\Phi \in \mathbf{A}$ of order $\langle k, n+m \rangle$ $(n, m \geqslant 1)$, $\Phi(\mathscr{P})$ is a relation satisfying weak replacement.

REMARK. We apologize for the rather ungainly term, semi-hyperregularity. It is reasonable, however. There is a term, hyperregularity, used in admissible set recursion theory, and, in a sense, the above notion is half of that.

We will show that, under quite reasonable conditions, if we insist that all operators should be monotone and $\mathscr{D}$-metacompact, then all allowable relations must be semi-hyperregular. And we will show there is a broad class of examples in which there must be allowable relations other than the semi-hyperregular, hence that non-metacompactness is not uncommon.

We note that, in *recursion* theories, taking $\mathscr{D}$-finite to mean finite, every enumeration operator is $\mathscr{D}$-compact, every enumeration operator is $\mathscr{D}$-metacompact, every relation is allowable, every relation satisfies weak replacement, and hence every relation is semi-hyperregular. Thus these notions are of no real interest for recursion theories. We must look at other production systems: ω-recursion theories now, and related production systems in the next chapter.

LEMMA 13.1. *Let* **A** *be a production system with a notion of* $\mathscr{D}$*-finiteness and an effective pairing function. Suppose the* $\mathscr{D}$*-finite relations are closed under* *projections. Then, if F is $\mathscr{D}$-finite and $F \subseteq \mathscr{P} \times \mathscr{Q}$ ($\mathscr{P} \times \mathscr{Q}$ allowable) then $F \subseteq G \times H$ where G and H are $\mathscr{D}$-finite, $G \subseteq \mathscr{P}$ and $H \subseteq \mathscr{Q}$.*

Proof. Say $\mathscr{P}$ is n-ary and $\mathscr{Q}$ is m-ary. Let

$$G = P^{n+1}_{n+1} \cdots P^{n+m-1}_{n+m-1} P^{n+m}_{n+m}(F) \quad \text{and} \quad H = P^{n+1}_1 \cdots P^{n+m-1}_1 P^{n+m}_1(F).$$

Then G and H are $\mathscr{D}$-finite, and meet the conditions.

Theorem 13.2. *Let **A** be a production system with a notion of $\mathscr{D}$-finiteness and an effective pairing function. Suppose the $\mathscr{D}$-finite relations are closed under projections. If every operator is monotone and $\mathscr{D}$-metacompact, then every allowable relation is semi-hyperregular.*

Proof. Suppose every operator is monotone and $\mathscr{D}$-metacompact. Let $\mathscr{P} \in [\mathscr{A}]^k$, let $\Phi \in \mathbf{A}$ be of order $\langle k, n+m \rangle$ and let $\mathscr{R} = \Phi(\mathscr{P})$. We show $\mathscr{R}$ satisfies weak replacement.

The proof divides into parallel arguments depending on whether $\mathscr{P} = \emptyset$ or $\mathscr{P} \neq \emptyset$. We present only the $\mathscr{P} \neq \emptyset$ argument; if $\mathscr{P} = \emptyset$, Corollary 2.5.2, part 1 is needed, and the argument is somewhat simpler.

Since $\mathscr{R} = \Phi(\mathscr{P})$ then $\mathscr{R} \leq \mathscr{P}$. Define S by $S(\mathbf{y}, \mathbf{x}) \Leftrightarrow \mathscr{R}(\mathbf{x}, \mathbf{y})$ ($\mathbf{y}$ is an m-tuple, $\mathbf{x}$ is an n-tuple). Then $S \leq \mathscr{P}$ too. Under the assumption that $\mathscr{P} \neq \emptyset$, by Corollary 2.5.2 part 2, there is an operator $\Psi \in \mathbf{A}$ of order $\langle k+m, n \rangle$ such that, whenever there is a relation $\mathscr{Q}$ for which $\mathscr{P} \times \mathscr{Q}$ is allowable,

$$\Psi(\mathscr{P} \times \mathscr{Q}) = S''(\mathscr{Q}).$$

Now, suppose F is $\mathscr{D}$-finite and $(\forall \mathbf{x} \in F)(\exists \mathbf{y})\mathscr{R}(\mathbf{x}, \mathbf{y})$. Define a relation $\mathscr{Q}_0$ by

$$\mathbf{y} \in \mathscr{Q}_0 \Leftrightarrow (\exists \mathbf{x})[\mathbf{x} \in F \wedge \mathscr{R}(\mathbf{x}, \mathbf{y})].$$

Since F, being $\mathscr{D}$-finite, is generated, and $\mathscr{R} \leq \mathscr{P}$, it follows that $\mathscr{Q}_0 \leq \mathscr{P}$. Since $\mathscr{P} \leq \mathscr{P}$, by Proposition 2.4.3 (2), $\mathscr{P} \times \mathscr{Q}_0$ is allowable. Then

$$\Psi(\mathscr{P} \times \mathscr{Q}_0) = S''(\mathscr{Q}_0).$$

We claim $F \subseteq \Psi(\mathscr{P} \times \mathscr{Q}_0)$. This is shown as follows. Let $\mathbf{x} \in F$. Then for some $\mathbf{y}$, say for $\mathbf{y}_0$, $\mathscr{R}(\mathbf{x}, \mathbf{y}_0)$. Then by definition $\mathbf{y}_0 \in \mathscr{Q}_0$. Also $S(\mathbf{y}_0, \mathbf{x})$, hence $\mathbf{x} \in S''(\mathscr{Q}_0)$ so $\mathbf{x} \in \Psi(\mathscr{P} \times \mathscr{Q}_0)$.

Since $F \subseteq \Psi(\mathscr{P} \times \mathscr{Q}_0)$ and all operators are $\mathscr{D}$-metacompact, $F \subseteq \Psi(F_0)$ for some $\mathscr{D}$-finite $F_0 \subseteq \mathscr{P} \times \mathscr{Q}_0$. Then by Lemma 13.1, $F_0 \subseteq G \times H$ for some

$\mathcal{D}$-finite $G \subseteq \mathcal{P}$ and $H \subseteq \mathcal{Q}_0$. Now, $G \times H \subseteq \mathcal{P} \times H$ and Ψ is monotone, so $F \subseteq \Psi(F_0) \subseteq \Psi(G \times H) \subseteq \Psi(\mathcal{P} \times H)$. Thus: there is a $\mathcal{D}$-finite $H \subseteq \mathcal{Q}_0$ such that $F \subseteq S''(H)$. We claim H is the relation we are looking for.

Suppose $\boldsymbol{x} \in F$. Then $\boldsymbol{x} \in S''(H)$ so for some $\boldsymbol{y} \in H$, $S(\boldsymbol{y}, \boldsymbol{x})$ or $\mathcal{R}(\boldsymbol{x}, \boldsymbol{y})$. Thus $(\forall \boldsymbol{x} \in F)(\exists \boldsymbol{y} \in H)\mathcal{R}(\boldsymbol{x}, \boldsymbol{y})$.

Suppose $\boldsymbol{y} \in H$. Then $\boldsymbol{y} \in \mathcal{Q}_0$ so for some $\boldsymbol{x} \in F$, $\mathcal{R}(\boldsymbol{x}, \boldsymbol{y})$. Thus $(\forall \boldsymbol{y} \in H)(\exists \boldsymbol{x} \in F)\mathcal{R}(\boldsymbol{x}, \boldsymbol{y})$.

Thus $\mathcal{R}$ satisfies weak replacement, and we are done.

Now we look at a few of the immediate consequences of semi-hyperregularity.

PROPOSITION 13.3. *Let* **A** *be a production system with a notion of* $\mathcal{D}$-*finiteness and an effective pairing function. Suppose* $\mathcal{P}$ *is semi-hyperregular and* $f \leqslant \mathcal{P}$ *where* f *is a partial function. If* $K \subseteq \operatorname{dom} f$ *where* K *is* $\mathcal{D}$-*finite then* $f''K$ *is also* $\mathcal{D}$-*finite.*

PROOF. Immediate.

DEFINITION. $\mathcal{P}$ is called *regular* if $\mathcal{P} \cap K$ is $\mathcal{D}$-finite for every $\mathcal{D}$-finite K.

LEMMA 13.4. *Let* **A** *be a production system with a notion of* $\mathcal{D}$-*finiteness. Suppose also that the empty set is* $\mathcal{D}$-*finite. Then, for* $\mathcal{P} \subseteq \mathcal{A}$, *if* $\mathcal{P} \times \bar{\mathcal{P}}$ *is semi-hyperregular then* $\mathcal{P}$ *is regular.*

PROOF. Let $\mathcal{P} \times \bar{\mathcal{P}}$ be semi-hyperregular, and let K be a $\mathcal{D}$-finite set; we show $\mathcal{P} \cap K$ is $\mathcal{D}$-finite.

If $\mathcal{P} \cap K = \emptyset$ the result is by hypothesis.

Now suppose $\mathcal{P} \cap K \neq \emptyset$. Choose $a \in \mathcal{P} \cap K$. Define a function f by

$$f(x) = \begin{cases} x & \text{if } x \in \mathcal{P}, \\ a & \text{if } x \in \bar{\mathcal{P}}. \end{cases}$$

Then $f \leqslant \mathcal{P} \times \bar{\mathcal{P}}$ since f is $[=_{\mathcal{A}} \cap (\mathcal{P} \times \mathcal{P})] \cup [\bar{\mathcal{P}} \times \{a\}]$. Also f is total, so $K \subseteq \operatorname{dom} f$. Then by Proposition 13.3, $f''K$ is $\mathcal{D}$-finite. But $f''K = \mathcal{P} \cap K$.

THEOREM 13.5. *Let* **A** *be an indexed production system with a notion of* $\mathcal{D}$-*finiteness. Also suppose that* $\emptyset$ *and* $\mathcal{A}$ *are* $\mathcal{D}$-*finite. Then there is a set which is not semi-hyperregular.*

PROOF. Let $\mathcal{U}^{\langle 1,1\rangle}$ be a universal operator of order $\langle 1, 1\rangle$, and define a set $\mathcal{P}$ by

$$x \in \mathcal{P} \iff \langle x, x\rangle \in \mathcal{U}^{\langle 1,1\rangle}(\emptyset).$$

We claim $\bar{\mathcal{P}}$ is not semi-hyperregular.

First we show $\bar{\mathcal{P}}$ is not regular. Well, if $\bar{\mathcal{P}}$ were regular, since $\mathcal{A}$ is $\mathcal{D}$-finite, $\bar{\mathcal{P}} \cap \mathcal{A} = \bar{\mathcal{P}}$ would be $\mathcal{D}$-finite, and hence generated. But a simple diagonal argument shows $\mathcal{P}$ is not generated.

Since $\bar{\mathcal{P}}$ is not regular, $\mathcal{P} \times \bar{\mathcal{P}}$ is not semi-hyperregular, by Lemma 13.4. Now, $\mathcal{P}$ is a generated set and hence for any $\mathcal{R}$

$$\mathcal{R} \leqslant \bar{\mathcal{P}} \iff \mathcal{R} \leqslant \mathcal{P} \times \bar{\mathcal{P}}.$$

It follows that $\bar{\mathcal{P}}$ is semi-hyperregular iff $\mathcal{P} \times \bar{\mathcal{P}}$ also is, hence $\bar{\mathcal{P}}$ is not semi-hyperregular.

REMARKS. ω-recursion theories are production systems and, if there is an effective pairing function, they are indexed. Further, if conditions (1)–(3) of Example IIB in Section 6 are met, the $\mathcal{D}$-finiteness conditions hold, taking $\mathcal{D}$-finite to mean ω-recursive. But then, by Theorem 13.5, there must be a set which is not semi-hyperregular. Now, the presence of a dual projection operator easily implies that the ω-recursive $= \mathcal{D}$-finite relations are closed under projections. Since ω-enumeration operators are total and montone, then by Theorem 13.2, in all such ω-recursion theories there must be operators which are not $\mathcal{D}$-metacompact.

PROPOSITION 13.6. *Let* **A** *be a production system with a notion of $\mathcal{D}$-finiteness and an effective pairing function. Suppose there are two sets, $\mathcal{P}, \mathcal{Q} \in [\mathcal{A}]^1$, one $\mathcal{D}$-finite, the other not, both of the same cardinality. Then there is a relation which is not semi-hyperregular.*

PROOF. Say $\mathcal{P}$ is $\mathcal{D}$-finite, $\mathcal{Q}$ is not, and they have the same cardinality. Let $f : \mathcal{P} \to \mathcal{Q}$ be 1–1 and onto. $\mathcal{P} \subseteq \operatorname{dom} f$ is $\mathcal{D}$-finite but $f''\mathcal{P} = \mathcal{Q}$ which is not $\mathcal{D}$-finite. Since (if f is allowable) $f \leqslant f$, then by Proposition 13.3, f is not semi-hyperregular. Of course, if f is not allowable, it can not be semi-hyperregular either.

REMARKS. This too can be used to show the existence of operators that are not $\mathcal{D}$-metacompact in many ω-recursion theories. We skip details. We will also use this result in the next chapter.

14. Compact cores

We have seen that $\mathscr{D}$-metacompactness and semi-hyperregularity play important roles. Here we show that many production systems can be "cut down" to subsystems in which all operators are $\mathscr{D}$-metacompact and all alowable relations are semi-hyperregular. Often such production systems will be partial, however. We begin with some results about restricting inputs.

DEFINITION. Let $\mathbf{A}$ be a production system with a notion of $\mathscr{D}$-finiteness. We call $\mathbf{A} = \langle \mathbf{A}, \mathscr{D} \rangle$ *full* if: $\mathbf{A}$ is indexed and has the input and the output place-fixing properties; there is an effective pairing function; there is a positive canonical coding of the $\mathscr{D}$-finite sets.

THEOREM 14.1. *Let* $\mathbf{A}$ *be a full production system. Let* $\mathfrak{Q}$ *be a collection of relations on* $\mathscr{A}$ *meeting these conditions: every generated relation is in* $\mathfrak{Q}$, *every relation in* $\mathfrak{Q}$ *is allowable, and* $\mathfrak{Q}$ *is closed under the operators of* $\mathbf{A}$. *Let* $\mathbf{A} \restriction \mathfrak{Q}$ *be like* $\mathbf{A}$ *but with operators restricted to members of* $\mathfrak{Q}$ *as inputs*; *that is, in* $\mathbf{A} \restriction \mathfrak{Q}$, *allowable = member of* $\mathfrak{Q}$. *Then* $\mathbf{A} \restriction \mathfrak{Q}$ *is also a full production system* (*using the same notion of* $\mathscr{D}$*-finiteness as in* $\mathbf{A}$) *in which the same relations are generated as in* $\mathbf{A}$.

PROOF. Suppose $\mathscr{R}$ is generated in $\mathbf{A}$. Then $\mathscr{R}$ is the value of some constant operator of $\mathbf{A}$. That operator, restricted to $\mathfrak{Q}$, is still constant, and is in $\mathbf{A} \restriction \mathfrak{Q}$, hence $\mathscr{R}$ is generated in $\mathbf{A} \restriction \mathfrak{Q}$. It follows that $\emptyset$, $=_{\mathscr{A}}$, and singletons are generated in $\mathbf{A} \restriction \mathfrak{Q}$.

That the operators of $\mathbf{A} \restriction \mathfrak{Q}$ are closed under $\cap$, $\cup$, $\times$ and composition and include transposition, projection and place-adding operators, is straightforward.

Thus $\mathbf{A} \restriction \mathfrak{Q}$ is a production system. Let $\mathscr{R}$ be generated in $\mathbf{A} \restriction \mathfrak{Q}$. Then (Proposition 2.3.1) $\mathscr{R} = \Phi(\emptyset)$ for some Φ. But this equation holds in $\mathbf{A}$ as well, hence $\mathscr{R}$ is generated in $\mathbf{A}$. Thus $\mathbf{A}$ and $\mathbf{A} \restriction \mathfrak{Q}$ have the same generated relations.

It is easy to see that the universal operators of $\mathbf{A}$, suitably restricted, become universal operators for $\mathbf{A} \restriction \mathfrak{Q}$, thus it is indexed.

The place-fixing and $\mathscr{D}$-finiteness features are straightforward.

REMARK. If $\mathfrak{Q}$ is exactly the generated relations of $\mathbf{A}$, the hypotheses of the above are satisfied, thus allowable need not mean more than generated.

LEMMA 14.2. *If* **A** *is a full production system, operators take semi-hyperregular inputs to semi-hyperregular outputs.*

PROOF. Let $\mathscr{P}$ be semi-hyperregular. We show $\Phi(\mathscr{P})$ also is. That is, we must show $\Psi(\Phi(\mathscr{P}))$ satisfies weak replacement. But this is $(\Psi\Phi)(\mathscr{P})$, and this satisfies weak replacement since $\mathscr{P}$ is semi-hyperregular.

LEMMA 14.3. *Let* **A** *be a full production system. If every generated relation satisfies weak replacement, then every generated relation is semi-*hyperregular.

PROOF. If $\mathscr{R}$ is generated, so is $\Phi(\mathscr{R})$ for every $\Phi \in \mathbf{A}$.

THEOREM 14.4. *Let* **A** *be a full production system. Suppose that, in it, every generated relation satisfies weak replacement. Let* $\mathfrak{Q}$ *consist of the semi-hyperregular relations of* **A**. *Then* $\mathbf{A} \restriction \mathfrak{Q}$ *is also a full production system (using the same notion of* $\mathscr{D}$*-finiteness) having the same relations generated as in* **A**, *and in which allowable = semi-hyperregular.*

PROOF. Theorem 14.1 and Lemmas 14.2 and 14.3.

Next, we follow up on the consequences of "allowable = semi-hyperregular", which appears is the conclusion of the above theorem.

DEFINITION. Let **A** be full. We say that, in **A**, the $\mathscr{D}$-finite union of $\mathscr{D}$-finite relations is $\mathscr{D}$-finite provided, for each $\mathscr{D}$-finite set F of $\mathscr{D}$-finite codes for $\mathscr{D}$-finite n-ary relations, $\bigcup_{a\in F} D^n_a$ is $\mathscr{D}$-finite.

THEOREM 14.5. *Let* **A** *be a full production system and suppose in it, the* $\mathscr{D}$*-finite union of* $\mathscr{D}$*-finite relations is* $\mathscr{D}$*-finite. If allowable = semi-hyperregular then every monotone,* $\mathscr{D}$*-compact operator is also* $\mathscr{D}$-metacompact.

PROOF. Let Φ be monotone and $\mathscr{D}$-compact; we show Φ is $\mathscr{D}$-metacompact. Thus, let $\mathscr{P}$ be allowable, and suppose $K \subseteq \Phi(\mathscr{P})$ where K is $\mathscr{D}$-finite; we show $K \subseteq \Phi(K')$ for some $\mathscr{D}$-finite $K' \subseteq \mathscr{P}$.

Since Φ is monotone and $\mathscr{D}$-compact, by Theorem 8.4, there is a generated $m+1$ place relation R such that $\Phi = R''F[J^n]$. Define an $m+1$ place relation Q by

$$Q(\mathbf{x}, y) \Leftrightarrow y \in F[J^n](\mathscr{P}) \wedge R(y, \mathbf{x}).$$

It is easy to see $Q \leqslant \mathcal{P}$. $\mathcal{P}$, being allowable, is semi-hyperregular, hence Q satisfies weak replacement. Now $K \subseteq \Phi(\mathcal{P}) = R''F[J^n](\mathcal{P})$, hence

$$(\forall \boldsymbol{x} \in K)(\exists y)[y \in F[J^n](\mathcal{P}) \wedge R(y, \boldsymbol{x})]$$

or

$$(\forall \boldsymbol{x} \in K)(\exists y)Q(\boldsymbol{x}, y).$$

Since Q satisfies weak replacement, there is a $\mathcal{D}$-finite set H such that

$$(\forall \boldsymbol{x} \in K)(\exists y \in H)Q(\boldsymbol{x}, y)$$

and

$$(\forall y \in H)(\exists \boldsymbol{x} \in K)Q(\boldsymbol{x}, y).$$

Note that, by the second of these, if $y \in H$ then, in particular, $y \in F[J^n](\mathcal{P})$ and hence $D_y^n \subseteq \mathcal{P}$.

Now, let $K' = \bigcup_{y \in H} D_y^n$. By hypothesis, K' is $\mathcal{D}$-finite. Also, by our previous remarks, $K' \subseteq \mathcal{P}$.

Finally, $K \subseteq \Phi(K')$ since: suppose $\boldsymbol{x} \in K$; then for some $y \in H$, $Q(\boldsymbol{x}, y)$. That is, $y \in F[J^n](\mathcal{P})$ and $R(y, \boldsymbol{x})$. But then $D_y^n \subseteq \mathcal{P}$. Now if $y \in H$, in fact $D_y^n \subseteq K'$, so $y \in F[J^n](K')$, and $R(y, \boldsymbol{x})$. Then $\boldsymbol{x} \in R''F[J^n](K') = \Phi(K')$.

Now we continue by looking at the consequences of every $\mathcal{D}$-compact operator being $\mathcal{D}$-metacompact, which appears in the conclusion of Theorem 14.5. First, a lemma.

LEMMA 14.6. *Let* $\mathbf{A}$ *be an indexed production system having the input place-fixing property, a notion of* $\mathcal{D}$*-finiteness, and an effective pairing function. Suppose also that the* $\mathcal{D}$*-finite relations are closed under projections. Let* Φ *be a monotone,* $\mathcal{D}$*-compact operator in* $\mathbf{A}$ *of order* $\langle q+k, n\rangle$. *Then there is a monotone,* $\mathcal{D}$*-compact* $\Psi \in \mathbf{A}$ *of order* $\langle k, q+n\rangle$ *such that, for* $\mathcal{P} \in [\mathcal{A}]^k$,

$$\langle y_1, \ldots, y_n\rangle \in \Phi(\{\langle x_1, \ldots, x_q\rangle\} \times \mathcal{P})$$
$$\Leftrightarrow \ \langle x_1, \ldots, x_q, y_1, \ldots, y_n\rangle \in \Psi(\mathcal{P}).$$

PROOF. By the input place-fixing property for $\mathbf{A}$, there is a basic function $f : \mathcal{A}^q \to \mathcal{A}$ such that

$$\boldsymbol{y} \in \Phi(\{\boldsymbol{x}\} \times \mathcal{P}) \ \Leftrightarrow \ \boldsymbol{y} \in \Phi_{f(\boldsymbol{x})}^{\langle k,n\rangle}(\mathcal{P}).$$

Since f is basic, hence generated, there is a constant operator, $I_f \in \mathbf{A}$ of order $\langle k, q+1\rangle$ and value f. Now, set

$$\Psi = P_{q+1}^{q+n+1} P_{q+1}^{q+n+2} E_{q+1,q+2}^{q+n+2}(I_f \times \mathcal{U}^{\langle k,n\rangle}).$$

It is straightforward to check that

$$y \in \Phi(\{x\} \times \mathscr{P}) \iff \langle x, y\rangle \in \Psi(\mathscr{P})$$

and from this the monotonicity of Ψ follows easily from that of Φ.

We show Ψ is $\mathscr{D}$-compact. Suppose $\langle x, y\rangle \in \Psi(\mathscr{P})$. Then $y \in \Phi(\{x\} \times \mathscr{P})$ and since Φ is $\mathscr{D}$-compact, $y \in \Phi(K)$ for some $\mathscr{D}$-finite $K \subseteq \{x\} \times \mathscr{P}$. Then by Lemma 13.1, $K \subseteq G \times H$ where G, H are $\mathscr{D}$-finite, $G \subseteq \{x\}$ and $H \subseteq \mathscr{P}$. If $K \neq \emptyset$ then $G \neq \emptyset$ which implies $G = \{x\}$ and hence $K \subseteq \{x\} \times H$. If $K = \emptyset$, trivially $K \subseteq \{x\} \times H$. So, in any case, $K \subseteq \{x\} \times H$. Since Φ is monotone, $\Phi(K) \subseteq \Phi(\{x\} \times H)$, so $y \in \Phi(\{x\} \times H)$. Then $\langle x, y\rangle \in \Psi(H)$ where H is $\mathscr{D}$-finite and $H \subseteq \mathscr{P}$. Hence Ψ is $\mathscr{D}$-compact.

THEOREM 14.7. *Let **A** be a full production system and suppose, in addition:*

(1) *the union of two $\mathscr{D}$-finite relations is $\mathscr{D}$-finite,*
(2) *the collection of $\mathscr{D}$-finite relations is closed under projections,*
(3) *singletons are $\mathscr{D}$-finite,*
(4) *$\emptyset$ is $\mathscr{D}$-finite,*
(5) *every monotone, $\mathscr{D}$-compact operator is $\mathscr{D}$-metacompact.*

*Let $\mathbf{A}^0$ consist of just the monotone, $\mathscr{D}$-compact operators of **A**. Then $\mathbf{A}^0$ is also a full production system, with the same relations generated as in **A** (using the same notions of basic and $\mathscr{D}$-finite as in **A**).*

PROOF. Since singletons are $\mathscr{D}$-finite, then, as observed earlier, $\mathscr{D}$-metacompactness implies $\mathscr{D}$-compactness. Hence, in **A**, for monotone operators, $\mathscr{D}$-compact $=$ $\mathscr{D}$-metacompact.

Constant operators are trivially monotone. Also, if Φ is constant it is $\mathscr{D}$-compact since, if $x \in \Phi(\mathscr{P})$ then $x \in \Phi(\emptyset)$ and $\emptyset \subseteq \mathscr{P}$ and is $\mathscr{D}$-finite. Thus $\mathbf{A}^0$ contains all the constant operators of **A**. It follows that the same relations are generated in **A** and in $\mathbf{A}^0$. In particular, $\emptyset$, $=_{\mathscr{A}}$ and singletons are generated in $\mathbf{A}^0$.

Projection operators are $\mathscr{D}$-compact: say $x \in P_1^2(\mathscr{R})$; then for some y, $\langle y, x\rangle \in \mathscr{R}$; but $\{\langle y, x\rangle\}$, being a singleton, is $\mathscr{D}$-finite, $x \in P_1^2(\{\langle y, x\rangle\})$ and $\{\langle y, x\rangle\} \subseteq \mathscr{R}$. They are trivially monotone. Hence they are in $\mathbf{A}^0$.

Similarly transpositions and place-adding operators are in $\mathbf{A}^0$.

Suppose $\Phi, \Psi \in \mathbf{A}^0$. Since they are in **A**, $\Phi\Psi \in \mathbf{A}$. $\Phi\Psi$ is easily checked to be monotone and $\mathscr{D}$-metacompact, using that these are true of Φ and Ψ separately. Hence $\Phi\Psi \in \mathbf{A}^0$. (It is here that we use $\mathscr{D}$-compact $=$ $\mathscr{D}$-metacompact.)

Suppose $\Phi, \Psi \in \mathbf{A}^0$. Trivially $\Phi \cap \Psi$ is monotone; we show it is $\mathscr{D}$-compact. Suppose $x \in (\Phi \cap \Psi)(\mathscr{P})$. Then $x \in \Phi(\mathscr{P})$ and $x \in \Psi(\mathscr{P})$.

Since both are $\mathscr{D}$-compact, we have $\boldsymbol{x} \in \Phi(K')$ and $\boldsymbol{x} \in \Psi(K'')$ for some $\mathscr{D}$-finite $K', K'' \subseteq \mathscr{P}$. But $K = K' \cup K''$ is $\mathscr{D}$-finite, and by monotonicity, $\boldsymbol{x} \in \Phi(K)$ and $\boldsymbol{x} \in \Psi(K)$. So $\boldsymbol{x} \in (\Phi \cap \Psi)(K)$ where $K \subseteq \mathscr{P}$. Thus $\Phi \cap \Psi$ is $\mathscr{D}$-compact, hence in $\boldsymbol{A}^0$.

That $\boldsymbol{A}^0$ is closed under $\times$ is a similar argument, while closure under $\cup$ is easier.

Thus $\boldsymbol{A}^0$ is a production system, with the same relations generated as $\boldsymbol{A}$.

Next we show $\boldsymbol{A}^0$ is indexed. We produce a universal operator for the collection of operators in $\boldsymbol{A}^0$ of order $\langle n, m\rangle$.

Since $\boldsymbol{A}$ is an indexed production system, its generated relations constitute an indexed relational system. Let U^{m+1} be a generated $m+2$-place relation which is universal for the generated $m+1$-place relations of $\boldsymbol{A}$. Define a relation S by $S = T^{m+2}_{1,2}(U^{m+1})$. Now, define an operator $\mathscr{U}^{\langle n,m\rangle}$ of order $\langle n, m+1\rangle$ by

$$\mathscr{U}^{\langle n,m\rangle} = S''F[J^n].$$

By Theorem 8.4, $\mathscr{U}^{\langle n,m\rangle}$ is a $\mathscr{D}$-compact, monotone operator of $\boldsymbol{A}$, hence it is in $\boldsymbol{A}^0$. We show it is universal.

Let $\Phi \in \boldsymbol{A}^0$ be of order $\langle n, m\rangle$. Then $\Phi \in \boldsymbol{A}$, is monotone and $\mathscr{D}$-compact. By Theorem 8.4, there is an $m+1$-place relation R, generated in $\boldsymbol{A}$, such that $\Phi = R''F[J^n]$. Let i be a relational index for R, so $R(y, \boldsymbol{x}) \Leftrightarrow U^{m+1}(i, y, \boldsymbol{x}) \Leftrightarrow S(y, i, \boldsymbol{x})$. Then, for allowable $\mathscr{P}$,

$$\begin{aligned}
\boldsymbol{x} \in \Phi(\mathscr{P}) &\Leftrightarrow \boldsymbol{x} \in R''F[J^n](\mathscr{P})\\
&\Leftrightarrow (\exists y)[R(y, \boldsymbol{x}) \wedge y \in F[J^n](\mathscr{P})]\\
&\Leftrightarrow (\exists y)[S(y, i, \boldsymbol{x}) \wedge y \in F[J^n](\mathscr{P})]\\
&\Leftrightarrow \langle i, \boldsymbol{x}\rangle \in S''F[J^n](\mathscr{P})\\
&\Leftrightarrow \langle i, \boldsymbol{x}\rangle \in \mathscr{U}^{\langle n,m\rangle}(\mathscr{P}).
\end{aligned}$$

Thus $\mathscr{U}^{\langle n,m\rangle}$ is universal, and $\boldsymbol{A}^0$ is indexed.

Next we show the output place-fixing property holds in $\boldsymbol{A}^0$, using the same notion of basic as in $\boldsymbol{A}$.

Let $\Phi \in \boldsymbol{A}^0$ be of order $\langle k, q+n\rangle$. Then Φ is monotone and $\mathscr{D}$-compact, so there is a generated $q+n+1$-place relation R such that $\Phi = R''F[J^k]$. Now, the iteration property holds for the generated relations of $\boldsymbol{A}$, so there is a basic function $f: \mathscr{A}^q \rightarrow \mathscr{A}$ such that

$$R(y, x_1, \ldots, x_q, y_1, \ldots, y_n) \Leftrightarrow U^{n+1}(f(x_1, \ldots, x_q), y, y_1, \ldots, y_n).$$

It then follows easily that, using the indexing of operators defined above,

$$\langle x_1,\ldots,x_q,y_1,\ldots,y_n\rangle\in\Phi(\mathscr{P})$$
$$\Leftrightarrow\ \langle f(x_1,\ldots,x_q),y_1,\ldots,y_n\rangle\in\mathscr{U}^{\langle k,n\rangle}(\mathscr{P}).$$

Thus we have the output place-fixing property in $\mathbf{A}^0$.

Next we turn to the input place-fixing property. Let $\Phi\in\mathbf{A}^0$ be of order $\langle q+k,n\rangle$. By Lemma 14.6, there is an operator $\Psi\in\mathbf{A}$ of order $\langle k,q+n\rangle$ which is monotone and $\mathscr{D}$-compact, hence in $\mathbf{A}^0$, such that

$$\boldsymbol{y}\in\Phi(\{\boldsymbol{x}\}\times\mathscr{P})\ \Leftrightarrow\ \langle\boldsymbol{x},\boldsymbol{y}\rangle\in\Psi(\mathscr{P}).$$

By the *output* place-fixing property applied to Ψ in $\mathbf{A}^0$, there is a basic $f:\mathscr{A}^q\to\mathscr{A}$ such that

$$\langle\boldsymbol{x},\boldsymbol{y}\rangle\in\Psi(\mathscr{P})\ \Leftrightarrow\ \langle f(\boldsymbol{x}),\boldsymbol{y}\rangle\in\mathscr{U}^{\langle k,n\rangle}(\mathscr{P})$$

and this immediately gives us the input place-fixing property for Φ.

Finally we turn to the $\mathscr{D}$-finiteness conditions. $\mathbf{A}^0$ has an effective pairing function since $\mathbf{A}$ does, and the two have the same generated relations. Similarly, the relation: y is a $\mathscr{D}$-finite code and $x\in D_y$, being generated in $\mathbf{A}$, is generated in $\mathbf{A}^0$. And the operator $F\in\mathbf{A}$ such that $F(\mathscr{P})=\{x\mid D_x\subseteq\mathscr{P}\}$ is easily seen to be monotone and $\mathscr{D}$-compact, hence it is in $\mathbf{A}^0$.

Thus $\mathbf{A}^0$ is a full production system. This ends the proof.

Finally, we introduce some terminology, then we combine the effects of several of the preceding theorems.

DEFINITION. Let $\mathbf{A}$ be a production system with a notion of $\mathscr{D}$-finiteness and an effective pairing function. Let $\mathfrak{Q}$ be the collection of semi-hyperregular relations of $\mathbf{A}$. By the *compact core* of $\mathbf{A}$ we mean $(\mathbf{A}\restriction\mathfrak{Q})^0$, using the notation of Theorems 14.1 and 14.7.

THEOREM 14.8. *Let* $\mathbf{A}$ *be a full production system, and suppose*:

(1) *every generated relation satisfies weak replacement,*

(2) *the $\mathscr{D}$-finite union of $\mathscr{D}$-finite relations is $\mathscr{D}$-finite,*

(3) *the $\mathscr{D}$-finite relations are closed under projections,*

(4) *every finite relation is also $\mathscr{D}$-finite.*

Then, the compact core of $\mathbf{A}$ *is also a full production system with the same relations generated as in* $\mathbf{A}$*, in which allowable = semi-hyperregular, and in which every operator is monotone and $\mathscr{D}$-metacompact.*

In addition, if the $\mathscr{D}$-finite coding of $\mathbf{A}$ *is canonical (not just positive canonical) the same is true of the compact core. And if the $\mathscr{D}$-finite approximability property holds in* $\mathbf{A}$*, it also holds in the compact core.*

PROOF. The hypotheses here allow the application of Theorems 14.4, 14.5 and 14.7, which give most of the conclusion. The other $\mathscr{D}$-finiteness conditions mentioned only refer to generated and to $\mathscr{D}$-finiteness, neither of which changes in passing from $\boldsymbol{A}$ to its compact core. This observation yields the rest of the conclusion.

EXAMPLE. ω-rec($\mathfrak{S}(\mathsf{N})$) = hyperarithmetic theory, satisfies the hypotheses of Theorem 14.8, hence its compact core is a full production system. This is useful since then we have Rogers' form available. But, using Theorem 13.5, there are sets which are not semi-hyperregular in ω-rec($\mathfrak{S}(\mathsf{N})$), hence which are not allowable inputs in the compact core. Thus it is a *partial* production system.

15. The least fixed point theorem, again

In Section 11 we proved a least fixed point theorem in an axiomatic setting. Now we do it again. The present proof involves different more restrictive hypotheses, but is, correspondingly, more elegant. The material is based on work of Scott and Park, see Scott [1976].

The proof below only applies to operators of order $\langle 1, 1\rangle$. At the end of the section we discuss how to reduce the $\langle n, n\rangle$ case to the $\langle 1, 1\rangle$ case, but the assumption of an *onto* pairing function seems to be essential.

Suppose $\boldsymbol{A}$ is a production system with a notion of $\mathscr{D}$-finiteness for which there is a positive canonical coding. Then, as in Section 7, certain operators Φ of order $\langle 1, 1\rangle$ can be expressed in Rogers' form, $\Phi = \mathscr{R}''F$, where $\mathscr{R}$ is a generated, binary relation.

If we also have an effective pairing function J, then every generated binary relation can be written in the form $[J^2]^{-1}(\mathscr{P})$ where $\mathscr{P}$ is a generated *set*. For, suppose $\mathscr{R} \in [\mathscr{A}]^2$ is generated. Let $\mathscr{P} = [J^2](\mathscr{R})$. Then $\mathscr{P}$ is also generated, is a set, and $[J^2]^{-1}(\mathscr{P}) = [J^2]^{-1}[J^2](\mathscr{R}) = \mathscr{R}$ by Proposition 8.2.

DEFINITION. Let $\boldsymbol{A}$ be a production system, with a notion of $\mathscr{D}$-finiteness for which there is a positive canonical coding, and an effective pairing function J.

For each generated set $\mathscr{P}$, by $\Phi_{\mathscr{P}}$ we mean the operator in $\boldsymbol{A}$ of order $\langle 1, 1\rangle$ given by $\Phi_{\mathscr{P}} = ([J^2]^{-1}(\mathscr{P}))''F$.

PROPOSITION 15.1 (Existence of a diagonal operator). *Let $\boldsymbol{A}$ be a production system with a notion of $\mathscr{D}$-finiteness for which there is a positive canonical coding, and with an effective pairing function J.*

There is an operator $\Psi \in \mathbf{A}$ of order $\langle 1, 1\rangle$ such that, for generated $\mathcal{P} \in [\mathcal{A}]^1$, $\Psi(\mathcal{P}) = \Phi_{\mathcal{P}}(\mathcal{P})$.

PROOF. Let

$$\Psi = P_1^2(T_{1,2}^2 A^1 F \cap [J^2]^{-1}).$$

Then, if $\mathcal{P} \in [\mathcal{A}]^1$ is generated,

$$\begin{aligned}
& x \in \Psi(\mathcal{P}) \\
\Leftrightarrow\ & x \in P_1^2(T_{1,2}^2 A^1 F \cap [J^2]^{-1})(\mathcal{P}) \\
\Leftrightarrow\ & \text{for some } y, \langle y, x\rangle \in T_{1,2}^2 A^1 F(\mathcal{P}) \text{ and } \langle y, x\rangle \in [J^2]^{-1}(\mathcal{P}) \\
\Leftrightarrow\ & \text{for some } y, \langle x, y\rangle \in A^1 F(\mathcal{P}) \text{ and } \langle y, x\rangle \in [J^2]^{-1}(\mathcal{P}) \\
\Leftrightarrow\ & \text{for some } y, y \in F(\mathcal{P}) \text{ and } \langle y, x\rangle \in [J^2]^{-1}(\mathcal{P}) \\
\Leftrightarrow\ & x \in ([J^2]^{-1}(\mathcal{P}))''(F(\mathcal{P})) \\
\Leftrightarrow\ & x \in \Phi_{\mathcal{P}}(\mathcal{P}).
\end{aligned}$$

REMARKS. Let $\mathbf{A}$ be a production system with a notion of $\mathcal{D}$-finiteness for which there is positive canonical coding, and with an effective pairing function J.

Let Φ be an operator of order $\langle 1, 1\rangle$ and suppose the composition $\Phi\Psi$ turns out to be $\Phi_{\mathcal{P}}$ for some generated set $\mathcal{P}$ (here Ψ is the diagonal operator of Proposition 15.1). Then Φ has a generated fixed point of a rather interesting form, namely $\Phi_{\mathcal{P}}(\mathcal{P})$. This, being an operator applied to a generated set, is generated. And it is a fixed point since

$$\begin{aligned}
\Phi(\Phi_{\mathcal{P}}(\mathcal{P})) &= \Phi\Psi(\mathcal{P}) && \text{(general behavior of } \Psi) \\
&= \Phi_{\mathcal{P}}(\mathcal{P}) && \text{(since } \Phi\Psi = \Phi_{\mathcal{P}}).
\end{aligned}$$

This construction of a generated fixed point for Φ is based on an argument in Combinatory Logic. Its application to recursion theory is due to Scott. See Scott [1976].

Now, one reason this particular fixed point for Φ is of interest is that Park has shown that under reasonable assumptions it will be the *least* fixed point of Φ. The following argument is essentially his. See Scott [1976], pp. 569–570 and p. 577. (Park, himself, never published this argument.)

THEOREM 15.2 (Least fixed point theorem). *Let $\mathbf{A}$ be an indexed production system having the input place-fixing property, with a notion of $\mathcal{D}$-finiteness for which there is a positive canonical coding, and with an effective pairing* *function, J. Suppose also that every operator in* **A** *is monotone and* $\mathscr{D}$-*compact (see Theorem* 14.8).

Suppose, in addition, there is a well-founded, transitive relation $<$ *on* $\mathscr{A}$ *(it need not be generated) meeting the two conditions:*

(1) $x < J(x, y)$,

(2) *if* $x \in D_y$ *then* $x < y$.

Let $\Phi \in \mathbf{A}$ *be of order* $\langle 1,1 \rangle$. *Then* Φ *has a generated least fixed point.*

PROOF. $\Phi \in \mathbf{A}$ is order $\langle 1,1 \rangle$. We produce a generated least fixed point for Φ.

Let Ψ be the diagonal operator of Proposition 15.1. $\Psi \in \mathbf{A}$, hence $\Phi\Psi \in \mathbf{A}$. Since we are supposing every operator in $\mathbf{A}$ is monotone and $\mathscr{D}$-compact, this applies to $\Phi\Psi$. Then by Theorem 7.3, $\Phi\Psi$ can be put in Rogers' form, hence by our remarks at the beginning of this section, $\Phi\Psi = \Phi_{\mathscr{P}}$ for some generated set $\mathscr{P}$.

By the remarks preceding this theorem, $\Phi_{\mathscr{P}}(\mathscr{P})$ is a generated fixed point for Φ. What we now show is that it is least among the allowable fixed points.

Let $\mathscr{B}$ be any allowable fixed point for Φ. We must show $\Phi_{\mathscr{P}}(\mathscr{P}) \subseteq \mathscr{B}$ or, equivalently,

$$\Psi(\mathscr{P}) \subseteq \mathscr{B}. \qquad (*)$$

Suppose we could show

$$D_y \subseteq \mathscr{P} \Rightarrow \Psi(D_y) \subseteq \mathscr{B}. \qquad (**)$$

This would suffice to establish (*), for, if $x \in \Psi(\mathscr{P})$ then, since Ψ is $\mathscr{D}$-compact, $x \in \Psi(D_y)$ for some $D_y \subseteq \mathscr{P}$. Then by (**), $\Psi(D_y) \subseteq \mathscr{B}$, hence $x \in \mathscr{B}$. Thus we show (**).

We suppose (**) is not true. Let k be a minimal member of $\mathscr{A}$, in the $<$ ordering, such that (**) fails, that is,

$$D_k \subseteq \mathscr{P} \quad \text{but not} \quad \Psi(D_k) \subseteq \mathscr{B}.$$

Choose some y_0 such that $y_0 \in \Psi(D_k)$ but $y_0 \notin \mathscr{B}$. Since $y_0 \in \Psi(D_k)$ and Ψ is a diagonal operator, then $y_0 \in \Phi_{D_k}(D_k)$ or

$$y_0 \in ([J^2]^{-1}(D_k))''F(D_k).$$

That is, there is some $D_t \subseteq D_k$ with $J(t, y_0) \in D_k$. Now, by our assumptions on $<$, this implies

$$J(t, y_0) < k.$$

But also $t < J(t, y_0)$. Hence $t < k$. Since k was minimal, (**) must hold for t, that is,

$$D_t \subseteq \mathcal{P} \Rightarrow \Psi(D_t) \subseteq \mathcal{B}.$$

But $D_t \subseteq D_k$ and $D_k \subseteq \mathcal{P}$ hence $D_t \subseteq \mathcal{P}$, and thus $\Psi(D_t) \subseteq \mathcal{B}$. Φ is monotone, hence

$$\Phi\Psi(D_t) \subseteq \Phi(\mathcal{B}).$$

Also, $\Phi\Psi = \Phi_{\mathcal{P}}$, and $\mathcal{B}$ is a fixed point for Φ, hence

$$\Phi_{\mathcal{P}}(D_t) \subseteq \mathcal{B}.$$

Now, $J(t, y_0) \in D_k \subseteq \mathcal{P}$ so $J(t, y_0) \in \mathcal{P}$. And $D_t \subseteq D_t$ hence $t \in F(D_t)$. Thus $y_0 \in ([J^2]^{-1}(\mathcal{P}))''F(D_t)$ or $y_0 \in \Phi_{\mathcal{P}}(D_t)$; and thus $y_0 \in \mathcal{B}$, a contradiction, since $y_0 \notin \mathcal{B}$. Thus (**) always holds, and the proof is complete.

EXAMPLES. I. Ordinary recursion theory, rec($\mathfrak{S}$(ℕ)), where we can take $<$ to be the usual $<$ relation. Indeed, the pairing function and finite coding of Rogers [1967] will do very nicely. This, after all, is the grand-daddy example.

II. rec($\mathfrak{S}(\mathrm{L}_\omega)$), where we can take $<$ to be the relation: is of lower rank than. For a pairing function, use the standard Kuratowski one. For a finite coding, let each finite subset of L_ω code itself.

III. rec($\mathfrak{S}(a_1, \ldots, a_n)$), where we can take $<$ to be the relation: "is shorter than". The pairing function and finite coding constructed in Chapter 6 are suitable.

We will give morc cxamples in the next chapter.

Finally we say how to extend the conclusion of Theorem 15.2 from operators of order $\langle 1, 1\rangle$ to operators of order $\langle n, n\rangle$. The following suffices; we leave details to the reader.

PROPOSITION 15.3. *Let* **A** *be a production system with a notion of $\mathcal{D}$-finiteness for which there is a $\mathcal{D}$-finite coding, and with an effective pairing function* which is onto. *Then*

(1) $[J^n][J^n]^{-1} = I^n$.

*Let $\Phi \in$ **A** be of order $\langle n, n\rangle$ and define Φ^* by $\Phi^* = [J^n]\Phi[J^n]^{-1}$, so Φ^* is of order $\langle 1, 1\rangle$. Then*

(2) *If Φ is monotone, so is Φ^*.*

(3) *If Φ is $\mathcal{D}$-compact, so is Φ^*.*

(4) *$\mathcal{P}$ is a fixed point of Φ^* iff $[J^n]^{-1}(\mathcal{P})$ is a fixed point of Φ.*

(5) *If $\mathcal{P}$ is the least fixed point of Φ^* then $[J^n]^{-1}(\mathcal{P})$ is the least fixed point of Φ.*

PROOF. Left to the reader.

CHAPTER NINE

ADMISSIBLE SET RECURSION THEORIES

1. Introduction

We return again to the informal description of our subject in terms of putting things into boxes. We have, so far, considered two rather extreme cases: (1) you are the one manipulating the boxes (Ch. 1, §1), and (2) some being that can live transfinitely long is doing it (Ch. 1, §8). Case (1) corresponds to recursion theory, case (2) to ω-recursion = hyperelementary theory. Now we consider intermediate cases. Suppose the boxes are under the management of a being whose lifespan is infinite, but not unbounded. He can, for example, count off infinitely many ordinals, but not all of them. The notion of an *admissible set* is what is needed here.

Consider the collection of all hereditarily finite sets (that its, sets which are finite, whose members are finite, with members of members finite, and so on). In a sense, you live within the collection of hereditarily finite sets. And the being that "does" ω-recursion theory lives within the universe of all sets. Well, an admissible set may be thought of as an intermediate collection, with enough structure for some being to "live" in it. The collection of hereditarily finite sets is the simplest admissible set, indeed, a part of all others; the class of all sets is the largest admissible "set". But there are many others in between.

Now imagine that it is not you putting items into boxes, you who live among the hereditarily finite sets, imagine instead that it is a being who lives within some other admissible set. What he can do properly includes what you could do, since your admissible set is part of his, but exactly what his abilities are depends on his admissible set.

Now, having indulged our fantasy, let us begin again, somewhat more soberly.

It is standard practice in set theory to replace *properties* by *sets* whenever possible, the set of things having the property in question. One cannot do this with the property of finiteness, however, since the collection of all finite sets constitutes a proper class. But suppose we choose a structure,

$\mathfrak{A} = \langle \mathscr{A}; \mathscr{R}_1, \ldots, \mathscr{R}_k \rangle$, and keep it fixed for the time being. The property of being finite, restricted to those sets we need to work with in "discussing" $\mathfrak{A}$, does determine a set. In fact, it may be described precisely as follows. Let

$$\mathscr{A}^0 = \mathscr{A},$$

$$\mathscr{A}^{n+1} = \text{the collection of all "finite" subsets of } \mathscr{A}^n \cup \mathscr{A}.$$

[The quotation marks are explained in Ch. 6, §2 under topic 3.] Let

$$\mathscr{A}^\omega = \bigcup_{n \in \omega} \mathscr{A}^n.$$

The members of $\mathscr{A}^\omega$ are the *hereditarily finite sets* built up from members of $\mathscr{A}$ as *urelements.* And, in our dealings with $\mathfrak{A}$, we can generally replace any talk about *being finite* with talk about *being a member of* $\mathscr{A}^\omega$. Thus, in working with $\mathfrak{A}$, it is reasonable to consider the structure $\langle \mathscr{A}; \mathscr{R}_1, \ldots, \mathscr{R}_k, \mathscr{A}^\omega, \in \rangle = \langle \mathfrak{A}, \mathscr{A}^\omega, \in \rangle$ [where $\in$ is the membership relation restricted to $\mathscr{A}^\omega$]. This is essentially the structure called $\mathfrak{A}^s$ in Chapter 6.

Elementary formal system derivations are finite objects; how do they fit in with the above discussion ? Well, a derivation is simply a word; a word is a finite sequence of symbols; but we never said what our formal symbols actually were. Suppose we say they are certain finite sets (are *coded* by certain finite sets, if you prefer). Then words are really functions from the finite ordinals to the collection consisting of these symbols and members of $\mathscr{A}$. In standard set-theoretic fashion, we may think of such functions as single-valued *sets* of ordered pairs. The finite ordinals are all members of $\mathscr{A}^\omega$, and it follows that elementary formal system derivations over $\mathfrak{A}$ simply are particular *members* of the collection $\mathscr{A}^\omega$ of the structure $\langle \mathfrak{A}, \mathscr{A}^\omega, \in \rangle$.

In fact, every informal use we made of *finite* in discussing rec($\mathfrak{A}$) in Chapter 1 could be replaced by a formal counterpart involving membership in $\mathscr{A}^\omega$. Then, a plausible approach to generalizing the role of finiteness in recursion theory is to try replacing the structure $\langle \mathfrak{A}, \mathscr{A}^\omega, \in \rangle$ by other, "similar" structures $\langle \mathfrak{A}, S, \in \rangle$ and see where that leads. More precisely, one should investigate which features of $\mathscr{A}^\omega$ play an essential role in developing the theory of rec($\mathfrak{A}$) using the structue $\langle \mathfrak{A}, \mathscr{A}^\omega, \in \rangle$ as our embodiment of finiteness, and see if there are other structures $\langle \mathfrak{A}, S, \in \rangle$ in which S also has these features.

One of the features of $\mathscr{A}^\omega$ is that it is in "layers", $\mathscr{A}^0$, $\mathscr{A}^1$, $\mathscr{A}^2$, etc. We might decide to keep this feature in S, but if we are to have a generalization, we might want to allow transfinite ordinal "levels".

Let us set up a first-order language to discuss $\langle \mathfrak{A}, S, \in \rangle$. We should have predicate symbols to represent $\mathcal{R}_1, \ldots, \mathcal{R}_k$, $\in$, $=$, and "being a set" and "being an urelement." We interpret formulas over $\langle \mathfrak{A}, S, \in \rangle$ in the obvious way. Now, consider formulas in which all the quantifiers are bounded, $(\forall x \in y)$ or $(\exists x \in y)$. Since the quantifiers are bounded by variables that range over S, and members of S are "generalized finite", we should be able to check the truth or falsity of any instances of such formulas by some "generalized finite" procedure. We needn't go into what this might mean now, but it should be clear that such formulas probably will play a key role in characterizing S.

Consider $\langle \mathfrak{A}, \mathcal{A}^{\omega}, \in \rangle$ again. Suppose $\mathcal{R}(x, y)$ is a relation on $\mathcal{A}^{\omega}$. If we restrict the domain to a finite collection a, it is not hard to see that a corresponding restriction of the range can be made. That is, if $(\forall x \in a)(\exists y)\mathcal{R}(x, y)$, then there is a finite b such that $(\forall x \in a)(\exists y \in b)\mathcal{R}(x, y)$. In fact, this is often used in developing recursion theory; we saw a variant in our axiomatic development in previous chapters, in the requirement of weak replacement. Now, is this a feature we want to have in a generalization $\langle \mathfrak{A}, S, \in \rangle$? One can make the case that it is much too strong, and would greatly restrict the range of possible generalizations. But investigation and experimentation has shown that many of the uses of this principle are actually applications of a very weak form of it, namely for those relations $\mathcal{R}$ definable by a formula with all quantifiers bounded. And this weak form seems much less open to objection as a requirement for "generalized finiteness" structures.

After much experimentation by several people, certain features of $\langle \mathfrak{A}, \mathcal{A}^{\omega}, \in \rangle$, including the ones discussed above, were selected as desirable for an interesting and useful generalization. Structures $\langle \mathfrak{A}, S, \in \rangle$ having these features are called *admissible sets.* We will define them properly in the next section. There is a large and rapidly growing literature about them. They have proved to be a successful generalization, and provide a common ground for recursion theoretic and set theoretic investigation.

In most treatments, admissible sets directly provide generalizations of both finiteness and of recursive enumerability. Thus, if $\mathbb{A}$ is an admissible set, $\mathbb{A}$-*finite* is taken to mean *member of* $\mathbb{A}$ while $\mathbb{A}$-*r.e.* is taken to mean *definable over* $\mathbb{A}$ *by a* Σ *formula.* We have chosen to separate these two roles. We take admissible sets as directly providing a generalization of finiteness only. We get our generalization of recursive enumerability by using a suitable generalization of elementary formal system (as might be expected after seeing the earlier chapters of the book). Still, it is shown that, for an interesting class of admissible sets, our generalization coincides with the customary one.

Having elementary formal systems around suggests a reasonable generalization of enumeration operator. This, in turn, provides us with many more production systems which satisfy the various assumptions of earlier chapters.

Admissible sets, as described above, are better termed admissible sets *with urelements.* Barwise [1975] is *the* reference on them; we sketch and summarize, but that book should be consulted for depth and details. Barwise [1974] is an earlier version of a portion of the material of Barwise [1975].

Originally, admissible sets were without urelements, and arose from the work of Platek [1966] and Kripke [1964]. And in those works much centered around ordinal numbers. An ordinal α is called *admissible* if the αth constructible set, L_α, is an admissible set (without urelements). But the admissible ordinals can be characterized directly, without reference to admissible sets, and the generalization of ordinary recursion theory, on the natural numbers, to admissible ordinals (called α-recursion theory) has turned out to be quite striking. Simpson [1974] and Simpson [1978] are two survey papers on this subject which reveal its beauty and coincidentally indicate how fast it is developing. The last few sections of this chapter discuss some relationships between our development of α-recursion theory and the one in the two Simpson papers.

Finally we note that this is largely a summary, descriptive chapter. The intent is to show how an admissible set generalization of the work in earlier chapters can be developed. A reader of Barwise [1975] should have no trouble supplying proofs of our assertions in this chapter, but a full rigorous development here would make for a book twice this size and half as interesting.

2. Admissible sets

In this section we give, in summary fashion, the definition and basic properties of admissible sets with urelements. The definition we use is a slightly restricted version of that in Barwise [1975], which should be read for proofs, discussions, beauty and truth.

Let $\mathfrak{A} = \langle \mathscr{A}; \mathscr{R}_1, \ldots, \mathscr{R}_k \rangle$ be a structure, fixed for the time being and, as usual, let $R_1, \ldots, R_k$ be predicate symbols assigned to $\mathscr{R}_1, \ldots, \mathscr{R}_k$. We have two additional predicate symbols, $\in$ and $=$, intended to represent membership and equality respectively.

We suppose available *three sorts of variables*: $p, q, p_1, \ldots,$ intended to range over members of $\mathscr{A}$ (urelements); $a, b, c, d, f, r, a_1, \ldots$, intended to range over sets built up from $\mathscr{A}$ (for short, sets); $x, y, z, \ldots$, intended to range over both sets and urelements.

Let $L(\mathfrak{A})$ be the three-sorted first order language built up out of the above materials.

The collection of Δ_0 *formulas* of $L(\mathfrak{A})$ is the smallest collection Y containing the atomic formulas (involving $R_1, \ldots, R_k$, $\in$ and $=$) and closed under:

(1) If φ is in Y so is $\sim\varphi$.

(2) If φ and Ψ are in Y so are $(\varphi \wedge \Psi)$ and $(\varphi \vee \Psi)$.

(3) If φ is in Y so are $(\forall u \in v)\varphi$ and $(\exists y \in v)\varphi$ for all variables u and v.

Notice this definition allows negation, unlike our use of Δ_0 in Chapter 6. In effect, we assume the *given relations and their complements*, are available for use.

The collection of Σ *formulas* is the smallest class Y containing the Δ_0 formulas and also meeting the condition

(4) If φ is in Y so is $(\exists u)\varphi$ for all variables u.

The collection of Π *formulas* is the smallest class Y containing the Δ_0 formulas and also meeting the condition

(5) If φ is in Y so is $(\forall u)\varphi$ for all variables u.

The theory KPU (Kripke–Platek set theory with urelements) consists of the universal closures of the following formulas (note that the formulas are in $L(\mathfrak{A})$ and hence depend on the structure $\mathfrak{A}$ chosen, or at least its type).

extensionality:

$$(\forall x)(x \in a \equiv x \in b) \supset a = b.$$

foundation:

$$(\exists x)\varphi(x) \supset (\exists x)[\varphi(x) \wedge (\forall y \in x) \sim \varphi(y)]$$

for all formulas $\varphi(x)$ in which y does not occur free.

pair:

$$(\exists a)(x \in a \wedge y \in a).$$

union:

$$(\exists b)(\forall y \in a)(\forall x \in y)(x \in b).$$

Δ_0 *separation*:

$$(\exists b)(\forall x)[x \in b \equiv (x \in a \wedge \varphi(x))]$$

for all Δ_0 formulas φ in which b does not occur free.

Δ_0 *collection*:

$$(\forall x \in a)(\exists y)\varphi(x, y) \supset (\exists b)(\forall x \in a)(\exists y \in b)\varphi(x, y)$$

for all Δ_0 formulas φ in which b does not occur free.

Next, we restrict our consideration to certain kinds of models for the above axioms. For each ordinal α a set $V_{\mathscr{A}}(\alpha)$ is defined as follows:

$$V_{\mathscr{A}}(0) = \emptyset,$$

$$V_{\mathscr{A}}(\alpha + 1) = \text{power set of } (\mathscr{A} \cup V_{\mathscr{A}}(\alpha)),$$

$$V_{\mathscr{A}}(\lambda) = \bigcup_{\alpha<\lambda} V_{\mathscr{A}}(\alpha) \quad \text{for limit } \lambda$$

and

$$V_{\mathscr{A}} = \bigcup_{\alpha} V_{\mathscr{A}}(\alpha).$$

By $\in_{\mathscr{A}}$ is meant the membership relation on $V_{\mathscr{A}}$.

Now, consider the following structure:

$$\langle \mathscr{A}; \mathscr{R}_1, \ldots, \mathscr{R}_k, S, \in \rangle = \langle \mathfrak{A}, S, \in \rangle$$

where $\mathscr{A} \cup S$ is transitive in $V_{\mathscr{A}}$ and $\in$ is $\in_{\mathscr{A}}$ restricted to $\mathscr{A} \cup S$. Suppose this structure turns out to be a model for KPU when we interpret urelement variables as ranging over $\mathscr{A}$, set variables as ranging over S and "mixed" variables as ranging over $\mathscr{A} \cup S$. Then the structure is called an *admissible set* (more properly, structure) *over* $\mathfrak{A}$. Admissible sets are denoted by $\mathbb{A}$, $\mathbb{B}$, $\mathbb{C}$, etc. If it is necessary to indicate the underlying structure, we write $\mathbb{A}_{\mathfrak{A}}$, etc.

EXAMPLES. Let TC(a) be the *transitive closure* of a.

(1) Let $\mathrm{HF}_{\mathscr{A}} = \{a \in V_{\mathscr{A}} \mid \mathrm{TC}(a) \text{ is finite}\}$. Let $\mathfrak{A}$ be a structure with domain $\mathscr{A}$, and set $\mathrm{HF}_{\mathfrak{A}} = \langle \mathfrak{A}, \mathrm{HF}_{\mathscr{A}}, \in \rangle$. $\mathrm{HF}_{\mathfrak{A}}$ is an admissible set. The "HF" stands for hereditarily finite. See Barwise [1975], p. 46. It should be clear that this is essentially what we called $\mathfrak{A}^s$ in Ch. 6, §2.

(2) Let κ be an infinite cardinal. Let $\mathrm{H}(\kappa)_{\mathscr{A}} = \{a \in V_{\mathscr{A}} \mid \mathrm{TC}(a)$ has cardinality $< \kappa\}$. Let $\mathfrak{A}$ be a structure with domain $\mathscr{A}$, and set $\mathrm{H}(\kappa)_{\mathfrak{A}} = \langle \mathfrak{A}, \mathrm{H}(\kappa)_{\mathscr{A}}, \in \rangle$. $\mathrm{H}(\kappa)_{\mathfrak{A}}$ is an admissible set. The "H(κ)" stands for hereditary cardinality less than κ. Note that $\mathrm{HF}_{\mathfrak{A}} = \mathrm{H}(\omega)_{\mathfrak{A}}$. See Barwise [1975], p. 52.

It is important to observe that the domains of $\mathrm{HF}_{\mathfrak{A}}$ and of $\mathrm{H}(\kappa)_{\mathfrak{A}}$ depend only on $\mathscr{A}$, and not on the given relations of $\mathfrak{A}$. This will play a role later on. It is, however, far from being a feature of all admissible sets.

We conclude with a list of basic theorems of KPU, and hence of true statements about admissible sets.

(1) There is a unique set $\emptyset$ with no elements.

(2) Given sets a and b, the following are sets: $\bigcup a$, $a \cup b$, $a \cap b$, $a \times b$, $\mathrm{TC}(a)$ (transitive closure).

(3) Σ reflection: For all Σ formulas φ, $\varphi \equiv (\exists a)\varphi^{(a)}$ [where $\varphi^{(a)}$ is φ with all its unbounded quantifiers restricted to a.]

(4) Σ collection: If φ is a Σ formula and if $(\forall x \in a)(\exists y)\varphi(x, y)$, then there is a set b such that $(\forall x \in a)(\exists y \in b)\varphi(x, y)$ and $(\forall y \in b)(\exists x \in a)\varphi(x, y)$.

(5) Δ separation: For any Σ formula $\varphi(x)$ and any Π formula $\Psi(x)$, if for all $x \in a$, $\varphi(x) \equiv \Psi(x)$, then there is a set $b = \{x \in a \mid \varphi(x)\}$.

(6) Σ replacement: If $\varphi(x, y)$ is a Σ formula and if $(\forall x \in a)(\exists! y)\varphi(x, y)$ then there is a function f with domain a such that $(\forall x \in a)\varphi(x, f(x))$.

(7) Strong Σ replacement: If $\varphi(x, y)$ is a Σ formula and if $(\forall x \in a)(\exists y)\varphi(x, y)$ then there is a function f with domain a such that $(\forall x \in a) f(a) \neq \emptyset$ and $(\forall x \in a)(\forall y \in f(x))\varphi(x, y)$.

(8) Proof by induction over TC: For any formula $\varphi(x)$, if, for each x, $(\forall y \in \mathrm{TC}(x))\varphi(y)$ implies $\varphi(x)$, then $(\forall x)\varphi(x)$.

3. Terminology

Let $\mathfrak{A} = \langle \mathscr{A}; \mathscr{R}_1, \ldots, \mathscr{R}_k \rangle$, and let $\mathbb{A}_{\mathfrak{A}} = \langle \mathfrak{A}, S, \in \rangle$ be admissible over $\mathfrak{A}$. The members of $\mathscr{A}$ are the ***urelements*** of $\mathbb{A}_{\mathfrak{A}}$; the members of S are the ***sets*** of $\mathbb{A}_{\mathfrak{A}}$.

An object x is said to be *in* $\mathbb{A}_{\mathfrak{A}}$ is $x \in \mathscr{A} \cup S$. An n-place relation $\mathscr{R}$ on $\mathbb{A}_{\mathfrak{A}}$ is $\mathbb{A}$-*finite* if $\mathscr{R}$ is in $\mathbb{A}_{\mathfrak{A}}$. $\mathscr{R}$ is Σ *on* $\mathbb{A}_{\mathfrak{A}}$ if there is a Σ formula φ, possibly having constants $y_1, \ldots, y_k$ from $\mathbb{A}_{\mathfrak{A}}$, such that $\mathscr{R}(x_1, \ldots, x_n) \Leftrightarrow \varphi(x_1, \ldots, x_n)$ is true in the structure $\mathbb{A}_{\mathfrak{A}}$. Similarly for a relation $\mathscr{R}$ being Π on $\mathbb{A}_{\mathfrak{A}}$. Also $\mathscr{R}$ is Δ *on* $\mathbb{A}_{\mathfrak{A}}$ if $\mathscr{R}$ is both Σ and Π on $\mathbb{A}_{\mathfrak{A}}$. A function F *on* $\mathbb{A}_{\mathfrak{A}}$ is a function with domain a subset of $(\mathscr{A} \cup S)^n$ for some n, and range a subset of $\mathscr{A} \cup S$. F is Σ *on* $\mathbb{A}_{\mathfrak{A}}$ if, considered as a relation, it is Σ on $\mathbb{A}_{\mathfrak{A}}$. Similarly for F being Π or Δ on $\mathbb{A}_{\mathfrak{A}}$:

We write $\mathrm{Ord}(\mathbb{A}_{\mathfrak{A}})$ for the collection of ordinals in $\mathbb{A}_{\mathfrak{A}}$. Since admissible sets are transitive, if an ordinal is in $\mathbb{A}_{\mathfrak{A}}$, so is every smaller ordinal. Ordinals are *sets*, that is, members of S. The property of being an ordinal is

Δ. We use $\alpha, \beta, \gamma, \ldots$ to range over ordinals. Note that $\mathrm{Ord}(\mathbb{A}_{\mathfrak{A}})$, though itself an ordinal, is not *in* $\mathbb{A}_{\mathfrak{A}}$.

We conclude by stating two more very powerful principles of proof for admissible sets.

(9) Definition by Σ recursion: Let G be an $n+2$-ary Σ function. There is an $n+1$-ary Σ function F so that

$$F(x_1, \ldots, x_n, y) = G(x_1, \ldots, x_n, y, \{\langle z, F(x_1, \ldots, x_n, z)\rangle \mid z \in \mathrm{TC}(y)\}).$$

(10) Δ relations defined by recursion: Let P and Q be Δ relations of $n+1$ and $n+2$ places respectively. There is a Δ relation R so that

$$R(x_1, \ldots, x_n, p) \Leftrightarrow P(x_1, \ldots, x_n, p),$$

$$R(x_1, \ldots, x_n, a) \Leftrightarrow Q(x_1, \ldots, x_n, a, \{b \in \mathrm{TC}(a) \mid R(x_1, \ldots, x_n, b)\}).$$

4. Elementary formal systems over admissible sets

In this section we define our notion of recursive enumerability with respect to an admissible set $\mathbb{A}$. Now, we want to allow derivations which are actually infinite, though they will be required to be finite in the sense of the admissible set $\mathbb{A}$. Since our usual notion of elementary formal system can not produce infinite derivations, something must be added. The ω-rule is too much to add; in a sense, it gives us no control on what is happening. Instead we add machinery which will allow us to take *limits* of certain chains, but only of those chains which are finite in the sense of the admissible set $\mathbb{A}$.

DEFINITION. Let $\mathbb{A}$ be an admissible set and let $<$ be a relation on $\mathbb{A}$. We call $<$ a *usable partial ordering on* $\mathbb{A}$ if

(1) $<$ is a partial ordering of the sets of $\mathbb{A}$,

(2) $<$ is Δ on $\mathbb{A}$,

(3) if c is an $\mathbb{A}$-finite chain of sets under $<$ then $\sup c$ exists and is in $\mathbb{A}$,

(4) the relation (c is an $\mathbb{A}$-finite chain of sets and $d = \sup c$) is Σ on $\mathbb{A}$.

EXAMPLE. Take $<$ to be $\subseteq$, in which case, if c is a chain of sets, $\sup c$ is simply $\bigcup c$. Other examples will occur in Section 11.

REMARKS. (1) (c is an $\mathbb{A}$-finite chain of sets) is Δ since it is equivalent to $(\forall x \in c)(\forall y \in c)[x < y \vee y < x]$, and $<$ is given to be Δ.

(2) (c is an $\mathbb{A}$-finite chain of sets and $d \neq \sup c$) is Σ since it is equivalent to (c is an $\mathbb{A}$-finite chain of sets) $\wedge (\exists e)(e = \sup c \wedge e \neq d)$.

(3) (c is an $\mathbb{A}$-finite chain of sets and $d = \sup c$) is actually Δ, by Remarks (1) and (2).

DEFINITION. Let $\mathbb{A}$ be an admissible set, let $\mathscr{B}$ be a set on $\mathbb{A}$, and let $\mathscr{S}_1, \ldots, \mathscr{S}_n$ be relations on $\mathscr{B}$. If each of $\mathscr{B}, \mathscr{S}_1, \ldots, \mathscr{S}_n$ is Σ on $\mathbb{A}$ we call the structure $\mathfrak{B} = \langle \mathscr{B}; \mathscr{S}_1, \ldots, \mathscr{S}_n \rangle$ an *$\mathbb{A}$-structure.*

Let $\mathfrak{B}$ be an $\mathbb{A}$-structure and let $<$ be a usable partial ordering on $\mathbb{A}$. We call the pair $\langle \mathfrak{B}, < \rangle$ an *augmented $\mathbb{A}$-structure* provided $\mathfrak{B}$ is closed under $\mathbb{A}$-finite sups, that is, provided whenever $c \subseteq \mathscr{B}$ is an $\mathbb{A}$-finite chain of sets, under $<$, then $\sup c \in \mathscr{B}$.

EXAMPLES. Let $\mathbb{A}$ be admissible. Then $\langle \mathbb{A}, \subseteq \rangle$ is an augmented $\mathbb{A}$-structure. Also, let ord($\mathbb{A}$) be the collection of ordinals of $\mathbb{A}$, and let s be the successor relation on ord($\mathbb{A}$). Then $\langle \text{ord}(\mathbb{A}); s \rangle$ is an $\mathbb{A}$-structure, and $\langle \langle \text{ord}(\mathbb{A}); s \rangle, \subseteq \rangle$ is an augmented $\mathbb{A}$-structure. Other examples result easily from the work in Section 11.

Now we want to define a notion of elementary formal system over $\langle \mathfrak{B}, < \rangle$, an augmented $\mathbb{A}$-structure, in which derivations, though possibly infinite, must be finite in the sense of $\mathbb{A}$. As defined in Chapter 1, derivations in recursion theories can never be infinite, so we begin by modifying the machinery.

We take over the definition, for $\mathfrak{B}$, of elementary formal system formula, atomic formula, pseudo-formula, and pseudo-atomic formula, all from Ch. 1, §3. (Note that $<$ plays no role here.) And we add the following. Let $\square$ be a new meta-linguistic symbol. Suppose $Px_1, \ldots, x_j$ is a pseudo-atomic formula. The result of replacing one or more of the x_i by $\square$ is called an *infinitary axiom.* The recursion theory axioms of Chapter 1, will now be called *ordinary axioms.* From now on we must specify both infinitary and ordinary axioms in giving elementary formal systems. The $\square$ is to be thought of as indicating the positions we are allowed to apply supremum operations on, as will be explained in a moment. If the x_i remain fixed during the course of a discussion, we may abbreviate $Px_1, \ldots, x_{i-1}, \square, x_{i+1}, \ldots, x_j$ by $\boldsymbol{Px} \square \boldsymbol{y}$, and similarly for two or more $\square$ occurrences. We extend the term *proper* so that an infinitary axiom is proper if the relation symbol of it is not one of those representing $\mathscr{S}_1, \ldots, \mathscr{S}_n$, the "given" relations of $\mathfrak{B}$.

Now we must say how the infinitary axioms are to be used in derivations. It is here that the partial ordering $<$ is used. First some terminology.

DEFINITION. $\langle \mathfrak{B}, < \rangle$ is our augmented $\mathbb{A}$ structure, so $<$ is a usable partial ordering. Let c be a collection of n-tuples. If $\langle x_1, \ldots, x_n \rangle \in c$ and $\langle y_1, \ldots, y_n \rangle \in c$, we write $\langle x_1, \ldots, x_n \rangle < \langle y_1, \ldots, y_n \rangle$ if $x_1 < y_1$ and ... and $x_n < y_n$. We call c a *chain* (under $<$) if any two members of c are comparable under this extended notion of $<$. If c is an $\mathbb{A}$-finite chain, by *sup* c we mean the n-tuple $\langle d_1, \ldots, d_n \rangle$ where, for each i,

$$d_i = \sup\{x_i \mid \langle x_1, \ldots, x_i, \ldots, x_n \rangle \in c\}.$$

Now, for each $n \geq 1$, we introduce an infinitary rule of derivation, I_n, as follows.

Rule I_n. Suppose $P\boldsymbol{x}_1 \square \boldsymbol{x}_2 \square \cdots \square \boldsymbol{x}_n \square \boldsymbol{x}_{n+1}$ is an instance of a proper infinitary axiom, and suppose there is an $\mathbb{A}$-finite chain c of n-tuples so that $P\boldsymbol{x}_1 y_1 \boldsymbol{x}_2 y_2 \cdots \boldsymbol{x}_n y_n \boldsymbol{x}_{n+1}$ is derivable for each $\langle y_1, \ldots, y_n \rangle \in c$. Then $P\boldsymbol{x}_1 s_1 \boldsymbol{x}_2 s_2 \cdots \boldsymbol{x}_n s_n \boldsymbol{x}_{n+1}$ follows, where $\langle s_1, \ldots, s_n \rangle = \sup c$.

Now we can define the notion of *derivable* from a set of axioms. Notice we now only define *derivable*, not *derivation*.

DEFINITION. $\langle \mathfrak{B}, < \rangle$ is our augmented $\mathbb{A}$-structure. Let E be a *finite* set of proper ordinary and infinitary elementary formal system axioms over $\mathfrak{B} = \langle \mathcal{B}; \mathcal{S}_1, \ldots, \mathcal{S}_n \rangle$.

(1) Each member of $\mathcal{S}_1^* \cup \cdots \cup \mathcal{S}_n^*$ is derivable from E.

(2) Each instance, over $\mathcal{B}$, of an ordinary axiom in E is derivable from E.

(3) If the two hypotheses of a rule MP application are derivable from E, so is the conclusion.

(4) If the hypotheses of a rule I_n application are derivable from E, so is the conclusion.

Next, we want to associate, with each derivable formula, something we can call a *derivation* of it, then we can add the restriction that derivations must be $\mathbb{A}$-finite. In order for this to make sense, formulas must be *sets* of some sort. The exact details are of little importance. Officially, we adopt the following. Our list of n-place predicate symbols is $\langle 0, n, 0 \rangle$, $\langle 0, n, 1 \rangle, \langle 0, n, 2 \rangle, \ldots$ (where $0, 1, \ldots$ are finite Von Neuman ordinals, and $\langle -, -, - \rangle$ is the usual set theory triple). We take the atomic formula $Pc_1, \ldots, c_k$ to be the $k+2$ tuple $\langle 1, P, c_1, \ldots, c_k \rangle$ (here each c_i is in $\mathcal{B}$). Finally we take $X \rightarrow Y$ to be $\langle 2, X, Y \rangle$. Then every elementary formal system formula over $\mathcal{B}$ is a set in $\mathbb{A}$. Still, we continue to write formulas as usual, for easy reading.

Incidentally, it would perhaps be of some interest to develop a notion of an admissible *word* structure, generalizing what we called $\mathfrak{A}^w$ in Chapter 6, thus avoiding this sort of business.

Finally, we define the notion of derivation. In Chapter 1, derivations were certain well-order sequences. But, building well-ordering into the definition now is going to get us all tangled up with the axiom of choice, which is not generally available in admissible sets. So instead we give a definition of derivation which is modelled on that of Barwise [1969] for infinitary logic, which avoids the problem by allowing a sort of "many-valuedness". The definition below can be shown to be equivalent to a well-ordering sort of definition in admissible sets in which a suitable form of the axiom of choice is available, specifically, in those that are *recursively listed* in the sense of Barwise [1975], p. 161.

DEFINITION. Let $\langle \mathfrak{B}, < \rangle$ be an augmented $\mathbb{A}$-structure, where $\mathfrak{B} = \langle \mathcal{B}, \mathcal{S}_1, \ldots, \mathcal{S}_n \rangle$, and let E be a finite set of proper ordinary and infinitary elementary formal system axioms over $\mathfrak{B}$. By a *derivation from* E *(with respect to* $\langle \mathfrak{B}, < \rangle$*) of the formula* Y we mean an ordered pair p such that one of the following holds.

(A1) $p = \langle 3, Y \rangle$ where Y is a member of $\mathcal{S}_1^* \cup \cdots \cup \mathcal{S}_n^*$.

(A2) $p = \langle 4, Y \rangle$ where Y is an instance, over $\mathcal{B}$, of one of the ordinary axioms in E.

(R0) (corresponding to rule MP) $p = \langle f, Y \rangle$ where: f is a function with domain $\{0, 1\}$, $f(0)$ is a *non-empty set* of derivations from E, each of the atomic formula X, and $f(1)$ is a *non-empty set* of derivations from E, each of the formula $X \to Y$.

(R_n) (corresponding to rule I_n). $p = \langle f, Y \rangle$ where: Y is $P\boldsymbol{x}_1 s_1 \boldsymbol{x}_2 s_2 \cdots \boldsymbol{x}_n s_n \boldsymbol{x}_{n+1}$; $P\boldsymbol{x}_1 \square \boldsymbol{x}_2 \square \cdots \boldsymbol{x}_n \square \boldsymbol{x}_{n+1}$ is an instance, over $\mathcal{B}$, of an infinitary axiom in E; f is a function with domain a set of n-tuples c; c is a $<$ chain; for each $\langle z_1, \ldots, z_n \rangle \in c$, $f(z_1, \ldots, z_n)$ is a non-empty set of derivations from E, each of $P\boldsymbol{x}_1 z_1 \boldsymbol{x}_2 z_2 \cdots \boldsymbol{x}_n z_n \boldsymbol{x}_{n+1}$; and $\langle s_1, \ldots, s_n \rangle = \sup c$.

DEFINITION. We call a derivation p an $\mathbb{A}$-*finite derivation* if $p \in \mathbb{A}$.

Note that, since $\mathcal{B}$ is required to be closed under $\mathbb{A}$-finite chain sups, if $Px_1, \ldots, x_n$ has an $\mathbb{A}$-finite derivation from E, then each $x_i \in \mathcal{B}$.

DEFINITION. Let P be an n-place predicate symbol and let $\mathcal{P} \subseteq \mathcal{B}^n$. We say P $\mathbb{A}$-*represents* $\mathcal{P}$ *over* $\langle \mathfrak{B}, < \rangle$ *in the elementary formal system* E if: $\boldsymbol{v} \in \mathcal{P}$ iff $P\boldsymbol{v}$ has an $\mathbb{A}$-finite derivation from E with respect to $\langle \mathfrak{B}, < \rangle$.

We say $\mathcal{P}$ *is* $\mathbb{A}$*-r.e. over* $\langle\mathfrak{B}, <\rangle$ if $\mathcal{P}$ is $\mathbb{A}$-represented over $\langle\mathfrak{B}, <\rangle$ by some predicate symbol in some elementary formal system.

EXAMPLE. Let $\mathbb{A}$ be any admissible set. Let $\mathfrak{B}$ be the structure $\langle \mathrm{Ord}(\mathbb{A}), s\rangle$ where s is the successor relation on $\mathrm{Ord}(\mathbb{A})$: $s(\alpha, \beta) \Leftrightarrow \beta = \alpha \cup \{\alpha\}$. Both $\mathrm{Ord}(\mathbb{A})$ and s are Σ (in fact, Δ) on $\mathbb{A}$, so $\mathfrak{B}$ is an $\mathbb{A}$-structure. Further, $\mathrm{Ord}(\mathbb{A})$ is closed under $\mathbb{A}$-finite chain unions, and it is easy to see $\langle\mathfrak{B}, \subseteq\rangle$ is then an augmented $\mathbb{A}$-structure. Now, ordinal addition is $\mathbb{A}$-r.e. over $\langle\mathfrak{B}, \subseteq\rangle$, having the following elementary formal system axioms. (For reading ease we have written $y = x^+$ for $s(x, y)$ and $x + y = z$ for $P(x, y, z)$.)

$$\left.\begin{array}{l} x + 0 = x \\ x + y = w \to z = y^+ \to u = w^+ \to x + z = u \end{array}\right\} \text{ordinary,}$$

$$x + \Box = \Box \qquad \text{infinitary.}$$

We are not yet in a position to prove that these axioms really do generate the addition relation; that will come in the next section.

5. Basic results

We set up some special tools for proving things about admissible set elementary formal systems.

For this section, $\mathbb{A}$ is a fixed admissible set, $\mathfrak{B} = \langle \mathcal{B}; \mathcal{S}_1, \dots, \mathcal{S}_n\rangle$ is a fixed $\mathbb{A}$-structure, and $\langle\mathfrak{B}, <\rangle$ is a fixed augmented $\mathbb{A}$-structure.

DEFINITION. Let $s_1, \dots, s_n$ be relations on b. If, for each $i = 1, 2, \dots, n$, s_i and $\mathcal{S}_i$ are both sets of k_i-tuples, we say $\langle b; s_1, \dots, s_n\rangle$ and $\langle \mathcal{B}; \mathcal{S}_1, \dots, \mathcal{S}_n\rangle$ are ***structures of the same type***. If each of $b, s_1, \dots, s_n$ is *in* $\mathbb{A}$, we call the structure $\langle b; s_1, \dots, s_n\rangle$ $\mathbb{A}$*-finite*. If $b \subseteq \mathcal{B}$, $s_1 \subseteq \mathcal{S}_1, \dots, s_n \subseteq \mathcal{S}_n$, then $\langle b; s_1, \dots, s_n\rangle$ is a *substructure* of $\langle \mathcal{B}; \mathcal{S}_1, \dots, \mathcal{S}_n\rangle$.

DEFINITION. Let E be an elementary formal system over $\mathfrak{B}$ (allowing infinitary axioms) and let p be an $\mathbb{A}$-finite derivation from E with respect to $\langle\mathfrak{B}, <\rangle$. Let $\langle b; s_1, \dots, s_n\rangle$ be an $\mathbb{A}$-finite structure of the same type as $\mathfrak{B}$. We say p is a *derivation using the* $\mathbb{A}$*-finite information* $\langle b; s_1, \dots, s_n\rangle$ if, in p, every elementary formal system formula in $\mathrm{TC}(p)$ [transitive closure of p] has all its constants in b, and every member of $\mathcal{S}_1^* \cup \dots \cup \mathcal{S}_n^*$ in $\mathrm{TC}(p)$ is actually a member of $s_1^* \cup \dots \cup s_n^*$.

LEMMA 5.1. *If p is an* $\mathbb{A}$*-finite derivation from E with respect to* $\langle \mathfrak{B}, < \rangle$*, then there is an* $\mathbb{A}$*-finite substructure* $\langle b; s_1, .., s_n \rangle$ *of* $\mathfrak{B}$ *so that p is a derivation using the* $\mathbb{A}$*-finite information* $\langle b; s_1, \dots, s_n \rangle$.

SKETCH OF PROOF. Since p is $\mathbb{A}$-finite, $p \in \mathbb{A}$, hence also $\mathrm{TC}(p) \in \mathbb{A}$. Now set $b = \{x \in \mathrm{TC}(p) \mid x$ occurs as a "constant" in a formula in $\mathrm{TC}(p)\}$. Using Δ-separation, $b \in \mathbb{A}$; $s_1, \dots, s_n$ are treated similarly.

LEMMA 5.2. *Let E be an elementary formal system over* $\mathfrak{B}$. *The following relation is* Δ *on* $\mathbb{A}$:

$$\mathrm{deriv}_E(p, Y, b, s_1, \dots, s_n) \Leftrightarrow \langle b; s_1, \dots, s_n \rangle \text{ is an } \mathbb{A}\text{-finite structure}$$

(*not necessarily* sub *structure of* $\mathfrak{B}$*, nor closed under* $\mathbb{A}$*-finite sups*), *of the same type as* $\mathfrak{B}$, *Y is an elementary formal system formula with all "constants" from b, and p is an* $\mathbb{A}$*-finite derivation from E of Y* (*with* $<$ *as the partial ordering*) *using the* $\mathbb{A}$*-finite information* $\langle b; s_1, \dots, s_n \rangle$.

SKETCH OF PROOF. Each of the clauses of a suitable recursive definition of "p is an $\mathbb{A}$-finite derivation from E of Y, a formula over b, using the $\mathbb{A}$-finite information $\langle b; s_1, \dots, s_n \rangle$" can be shown to be Δ. The result follows using item (10) in Section 3, Δ relations defined by recursion.

THEOREM 5.3. *Let E be an elementary formal system over* $\mathfrak{B}$. *The following relation is* Σ *on* $\mathbb{A}$:

$$\mathrm{deriv}_E(p, Y) \Leftrightarrow p \text{ is an } \mathbb{A}\text{-finite derivation from } E \text{ of } Y \text{ (w.r.t. } \langle \mathfrak{B}, < \rangle).$$

PROOF. Let $\mathscr{B}(x)$ be a Σ formula for $\mathscr{B}$, similarly for $\mathscr{S}_1(\boldsymbol{x}), \dots, \mathscr{S}_n(\boldsymbol{x})$. Using Lemma 5.1,

$$\mathrm{deriv}_E(p, Y) \Leftrightarrow (\exists b, s_1, \dots, s_n)[(\forall x \in b)\mathscr{B}(x) \wedge (\forall \boldsymbol{x} \in s_1)\mathscr{S}_1(\boldsymbol{x}) \wedge \cdots \wedge (\forall \boldsymbol{x} \in s_n)\mathscr{S}_n(\boldsymbol{x}) \wedge \mathrm{deriv}_E(p, Y, b, s_1, \dots, s_n)].$$

The result follows by Lemma 5.2.

COROLLARY 5.4. *If* $\mathscr{R}$ *is* $\mathbb{A}$*-r.e. over* $\langle \mathfrak{B}, < \rangle$, $\mathscr{R}$ *is* Σ *on* $\mathbb{A}$.

PROOF. Say $\mathscr{R}$ is $\mathbb{A}$-represented by R using axioms E. Then

$$\mathscr{R}\boldsymbol{x} \Leftrightarrow (\exists p)(\exists Y)[\mathrm{deriv}_E(p, Y) \wedge Y = R\boldsymbol{x}].$$

Next, we set up the machinery for proving things about derivations by a sort of induction on their complexity.

DEFINITION. Let E be an elementary formal system over $\mathfrak{B}$, and let p be an $\mathbb{A}$-finite derivation from E with respect to $\langle \mathfrak{B}, < \rangle$. By a *subderivation* of p we mean any member $q \in \mathrm{TC}(p)$ such that $q \neq p$ and and q is also an $\mathbb{A}$-finite derivation from E with respect to $\langle \mathfrak{B}, < \rangle$.

THEOREM 5.5 (Induction on the complexity of derivations). *Let E be an elementary formal system over $\mathfrak{B}$, and let $\varphi(x)$ be a formula. Suppose, for any $\mathbb{A}$-finite derivation p from E with respect to $\langle \mathfrak{B}, < \rangle$, $\varphi(p)$ holds in $\mathbb{A}$ whenever $\varphi(q)$ holds in $\mathbb{A}$ for every subderivation q of p. Then $\varphi(p)$ holds in $\mathbb{A}$ for* every *$\mathbb{A}$-finite derivation from E with respect to $\langle \mathfrak{B}, < \rangle$.*

PROOF. Let $\mathrm{deriv}_E(p)$ be a Σ formula for $(\exists Y)\,\mathrm{deriv}_E(p, Y)$. Now let $\Psi(x)$ be the formula $\sim \mathrm{deriv}_E(x) \vee [\mathrm{deriv}_E(x) \wedge \varphi(x)]$. The hypothesis of the theorem gives us that $(\forall y \in \mathrm{TC}(x))\Psi(y)$ implies $\Psi(x)$. Then by item (8) in Section 2 (proof by induction over TC), $(\forall x)\Psi(x)$, which gives the desired conclusion.

DEFINITION. Let E be an elementary formal system over $\mathfrak{B}$ and let p be an $\mathbb{A}$-finite derivation from E with respect to $\langle \mathfrak{B}, < \rangle$. We define the notion of *immediate subderivation* of p as follows:

If p is $\langle 3, Y\rangle$ or $\langle 4, Y\rangle$, there are no immediate subderivations of p.

If p is $\langle f, Y\rangle$ where f is a function, and if $q \in f(x)$ for some x, then q is an immediate subderivation of p.

NOTE. An immediate subderivation of p is also a subderivation of p.

THEOREM 5.6 (Strong induction on the complexity of derivations). *Let E be an elementary formal system over $\mathfrak{B}$, and let $\varphi(x)$ be a formula. Suppose, for any $\mathbb{A}$-finite derivation p from E with respect to $\langle \mathfrak{B}, < \rangle$, $\varphi(p)$ holds in $\mathbb{A}$ whenever $\varphi(q)$ holds in $\mathbb{A}$ for every* immediate *subderivation q of p. Then $\varphi(p)$ holds in $\mathbb{A}$ for* every *$\mathbb{A}$-finite derivation from E with respect to $\langle \mathfrak{B}, < \rangle$.*

PROOF. Let $\Psi(x)$ be the formula

$$\mathrm{deriv}_E(x) \wedge \varphi(x) \wedge (\forall y \in \mathrm{TC}(x))[(\mathrm{deriv}_E(y) \wedge y \neq x) \supset \varphi(y)].$$

Then $\Psi(x) \equiv (x$ is a derivation, and it and all its subderivations have the property φ). That every derivation has property Ψ follows by Theorem 5.5, and this implies every derivation has the property φ.

EXAMPLE. We continue, sketchily, the ordinal addition example of Section 4. Let $\mathbb{A}$ be admissible. Let $\mathfrak{B}$ be $\langle \mathrm{Ord}(\mathbb{A}); \text{successor}\rangle$, and let E be the axioms for addition of Section 4. The problem is to show: $\alpha + \beta = \gamma$ is true in $\mathbb{A}$ iff $\alpha + \beta = \gamma$ has an $\mathbb{A}$-finite derivation from E with respect to $\langle \mathfrak{B}, \subseteq \rangle$.

First, define a notion of "truth" for elementary formal system formulas. For atomic formulas, $\alpha = \beta^+$ is true if, in fact, $\alpha = \beta^+$; $\alpha + \beta = \gamma$ is true if, in fact, $\alpha + \beta = \gamma$. Next, if $X_1, \ldots, X_n, Y$ are atomic, call $X_1 \to \cdots \to X_n \to Y$ true if one of $X_1, \ldots, X_n$ is not true, or if Y is true.

Now, we only need concern ourselves with truth for formulas that can be derived from E. This means that, in the clause above, covering $X_1 \to \cdots \to X_n \to Y$, we need only consider $n = 1, 2$ or 3, and in each case, we can specify exactly what sorts of atomic subformulas are involved. Thus a *formula* $\varphi(x)$ can be produced that expresses the content of "truth" for the formulas that are (potentially) derivable from E. Now, using Theorem 5.5, it can easily be shown that if Y actually has an $\mathbb{A}$-finite derivation from E with respect to $\langle \mathfrak{B}, \subseteq \rangle$, then Y is true. It follows that if $\alpha + \beta = \gamma$ has an $\mathbb{A}$-finite derivation from E with repect to $\langle \mathfrak{B}, \subseteq \rangle$, then, in fact, $\alpha + \beta = \gamma$.

Conversely, if $\alpha + \beta = \gamma$ where $\alpha, \beta, \gamma \in \mathrm{Ord}(\mathbb{A})$, one may show $\alpha + \beta = \gamma$ has an $\mathbb{A}$-finite derivation from E with respect to $\langle \mathfrak{B}, \subseteq \rangle$ by induction on β, using item (8) in Section 2, proof by induction over TC. (Strong Σ replacement is needed to handle limit ordinals.)

THEOREM 5.7. *If $\langle \mathfrak{B}, < \rangle$ is an augmented $\mathbb{A}$-structure, where $\mathbb{A}$ is admissible, the following are $\mathbb{A}$-r.e. over $\langle \mathfrak{B}, < \rangle$:*

(1) *each of the given relations of $\mathfrak{B}$,*

(2) *the empty set,*

(3) $=_{\mathscr{B}}$, *the equality relation on $\mathscr{B}$,*

(4) *for each $b \in \mathscr{B}$, the set $\{b\}$.*

PROOF. See Proposition 1.10.1.

6. Examples

We give examples which relate our present notions with those of Chapter 1, and with those in the literature.

EXAMPLE 1. Let $\mathfrak{A}$ be a structure. Let $\mathrm{HF}_{\mathfrak{A}}$ be defined as in Section 2. Then $\langle \mathfrak{A}, \subseteq \rangle$ is an augmented $\mathrm{HF}_{\mathfrak{A}}$ structure, and the relations $\mathrm{HF}_{\mathfrak{A}}$-r.e. over $\langle \mathfrak{A}, \subseteq \rangle$ are precisely the relations r.e. over $\mathfrak{A}$ in the sense of Chapter 1. Properly speaking, this is a degenerate example, since the infinitary rules of derivation are not needed, all $\mathrm{HF}_{\mathfrak{A}}$-finite chains being truly finite.

DEFINITION. Let $\mathbb{A} = \langle \mathfrak{A}, S, \in \rangle$ be an admissible set. If there is a mapping $F : \mathcal{A} \cup S \rightarrow \mathrm{Ord}(\mathbb{A})$ which is 1–1, onto, and Σ on $\mathbb{A}$, then $\mathbb{A}$ is said to be *recursively listed.* See Barwise [1975], p. 161.

DEFINITION. Let $\mathbb{A} = \langle \mathfrak{A}, S, \in \rangle = \langle \mathcal{A}, \mathcal{R}_1, \ldots, \mathcal{R}_k, S, \in \rangle$ be an admissible set. By $\mathbb{A}^*$ we mean the structure

$$\langle \mathcal{A}, \mathcal{R}_1, \ldots, \mathcal{R}_k, \bar{\mathcal{R}}_1, \ldots, \bar{\mathcal{R}}_k, S, x \cup \{y\} = z, \neq_{\mathcal{A}} \rangle.$$

Where: $x \cup \{y\} = z$ is the relation on $\mathcal{A} \in S$ which holds when set x with y added as member is set z; $\neq_{\mathcal{A}}$ is the unequals relation on urelements; and each $\bar{\mathcal{R}}_i$ is the complement of $\mathcal{R}_i$.

EXAMPLE 2. Let $\mathbb{A}$ be a recursively listed admissible set. Then $\langle \mathbb{A}^*, \subseteq \rangle$ is an augmented $\mathbb{A}$-structure, and the relations $\mathbb{A}$-r.e. over $\langle \mathbb{A}^*, \subseteq \rangle$ are exactly the relations Σ on $\mathbb{A}$.

We check this as follows. Half is by Corollary 5.4. The other half follows using the closure properties to be shown in Section 8, once two items have been established. First, $\notin$ is $\mathbb{A}$-r.e. over $\langle \mathbb{A}^*, \subseteq \rangle$ (that $\in$ is, is trivial). Second, the relations $\mathbb{A}$-r.e. over $\langle \mathbb{A}^*, \subseteq \rangle$ are closed under $(\forall x \in y)$. Appropriate elementary formal systems for these are not hard to come by (modify Example VII in Ch. 1, §4, for $\notin$, for instance). It is proving that they do the job that is tedious, and it is here that the recursive listing of $\mathbb{A}$ is used.

We note that in $\mathbb{A}^*$ we threw in complements of the given relations of $\mathbb{A}$. This is forced on us since the definition of Δ_0 formula used with admissible sets includes closure under negation, and hence each of the given relations of $\mathbb{A}$, *and its complement,* is Σ on $\mathbb{A}$.

In the literature on admissible sets, $\mathbb{A}$-*r.e.* is used to mean Σ *on* $\mathbb{A}$. Consequently this example relates our notion with those standard in the field, for recursively listed admissible sets. This includes those admissible sets of the form L_α, the kind first studied. For non-recursively listed admissible sets we do not know what happens.

DEFINITION. Let $x_1, \ldots, x_n \in \mathbb{A}$. By an $\mathbb{A}$-*word over* $\{x_1, \ldots, x_n\}$ we mean a function $f \in \mathbb{A}$ with domain an ordinal in $\mathbb{A}$ and range $\{x_1, \ldots, x_n\}$. By $\mathbb{A}^w(x_1, \ldots, x_n)$ we mean the collection of all $\mathbb{A}$-words over $\{x_1, \ldots, x_n\}$. And by $\mathbb{A}(x_1, \ldots, x_n)$ we mean the structure $\langle \mathbb{A}^w(x_1, \ldots, x_n), * \rangle$ where $*$ is the concatenation relation on $\mathbb{A}^w(x_1, \ldots, x_n)$.

EXAMPLE 3. Let $\mathbb{A}$ be a *pure* admissible set (that is, no urelements). Suppose also that $\mathbb{A}$ is recursively listed. Let $x_1, \ldots, x_n \in \mathbb{A}$, where $n \geqslant 2$. The structure $\mathbb{A}(x_1, \ldots, x_n)$ is an $\mathbb{A}$-structure. As it happens, the relation $\subseteq$ on $\mathbb{A}(x_1, \ldots, x_n)$ amounts to the relation of "being an initial subword of". $\langle \mathbb{A}(x_1, \ldots, x_n), \subseteq \rangle$ is an augmented $\mathbb{A}$-structure, and the relations $\mathbb{A}$-r.e. over it are precisely the relations on $\mathbb{A}^w(x_1, \ldots, x_n)$ that are Σ on $\mathbb{A}$.

This example will be continued in Section 13, where a sketch of the correctness of our claims will be given. By the way, we do not know the effect of allowing urelements, or of allowing an infinite but $\mathbb{A}$-finite set of "letters."

EXAMPLE 4. Consider an admissible set of the form L_α, the αth collection of constructible sets, using the definition of Gödel [1939] (where it was denoted M_α). L_α is a pure admissible set, there are no urelements (nor are there any given relations, only $\in$ and $=$ appear in Δ_0 and Σ formulas). These are the first admissible sets anyone looked at. See Plateck [1966] and Kripke [1964].

Now $\mathrm{Ord}(L_\alpha) = \alpha$. Let us write α^* for the structure $\langle \alpha$, successor on $\alpha \rangle$. Then $\langle \alpha^*, \subseteq \rangle$ is an augmented L_α-structure. Again, the relations L_α-r.e. over $\langle \alpha^*, \subseteq \rangle$ are precisely the relations on α which are Σ on L_α.

The correctness of this claim will be sketched in Section 13.

An *ordinal* α is called *admissible* if the set L_α is an admissible set. There are many admissible ordinals: the smallest is ω, the set of natural numbers; the second smallest is ω_1, the set of recursive ordinals; every cardinal is an admissible ordinal; and for every admissible ordinal, there is a bigger one of the same cardinality. Admissible ordinal theory, usually called α-recursion theory, can be developed directly, apart from the theory of admissible sets; see Simpson [1974] and Simpson [1978]. A direct, natural, definition of α-r.e. can be given. For the admissible set ω, the notion of ω-r.e. coincides with the usual notion of r.e. for ordinary recursion theory. In general, for an admissible ordinal α, the notion of α-r.e., as used in the literature, is the equivalent of being Σ over L_α, hence by the above, it is equivalent to being L_α-r.e. over $\langle \alpha^*, \subseteq \rangle$ using our elementary formal system approach.

7. Enumeration operators generalized

In the recursion theories of Chapter 1, enumeration operators could be given *any* set of n-tuples (for the appropriate n) as input. But for admissible sets, to get the closure of operators under composition, we are going to need (in effect) metacompactness. But then (for many structures) Theorem 8.13.2, will insist that only semi-hyperregular relations be allowable, and further, Proposition 8.13.6 will generally apply, so generally there will be relations not allowable. In other words, we can expect *partial* production systems to come up here.

If we define a generalization of enumeration operator in the most natural way, it is not hard to discover that things work very smoothly if only sets and relations that are Σ are used as allowable inputs. It is the case, however, that different admissible sets can have the same notions of generalized finiteness, yet differ on their Σ-relations. This leads us to work, not with single admissible sets, but with families of them.

DEFINITION. We say two admissible sets are *related* if they have the same urelements and the same sets. Thus $\langle \mathfrak{A}, S, \in \rangle$ and $\langle \mathfrak{B}, S', \in \rangle$ are related if $\mathcal{A} = \mathcal{B}$ and $S = S'$. Let $\mathbb{F}$ be a collection of admissible sets; we call $\mathbb{F}$ a *family* if any two members of $\mathbb{F}$ are related.

EXAMPLES. (1) Let $\mathfrak{A}$ be a structure. Let $\mathfrak{A}_1 = \langle \mathfrak{A}, \mathcal{R}_1, \ldots, \mathcal{R}_k \rangle$ and let $\mathfrak{A}_2 = \langle \mathfrak{A}, \mathcal{S}_1, \ldots, \mathcal{S}_n \rangle$. For any infinite cardinal κ, $\mathrm{H}(\kappa)_{\mathfrak{A}_1}$ and $\mathrm{H}(\kappa)_{\mathfrak{A}_2}$ are related. Let $\mathbb{F}_\kappa(\mathfrak{A}) = \{\mathrm{H}(\kappa)_{\mathfrak{B}} \mid \mathfrak{B} = \langle \mathfrak{A}, \mathcal{R}_1, \ldots, \mathcal{R}_j \rangle$ for some relations $\mathcal{R}_1, \ldots, \mathcal{R}_j$ on $\mathcal{A}\}$. Then $\mathbb{F}_\kappa(\mathfrak{A})$ is a family. In particular, $\mathbb{F}_\omega(\mathfrak{A})$ is a family, one which plays an important role.

(2) Let $\mathbb{F} = \{\mathbb{A}\}$ where $\mathbb{A}$ is an admissible set. $\mathbb{F}$ is trivially a family.

DEFINITION. Let $\mathbb{F}$ be a family of admissible sets. We say $\langle \mathfrak{B}, < \rangle$ is an *augmented* $\mathbb{F}$*-structure* if $\langle \mathfrak{B}, < \rangle$ is an augmented $\mathbb{A}$-structure for each $\mathbb{A} \in \mathbb{F}$.

EXAMPLE. Let $\mathbb{F}$ be a family of admissible sets. $\mathrm{Ord}(\mathbb{A}) = \alpha$ is the same for each $\mathbb{A} \in \mathbb{F}$ since the members of $\mathbb{F}$ contain the same sets. Let $\alpha^* = \langle \alpha$, successor on $\alpha\rangle$. Then it is easy to see that $\langle \alpha^*, \subseteq \rangle$ is an augmented $\mathbb{F}$-structure.

DEFINITION. Let $\mathbb{F}$ be a family of admissible sets, and $\langle \mathfrak{B}, < \rangle$ be an augmented $\mathbb{F}$-structure. We say $\mathcal{P}$ is $\mathbb{F}$*-r.e. over* $\langle \mathfrak{B}, < \rangle$ if $\mathcal{P}$ is $\mathbb{A}$-r.e. over

$\langle \mathfrak{B}, < \rangle$. For every $\mathbb{A} \in \mathbb{F}$ (equivalently, for any $\mathbb{A} \in \mathbb{F}$, since all members of $\mathbb{F}$ have the same notion of "finite", hence of "finite" derivation).

We spend the rest of this section generalizing, to families of admissible sets, the notion of enumeration operator from Chapter 1. The reader should review the material from that chapter before reading this.

DEFINITION. Let $\mathbb{F}$ be a family of admissible sets. A relation $\mathcal{I}$ is *allowable in* $\mathbb{F}$ if $\mathcal{I}$ is Σ on at least one member of $\mathbb{F}$.

Now, let $\langle \mathfrak{B}, < \rangle$ be an augmented $\mathbb{F}$-structure, where $\mathbb{F}$ is a family of admissible sets. Let E be an elementary formal system over $\mathfrak{B}$, possibly with infinitary axioms, in which the n-place predicate symbol I and the m-place predicate symbol O may occur, but in which I does not occur in the conclusion of any axiom. Let $\mathcal{I}$ be an n-ary relation on $\mathcal{B}$ which is allowable in $\mathbb{F}$, say $\mathcal{I}$ turns out to be Σ on both $\mathbb{A}$ and $\mathbb{B}$ in $\mathbb{F}$. Then $\langle\langle \mathfrak{B}, \mathcal{I}\rangle, < \rangle$ is both an augmented $\mathbb{A}$-structure and an augmented $\mathbb{B}$-structure. $\mathbb{A}$ and $\mathbb{B}$ are both in the family $\mathbb{F}$, so they have the same notions of derivation from E, since they have the same notions of generalized finiteness. Hence $O\boldsymbol{v}$ has an $\mathbb{A}$-derivation over $\langle\langle \mathfrak{B}, \mathcal{I}\rangle, < \rangle$ from E if and only if $O\boldsymbol{v}$ has a $\mathbb{B}$-derivation over $\langle\langle \mathfrak{B}, \mathcal{I}\rangle, < \rangle$ from E.

Finally, we are ready to define our generalization of enumeration operators.

DEFINITION. Let $\mathbb{F}$ be a family of admissible sets. Let $\langle \mathfrak{B}, < \rangle$ be an augmented $\mathbb{F}$-structure. Let E be an elementary formal system over $\mathfrak{B}$, possibly involving the predicate symbols I (n-place) and O(m-place), but in which I does not occur in the conclusion of any axiom. By $[E^I_O]$ we mean the operator of order $\langle n, m \rangle$, with domain the collection of all n-ary relations $\mathcal{R}$ which are allowable in $\mathbb{F}$, and such that, if $\mathcal{I}$ is in the domain of $[E^I_O]$, then $[E^I_O](\mathcal{I}) = \{\boldsymbol{v} \in \mathcal{B}^m \mid O\boldsymbol{v}$ has an $\mathbb{A}$-derivation over $\langle\langle \mathfrak{B}, \mathcal{I}\rangle, < \rangle$ from E, where $\mathbb{A} \in \mathbb{F}$ is any admissible set on which $\mathcal{I}$ is $\Sigma\}$. (It is understood that I represents $\mathcal{I}$.) We call $[E^I_O]$ an $\mathbb{F}$-*enumeration operator over* $\langle \mathfrak{B}, < \rangle$.

By $\mathbb{F}$-rec($\langle \mathfrak{B}, < \rangle$) we mean the pair $\langle \mathscr{C}, \mathscr{O} \rangle$, where $\mathscr{C}$ is the collection of all sets and relations allowable in $\mathbb{F}$, and $\mathscr{O}$ is the collection of all $\mathbb{F}$-enumeration operators over $\langle \mathfrak{B}, < \rangle$. We refer to this as the $\mathbb{F}$-*recursion theory over* $\langle \mathfrak{B}, < \rangle$. It is the main object of study for the rest of this chapter.

8. Structural properties

For this section, $\mathbb{F}$ is a family of admissible sets and $\langle \mathfrak{B}, < \rangle$ is an augmented $\mathbb{F}$-structure. Below we verify that $\mathbb{F}$-rec$(\langle \mathfrak{B}, < \rangle)$ is a production system, as defined in Ch. 2, §2. This is the first of several sections in which we consider the application of our earlier, abstract development in the present setting.

THEOREM 8.1. *Every $\mathbb{F}$-enumeration operator over $\langle \mathfrak{B}, < \rangle$ takes relations allowable in $\mathbb{F}$ to relations allowable in $\mathbb{F}$.*

PROOF. Let $[E^I_O]$ be an $\mathbb{F}$-enumeration operator over $\langle \mathfrak{B}, < \rangle$, and let $\mathscr{J}$ be allowable in $\mathbb{F}$, say $\mathscr{J}$ is Σ on $\mathbb{A} \in \mathbb{F}$. Then $[E^I_O](\mathscr{J}) = \{\boldsymbol{Ov} \mid \boldsymbol{Ov}$ has an $\mathbb{A}$-derivation over $\langle\langle \mathfrak{B}, \mathscr{J}\rangle, < \rangle$ from $E\}$. It follows that $[E^I_O](\mathscr{J})$ is $\mathbb{A}$-r.e. over $\langle\langle \mathfrak{B}, \mathscr{J}\rangle, < \rangle$, hence is Σ over $\mathbb{A}$ by Corollary 5.4, and hence is allowable in $\mathbb{F}$.

REMARK. It should be obvious that every $\mathbb{F}$-enumeration operator over $\langle \mathfrak{B}, < \rangle$ is *monotone*; we take up compactness later.

THEOREM 8.2. $\mathbb{F}$-rec$(\langle \mathfrak{B}, < \rangle)$ *is closed under composition.*

PROOF. Our proof is an admissible set version of that of Proposition 1.10.5.

Let $[E^A_B]$, of order $\langle n, m\rangle$ and $[F^C_D]$ of order $\langle m, p\rangle$ both be $\mathbb{F}$-enumeration operators over $\langle \mathfrak{B}, < \rangle$. We show $[F^C_D][E^A_B]$ is also an $\mathbb{F}$-enumeration operator over $\langle \mathfrak{B}, < \rangle$. We may suppose E and F are disjoint, that is, they contain no common predicate symbols except those representing the given relations of $\mathfrak{B}$. Let H consist of

the axioms of E,

the axioms of f, $\quad \boldsymbol{Bx} \to \boldsymbol{Cx}$.

Then $[H^A_D]$ is an $\mathbb{F}$-enumeration operator over $\langle \mathfrak{B}, < \rangle$. We claim $[H^A_D] = [F^C_D][E^A_B]$. That is, if $\mathscr{J}$ is allowable in $\mathbb{F}$, $[H^A_D][\mathscr{J}] = [F^C_D][E^A_B](\mathscr{J})$. We show this in two parts. We suppose $\mathscr{J}$ is Σ on $\mathbb{A} \in \mathbb{F}$.

(1) Suppose $\boldsymbol{v} \in [F^C_D][E^A_B](\mathscr{J})$; we show $\boldsymbol{v} \in [H^A_D](\mathscr{J})$.

Let $[E^A_B][\mathscr{J}] = \mathcal{J}$. Then of course $\boldsymbol{v} \in [F^C_D](\mathcal{J})$. Then $\boldsymbol{Dv}$ has an $\mathbb{A}$-derivation over $\langle\langle \mathfrak{B}, \mathcal{J}\rangle, < \rangle$ from F, in which C represents $\mathcal{J}$. Let p be one such $\mathbb{A}$-derivation. Let input$_p = \{\boldsymbol{x} \in \mathrm{TC}(p) \mid \boldsymbol{Cx} \in \mathrm{TC}(p)\}$. Since p is an $\mathbb{A}$-derivation, $p \in \mathbb{A}$, and it follows, by Δ-separation, that input$_p \in \mathbb{A}$. Now, if $\boldsymbol{x} \in \text{input}_p$, $\boldsymbol{x} \in \mathscr{J}$, so $\boldsymbol{x} \in [E^{A}_{B}](\mathscr{I})$ and so $B\boldsymbol{x}$ has an $\mathbb{A}$-derivation over $\langle\langle \mathfrak{B}, \mathscr{I}\rangle, <\rangle$ from E, in which A represents $\mathscr{I}$. Also, by Theorem 5.3, the relation: q is an $\mathbb{A}$-derivation of Y over $\langle\langle \mathfrak{B}, \mathscr{I}\rangle, <\rangle$ from E, is Σ on $\mathbb{A}$. (*Note*: here we use that $\mathscr{I}$ is Σ on $\mathbb{A}$, that is, we are using that $\mathscr{I}$ is allowable.) It follows, by the Σ collection principle, that there is a set $b \in \mathbb{A}$ such that, if $\boldsymbol{x} \in \text{input}_p$, some member of b is an $\mathbb{A}$-derivation of $B\boldsymbol{x}$ over $\langle\langle \mathfrak{B}, \mathscr{I}\rangle, <\rangle$ from E; and each member of b is an $\mathbb{A}$-derivation over $\langle\langle \mathfrak{B}, \mathscr{I}\rangle, <\rangle$ from E, of $B\boldsymbol{x}$ for some $\boldsymbol{x} \in \text{input}\ p$. Note that, if $\boldsymbol{x} \in \text{input}_p$ is fixed, then $\{y \in b \mid y \text{ is an } \mathbb{A}\text{-derivation of } B\boldsymbol{x} \text{ over } \langle\langle \mathfrak{B}, \mathscr{I}\rangle, <\rangle \text{ from } E\}$ is a non-empty *member* of $\mathbb{A}$. It is in $\mathbb{A}$ since it is the same as $\{y \in b \mid y = \langle u, B\boldsymbol{x}\rangle \text{ for some } u\}$ and we have Δ_0 separation.

Now we define a function K on $\mathbb{A}$, by recursion, as follows.

If $r = \langle 3, C\boldsymbol{x}\rangle$ then $K(r) = \langle f, C\boldsymbol{x}\rangle$ where f is a function with domain $\{0, 1\}$, and $f(0) = \{y \in b \mid y \text{ is an } \mathbb{A}\text{-derivation of } B\boldsymbol{x} \text{ over } \langle\langle \mathfrak{B}, \mathscr{I}\rangle, <\rangle \text{ from } E\}$ and $f(1) = \{\langle 4, B\boldsymbol{x} \to C\boldsymbol{x}\rangle\}$.

If $r = \langle f, Y\rangle$ where f is a function, then $K(r) = \langle f', Y\rangle$ where f' is a function with the same domain as f, and for each $x \in \text{domain } f$, $f'(x) = K(f(x))$.

$K(r) = r$ otherwise.

Using Definition by Σ recursion, K is a Σ function symbol. It follows, by Σ replacement, that, since $p \in \mathbb{A}$, $K(p) \in \mathbb{A}$. Now from the definition of K it should be clear that $K(p)$ will be like p, except that each occurrence of $C\boldsymbol{x}$ will have been replaced by (schematically):

derivation of $B\boldsymbol{x}$ from E over $\langle\langle \mathfrak{B}, \mathscr{I}\rangle, <\rangle$	$B\boldsymbol{x} \to C\boldsymbol{x}$

$$C\boldsymbol{x}$$

It follows that $K(p)$ is also an $\mathbb{A}$-derivation, but over $\langle\langle \mathfrak{B}, \mathscr{I}\rangle, <\rangle$ from the axioms H. p was a derivation of $D\boldsymbol{v}$ and so is $K(p)$. Hence $\boldsymbol{v} \in [H^{A}_{D}](\mathscr{I})$. This completes (1).

(2) Suppose $\boldsymbol{v} \in [H^{A}_{D}](\mathscr{I})$. We show $\boldsymbol{v} \in [F^{C}_{D}][E^{A}_{B}](\mathscr{I})$. This time we are quite sketchy.

Since $\boldsymbol{v} \in [H^{A}_{D}](\mathscr{I})$, there is an $\mathbb{A}$-derivation q, of $D\boldsymbol{v}$ over $\langle\langle \mathfrak{B}, \mathscr{I}\rangle, <\rangle$ from H, in which A represents $\mathscr{I}$. Let $\mathscr{J} = \{\boldsymbol{y} \in \text{TC}(q) \mid \text{there is a subderivation of } q, \text{ of } C\boldsymbol{y}\}$. We claim $\boldsymbol{v} \in [F^{C}_{D}](\mathscr{J})$ and $\mathscr{J} \subseteq [E^{A}_{B}](\mathscr{I})$ which will establish our desired result. Let T be some (otherwise irrelevant) instance of one of the given relations of $\mathfrak{B}$.

Using Σ recursion and Σ replacement, one can show that, since $q \in \mathbb{A}$, so does q', where q' is like q except that every "line" in q which contains a predicate symbol peculiar to E, has been replaced by T. It is not hard to see that q' is an $\mathbb{A}$-derivation, also of $D\boldsymbol{v}$, but over $\langle\langle\mathfrak{B}, \mathscr{J}\rangle, <\rangle$ from E', in which C represents $\mathscr{J}$. Thus $\boldsymbol{v} \in [F^C_D](\mathscr{J})$.

Suppose $\boldsymbol{y} \in \mathscr{J}$. Then there is a subderivation, say r, of q, of $C\boldsymbol{y}$. Then there must also be a subderivation s, of r, of $B\boldsymbol{y}$. Again, using Σ recursion and Σ replacement one can show that, since $s \in \mathbb{A}$, so does s', where s' is like s except that every "line" in s which contains a predicate symbol peculiar to F has been replaced by T. Again, s' is an $\mathbb{A}$-derivation, also of $B\boldsymbol{y}$, over $\langle\langle\mathfrak{B}, \mathscr{J}\rangle, <\rangle$, from E, in which A represents $\mathscr{J}$. Thus $\boldsymbol{y} \in [E^A_B](\mathscr{J})$. So $\mathscr{J} \subseteq [E^A_B](\mathscr{J})$, which completes the proof.

THEOREM 8.3. $\mathbb{F}$-rec($\langle\mathfrak{B}, <\rangle$) *is closed under intersection, union and Cartesian products.*

PROOF. Exactly like the proofs of Propositions 1.10.3 and 1.10.4, except that a little more care is needed now to ensure that the combined axiom systems behave in the expected way. We leave the details to the reader.

THEOREM 8.4. $\mathbb{F}$-rec($\langle\mathfrak{B}, <\rangle$) *contains the transposition, projection, and place-adding operators.*

PROOF. See Proposition 1.10.2.

REMARK. We have thus verified the following: $\mathbb{F}$-rec($\langle\mathfrak{B}, <\rangle$) *is a partial production system, where allowable means allowable in* $\mathbb{F}$.

9. More examples

EXAMPLE 1. Let $\mathfrak{A}$ be any structure, and let $\mathbb{F}_\omega(\mathfrak{A})$ be defined as in Example 1 of Section 7, that is $\mathbb{F}_\omega(\mathfrak{A}) = \{\mathrm{HF}_{\mathfrak{B}} \mid \mathscr{B} = \langle\mathfrak{A}, \mathscr{R}_1, \ldots, \mathscr{R}_j\rangle$ for some $\mathscr{R}_1, \ldots, \mathscr{R}_j$ on $\mathscr{A}\}$. $\langle\mathfrak{A}, \subseteq\rangle$ is an augmented $\mathbb{F}_\omega(\mathfrak{A})$-structure, and in fact, *every* relation on $\mathscr{A}$ is allowable in $\mathbb{F}_\omega(\mathfrak{A})$. Very simply, let $\mathscr{R}$ be any relation on $\mathscr{A}$; then if $\mathfrak{B} = \langle\mathfrak{A}, \mathscr{R}\rangle$ we have $\mathrm{HF}_{\mathfrak{B}} \in F_\omega(\mathfrak{A})$ and trivially, $\mathscr{R}$ is Σ on $\mathrm{HF}_{\mathfrak{B}}$. By the results of Section 8, $\mathbb{F}_\omega$-rec($\langle\mathfrak{A}, \subseteq\rangle$) is a *total* production system. In fact, it coincides with rec($\mathfrak{A}$) as defined in Chapter 1.

EXAMPLE 2. Again, let $\mathfrak{A}$ be any structure, and let κ be some infinite cardinal. Define $\mathbb{F}_\kappa(\mathfrak{A})$ as in Example 1 of Section 7. If $\langle \mathfrak{A}, < \rangle$ is an $\mathbb{F}_\kappa(\mathfrak{A})$-recursion theory structure (one must check closure under appropriate sups), then, as in the previous example, every relation on $\mathscr{A}$ will be allowable in $\mathbb{F}_\kappa(\mathfrak{A})$, and thus $\mathscr{F}_\kappa(\mathfrak{A})$-rec($\langle \mathfrak{A}, < \rangle$) will be a total production system. In particular, this will be the case if $\mathfrak{A} = \kappa^* = \langle \kappa$, successor on $\kappa \rangle$ and $<$ is $\subseteq$.

EXAMPLE 3. Let α be an admissible ordinal (see Example 4 in Section 6). Let $\mathbb{F}(L_\alpha)$ be the family of all admissible sets related to $\langle L_\alpha, \in \rangle$, and let $\alpha^* = \langle \alpha$, successor on $\alpha \rangle$. Then $\langle \alpha^*, \subseteq \rangle$ is an augmented $\mathbb{F}(L_\alpha)$-structure, thus we may consider $\mathbb{F}(L_\alpha)$-rec($\langle \alpha^*, \subseteq \rangle$). According to Example 4 of Section 6, the relations $\mathbb{F}(L_\alpha)$-r.e. over $\langle \alpha^*, \subseteq \rangle$ are those usually called α-r.e. For most admissible ordinals α, however, $\mathbb{F}(L_\alpha)$-rec($\langle \alpha^*, \subseteq \rangle$) will be a *partial* production system, as we will see in Section 16.

10. Pre-embeddings and embeddings

Let $\mathbb{F}$ be a family of admissible sets, and let $\langle \mathfrak{A}, <_{\mathfrak{A}} \rangle$ and $\langle \mathfrak{B}, <_{\mathfrak{B}} \rangle$ be two augmented $\mathbb{F}$-structures, fixed for this section. We consider to what extent the material developed in Chapter 3, can be applied.

DEFINITION. Let $\theta : \mathscr{A} \to \mathscr{B}$ be a coding, as defined in Ch. 3, §2. We call θ a *Σ coding in* $\mathbb{F}$ if the relation $y \in \theta(x)$ is Σ on every admissible set in $\mathbb{F}$.

PROPOSITION 10.1. *Let $\theta : \mathscr{A} \to \mathscr{B}$ be a Σ coding in* $\mathbb{F}$. *Then θ is a pre-embedding from* $\mathbb{F}$-rec($\langle \mathfrak{A}, <_{\mathfrak{A}} \rangle$) *to* $\mathbb{F}$-rec($\langle \mathfrak{B}, <_{\mathfrak{B}} \rangle$).

PROOF. Suppose $\mathscr{P} \subseteq \mathscr{A}^n$ is allowable in $\mathbb{F}$. Then $\mathscr{P}$ is Σ on some $\mathbb{A} \in \mathbb{F}$. $\mathscr{B}$ is Σ on $\mathbb{A}$ since $\mathfrak{B}$ is an $\mathbb{F}$-structure. Also θ is Σ on $\mathbb{A}$. But then,

$$\begin{aligned} &\langle y_1, \dots, y_n \rangle \in \theta_n(\mathscr{P}) \\ &\quad \Leftrightarrow \; y_1 \in \mathscr{B} \wedge \cdots \wedge y_n \in \mathscr{B} \wedge (\exists x_1, \dots, x_n)(\langle x_1, \dots, x_n \rangle \in \mathscr{P} \\ &\qquad \wedge y_1 \in \theta(x_1) \wedge \cdots \wedge y_n \in \theta(x_n)) \end{aligned}$$

and this is easily seen to be Σ on $\mathbb{A}$. Thus $\theta_n(\mathscr{P})$ is allowable in $\mathbb{F}$.

Similarly θ_n^{-1} takes allowable relations to allowable relations, so θ is a pre-embedding.

Now the question is, when can we be sure that a given pre-embedding is actually an *embedding* of one $\mathbb{F}$-recursion theory in another. The terminology of theory assignment and elementary theory assignment, from Ch. 3, §5, is not quite appropriate, since they applied to *structures*, while $\mathbb{F}$-rec(-) applies to *augmented* structures. Thus any analog of Theorem 3.5.1 must take the usable partial orderings into account. The following definition covers things; in it we use $\mathbb{F}$-*finite* to mean $\mathbb{A}$-finite for any (every) $\mathbb{A} \in \mathbb{F}$.

DEFINITION. Let $\theta : \mathcal{A} \to \mathcal{B}$ be a coding. We say θ *preserves* $\mathbb{F}$-*finite chain sups* from $\langle \mathfrak{A}, <_{\mathfrak{A}} \rangle$ to $\langle \mathfrak{B}, <_{\mathfrak{B}} \rangle$ if, for each $\mathbb{F}$-finite $a \subseteq \mathcal{A}$ which is a chain under $<_{\mathfrak{A}}$, and for each $s \in \theta(\sup a)$, there is an $\mathbb{F}$-finite $b \subseteq \mathcal{B}$ which is a chain under $<_{\mathfrak{B}}$, such that

(1) $\sup b = s$,

(2) if $y \in b$ then $y \in \theta(x)$ for some $x \in a$.

PROPOSITION 10.2. *The composition of two codings, each of which preserves $\mathbb{F}$-finite chain sups, is another coding which preserves $\mathbb{F}$-finite chain sups.*

Now we can state an analog of Theorem 3.5.1.

THEOREM 10.3. *Let $\mathfrak{A} = \langle \mathcal{A}; \mathcal{R}_1, \dots, \mathcal{R}_k \rangle$. Let $\theta : \mathcal{A} \to \mathcal{B}$ be a Σ-coding in $\mathbb{F}$ which preserves $\mathbb{F}$-finite sups from $\langle \mathfrak{A}, <_{\mathfrak{A}} \rangle$ to $\langle \mathfrak{B}, <_{\mathfrak{B}} \rangle$, and suppose that each of $(=_{\mathcal{A}})^{\theta}$, $\mathcal{R}_1^{\theta}, \dots, \mathcal{R}_k^{\theta}$ is $\mathbb{F}$-r.e. over $\langle \mathfrak{B}, <_{\mathfrak{B}} \rangle$, [that is, each is generated in $\mathbb{F}$-rec($\langle \mathfrak{B}, <_{\mathfrak{B}} \rangle$)]. Then θ is an* embedding, *θ : $\mathbb{F}$-rec($\langle \mathfrak{A}, <_{\mathfrak{A}} \rangle$) $\to$ $\mathbb{F}$-rec($\langle \mathfrak{B}, <_{\mathfrak{B}} \rangle$).*

REMARK. The proof is much like that of Theorem 3.5.1, but now Σ-replacement must be brought in, to ensure that $\mathbb{F}$-finite derivations arise. We leave details to the dedicated reader.

11. Combining augmented $\mathbb{F}$-strucures

The material in Chapter 4, does not apply directly in the present setting, since we must now work with *augmented* structures. However, only minor adjustments are needed. We begin in this section by defining $\dot{\cup}$ and $\times$ to take usable partial orderings into account.

Recall, in Chapter 4, $\dot{\cup}$ and $\times$ were defined for domains in Section 2, certain maps i and j were defined in Section 4, and finally, $\dot{\cup}$ and $\times$ were defined for structures, in Section 5. We make use of all this, below.

DEFINITION. Let $\mathbb{F}$ be a family of admissible sets, and let $\langle \mathfrak{A}, <_{\mathfrak{A}} \rangle$ and $\langle \mathfrak{B}, <_{\mathfrak{B}} \rangle$ be two augmented $\mathbb{F}$-structures. Then, if $\square$ is either $\dot{\cup}$ or $\times$, by $\langle \mathfrak{A}, <_{\mathfrak{A}} \rangle \square \langle \mathfrak{B}, <_{\mathfrak{B}} \rangle$ we mean $\langle \mathfrak{A} \square \mathfrak{B}, < \rangle$ where

$$< \text{ is } \begin{cases} (<_{\mathfrak{A}})^i \cup (<_{\mathfrak{B}})^j & \text{if } \square \text{ is } \dot{\cup}, \\ (<_{\mathfrak{A}})^i \cap (<_{\mathfrak{B}})^j & \text{if } \square \text{ is } \times. \end{cases}$$

There are several direct consequences of this definition, which we summarize in the following proposition, whose proof we leave to the reader.

PROPOSITION 11.1. *$\mathbb{F}$ is a family of admissible sets, $\langle \mathfrak{A}, <_{\mathfrak{A}} \rangle$ and $\langle \mathfrak{B}, <_{\mathfrak{B}} \rangle$ are augmented $\mathbb{F}$-structures, and $\square$ is either $\dot{\cup}$ or $\times$. Then:*

(1) *$\langle \mathfrak{A}, <_{\mathfrak{A}} \rangle \square \langle \mathfrak{B}, <_{\mathfrak{B}} \rangle$ is also an augmented $\mathbb{F}$-structure.*

(2) *Let the codings i and j be defined as in Ch. 4, §4. Then $i : \mathcal{A} \to \mathcal{A} \square \mathcal{B}$ preserves $\mathbb{F}$-finite chain sups from $\langle \mathfrak{A}, <_{\mathfrak{A}} \rangle$ to $\langle \mathfrak{A}, <_{\mathfrak{A}} \rangle \square \langle \mathfrak{B}, <_{\mathfrak{B}} \rangle$. Similarly for $j : \mathcal{B} \to \mathcal{A} \square \mathcal{B}$.*

(3) *Let $\square$ be defined on codings as in* Ch. 4, §2. *Let $\langle \mathfrak{A}', <_{\mathfrak{A}'} \rangle$ and $\langle \mathfrak{B}', <_{\mathfrak{B}'} \rangle$ be two more augmented $\mathbb{F}$-structures, and let $\alpha : \mathcal{A} \to \mathcal{A}'$ and $\beta : \mathcal{B} \to \mathcal{B}'$ be codings. If α preserves $\mathbb{F}$-finite chain sups from $\langle \mathfrak{A}, <_{\mathfrak{A}} \rangle$ to $\langle \mathfrak{A}', <_{\mathfrak{A}'} \rangle$ and β preserves $\mathbb{F}$-finite chain sups from $\langle \mathfrak{B}, <_{\mathfrak{B}} \rangle$ to $\langle \mathfrak{B}', <_{\mathfrak{B}'} \rangle$, then $\alpha \square \beta$ also preserves $\mathbb{F}$-finite chain sups from $\langle \mathfrak{A}, <_{\mathfrak{A}} \rangle \square \langle \mathfrak{B}, <_{\mathfrak{B}} \rangle$ to $\langle \mathfrak{A}', <_{\mathfrak{A}'} \rangle \square \langle \mathfrak{B}', <_{\mathfrak{B}'} \rangle$.*

12. More monoidal subcategories in Prod

As we have remarked, much of the work of Chapter 4 does not apply directly to the present situation, since we now must deal with *augmented* structures. In this section we sketch how certain modifications of the development in Ch. 4, §9 will yield currently applicable results.

DEFINITION. Let $\mathbb{F}$ be a family of admissible sets. We write $\mathcal{D}_{\mathbb{F}}$ for the collection of all augmented $\mathbb{F}$-structures. And we define a collection $\mathcal{E}_{\mathbb{F}}$ of codings as follows. Suppose $\langle \mathfrak{A}, <_{\mathfrak{A}} \rangle$ and $\langle \mathfrak{B}, <_{\mathfrak{B}} \rangle$ are in $\mathcal{D}_{\mathbb{F}}$. Let $\theta : \mathcal{A} \to \mathcal{B}$ be a Σ-coding in $\mathbb{F}$ which preserves $\mathbb{F}$-finite chain sups from $\langle \mathfrak{A}, <_{\mathfrak{A}} \rangle$ to $\langle \mathfrak{B}, <_{\mathfrak{B}} \rangle$. Then θ is to be one of the members of $\mathcal{E}_{\mathbb{F}}$.

PROPOSITION 12.1. *Whether $\square$ is $\dot{\cup}$ or $\times$, $\mathcal{D}_{\mathbb{F}}$ is closed under $\square$, and $\mathcal{E}_{\mathbb{F}}$ is closed under both $\square$ and composition, and contains the identity codings,* inj, *and the i and j codings.*

PROOF. By Propositions 11.1, 10.2, and a little work.

PROPOSITION 12.2. *Let* $\langle\mathfrak{A},<_{\mathfrak{A}}\rangle$, $\langle\mathfrak{A}',<_{\mathfrak{A}}\rangle$, $\langle\mathfrak{B},<_{\mathfrak{B}}\rangle$ *and* $\langle\mathfrak{B}',<_{\mathfrak{B}}\rangle$ *be from* $\mathscr{D}_{\mathbb{F}}$. *Suppose* $\mathbb{F}$-rec$(\langle\mathfrak{A},<_{\mathfrak{A}}\rangle)=\mathbb{F}$-rec$(\langle\mathfrak{A}',<_{\mathfrak{A}}\rangle)$ *and* $\mathbb{F}$-rec$(\langle\mathfrak{B},<_{\mathfrak{B}}\rangle)=\mathbb{F}$-rec$(\langle\mathfrak{B}',<_{\mathfrak{B}}\rangle)$. *Then, for* $\square$ *either* $\dot{\cup}$ *or* $\times$, $\mathbb{F}$-rec$(\langle\mathfrak{A},<_{\mathfrak{A}}\rangle\square\langle\mathfrak{B},<_{\mathfrak{B}}\rangle)=\mathbb{F}$-rec$(\langle\mathfrak{A}',<_{\mathfrak{A}}\rangle\square\langle\mathfrak{B}',<_{\mathfrak{B}}\rangle)$.

PROOF. An elaboration of that of Lemma 4.7.2, making use of Theorem 10.3.

DEFINITION. Let $\mathbb{F}$ be a family of admissible sets, and let $\mathscr{D}_{\mathbb{F}}$ and $\mathscr{E}_{\mathbb{F}}$ be as above. By $C(\mathbb{F}\text{-rec}(\text{-}),\mathscr{D}_{\mathbb{F}},\mathscr{E}_{\mathbb{F}})$ we mean the category formed as follows.

(1) The objects are all pairs $\langle\mathbb{F}\text{-rec}(\langle\mathfrak{A},<_{\mathfrak{A}}\rangle),<_{\mathfrak{A}}\rangle$ where $\langle\mathfrak{A},<_{\mathfrak{A}}\rangle$ is from $\mathscr{D}_{\mathbb{F}}$.

(2) $\theta:\langle\mathbb{F}\text{-rec}(\langle\mathfrak{A},<_{\mathfrak{A}}\rangle),<_{\mathfrak{A}}\rangle\rightarrow\langle\mathbb{F}\text{-rec}(\langle\mathfrak{B},<_{\mathfrak{B}}\rangle),<_{\mathfrak{B}}\rangle$ is a morphism provided θ is an embedding of $\mathbb{F}$-rec$(\langle\mathfrak{A},<_{\mathfrak{A}}\rangle)$ in $\mathbb{F}$-rec$(\langle\mathfrak{B},<_{\mathfrak{B}}\rangle)$ which is in $\mathscr{E}_{\mathbb{F}}$, and preserves $\mathbb{F}$-finite chain sups from $\langle\mathfrak{A},<_{\mathfrak{A}}\rangle$ to $\langle\mathfrak{B},<_{\mathfrak{B}}\rangle$.

It is straightforward to verify that $C(\mathbb{F}\text{-rec}(\text{-}),\mathscr{D}_{\mathbb{F}},\mathscr{E}_{\mathscr{F}})$ is a category. Then $\square$ may be defined on it, using Proposition 12.2, in the obvious way. A symmetric monoidal category results, which plays a role much like that of $C(T,\mathscr{D},\mathscr{E})$ in Chapters 4 and 5. Thus, notions of separability, reflexivity and effective embedding may be defined and investigated. We do not follow up on this here.

13. Extensions, again

In Ch. 5, §6, notions of *extension, conservative extension* and *inessential extension* were developed. They modify quite readily to the present setting, and we take a short space to discuss this.

DEFINITION. Let $\mathbb{F}$ be a family of admissible sets, let $\langle\mathfrak{A},<\rangle$ and $\langle\mathfrak{B},<\rangle$ be two augmented $\mathbb{F}$-structures (note: the *same* partial ordering occurs in both), and suppose $\mathscr{A}\subseteq\mathscr{B}$. Let $\text{inj}:\mathscr{A}\rightarrow\mathscr{B}$ be defined by $\text{inj}(x)=\{x\}$ for $x\in\mathscr{A}$. If inj is an *embedding* of $\mathbb{F}$-rec$(\langle\mathfrak{A},<\rangle)$ in $\mathbb{F}$-rec$(\langle\mathfrak{B},<\rangle)$ we call $\mathbb{F}$-rec$(\langle\mathfrak{B},<\rangle)$ an *extension* of $\mathbb{F}$-rec$(\langle\mathfrak{A},<\rangle)$. Further, if inj^{-1} is a *co-embedding* we say the extension is *conservative.*

Now Proposition 5.6.2, carries over directly. We state it, revised, as

PROPOSITION 13.1. Suppose $\mathbb{F}$-rec($\langle\mathfrak{B}, <\rangle$) is a conservative extension of $\mathbb{F}$-rec($\langle\mathfrak{A}, <\rangle$), and $\mathcal{R}$ is a relation on $\mathcal{A}$. Then $\mathcal{R}$ is $\mathbb{F}$-r.e. over $\langle\mathfrak{A}, <\rangle$ if and only if $\mathcal{R}$ is $\mathbb{F}$-r.e. over $\langle\mathfrak{B}, <\rangle$.

PROOF. By Proposition 3.4.2.

DEFINITION. Let $\mathbb{F}$-rec($\langle\mathfrak{B}, <\rangle$) be an extension of $\mathbb{F}$-rec($\langle\mathfrak{A}, <\rangle$), and let h be an embedding, $h : \mathbb{F}\text{-rec}(\langle\mathfrak{B}, <\rangle) \to \mathbb{F}\text{-rec}(\langle\mathfrak{A}, <\rangle)$. If the relation $y \in (h \restriction \mathcal{A})(x)$ is $\mathbb{F}$-r.e. over $\langle\mathfrak{A}, <\rangle$ we call $\mathbb{F}$-rec($\langle\mathfrak{B}, <\rangle$) an *inessential extension* of $\mathbb{F}$-rec($\langle\mathfrak{A}, <\rangle$) via h.

PROPOSITION 13.2. *If* $\mathbb{F}$-rec($\langle\mathfrak{B}, <\rangle$) *is an inessential extension of* $\mathbb{F}$-rec($\langle\mathfrak{A}, <\rangle$) *then it is also a conservative extension.*

PROOF. The same as that of Theorem 5.6.5.

Now we sketch the correctness of our claims in Examples 3 and 4 of Section 6.

EXAMPLE 3 (continued). Let $\mathbb{A}$ be a pure admissible set which is recursively listed. $\{\mathbb{A}\}$ is trivially a family, so all the present material can be applied. In particular, $\langle\mathbb{A}(x_1, \ldots, x_n), \subseteq\rangle$ is an augmented $\{\mathbb{A}\}$-structure.

Now, let us suppose } and { are among the "letters" $x_1, \ldots, x_n$. Using them, every set in $\mathbb{A}$ can be shown to have "names" in a natural sense, though generally the names will be infinitely long words. For example, { } "names" 0, {{ }} names {0} or 1, etc., and if 0*, 1*, etc. are "names" for 0, 1, etc., then {0*1* $\cdots$} is one "name" for ω. The recursive listing of $\mathbb{A}$ is used in showing that every set in $\mathbb{A}$ has "names" in $\mathbb{A}^w(x_1, \ldots, x_n)$.

Now, let $\mathbb{A}^*$ be the structure of Example 2 of Section 6. Then $\langle\mathbb{A}^*, \subseteq\rangle$ is also an augmented $\{\mathbb{A}\}$-structure. Define a *coding* $\theta : \mathbb{A}^* \to \mathbb{A}(x_1, \ldots, x_n)$ so that $\theta(x)$ is the collection of all "names" for x.

It can be shown that (1) θ is a Σ coding in $\{\mathbb{A}\}$; (2) θ preserves $\{\mathbb{A}\}$-finite sups from $\langle\mathbb{A}^*, \subseteq\rangle$ to $\langle\mathbb{A}(x_1, \ldots, x_n), \subseteq\rangle$; and (3) Each of the "given" relations of $\mathbb{A}^*$, as well as $=_{\mathcal{A}}$, is mapped by θ to a relation that is $\{\mathbb{A}\}$-r.e. over $\langle\mathbb{A}(x_1, \ldots, x_n), \subseteq\rangle$. Hence by Theorem 10.3, $\theta : \{\mathbb{A}\}\text{-rec}(\langle\mathbb{A}^*, \subseteq\rangle) \to \{\mathbb{A}\}\text{-rec}(\langle\mathbb{A}(x_1, \ldots, x_n), \subseteq\rangle)$ is an *embedding*.

It is not hard to see that $\{\mathbb{A}\}$-rec($\langle\mathbb{A}^*, \subseteq\rangle$) is an *extension* of $\{\mathbb{A}\}$-rec($\langle\mathbb{A}(x_1, \ldots, x_n), \subseteq\rangle$). Also (and this takes more work) the relation $y \in (\theta \restriction \mathbb{A}^w(x_1, \ldots, x_n))(x)$ is $\{\mathbb{A}\}$-r.e. in $\{\mathbb{A}\}$-rec($\langle\mathbb{A}(x_1, \ldots, x_n), \subseteq\rangle$). Thus the extension is an *inessential* one, so by Proposition 13.2, it is also *conservative*. Finally, by Proposition 13.1, we have: *If* $\mathcal{R}$ *is a relation on* $\mathbb{A}^w(x_1, \ldots, x_n)$, $\mathcal{R}$ *is* $\{\mathbb{A}\}$*-r.e. over* $\langle \mathbb{A}(x_1, \ldots, x_n), \subseteq \rangle$, *iff* $\mathcal{R}$ *is* $\{\mathbb{A}\}$*-r.e. over* $\langle \mathbb{A}^*, \subseteq \rangle$.

It follows from Example 2 in Section 6, that the relations $\mathbb{A}$-r.e. over $\langle \mathbb{A}(x_1, \ldots, x_n), \subseteq \rangle$ are precisely the relations on $\mathbb{A}^w(x_1, \ldots, x_n)$ that are Σ on $\mathbb{A}$.

EXAMPLE 4 (continued). Let L_α be the αth constructible set, defined as in Gödel [1939] (where it is denoted M_α), and suppose L_α is admissible. (Then α is an admissible ordinal.) In Section 6, we asserted that the relations L_α-r.e. over $\langle \alpha^*, \subseteq \rangle$ were just the relations on α which were Σ on L_α. The correctness of this assertion may be checked along lines similar to the above. In outline, one proceeds as follows.

$\{L_\alpha\}$ is, of course, a family, and it is not hard to see that $\{L_\alpha\}$-rec($\langle L_\alpha^*, \subseteq \rangle$) is an extension of $\{L_\alpha\}$-rec($\langle \alpha^*, \subseteq \rangle$).

Now, for each ordinal β, let F_β be the βth constructible set using the definition of Gödel [1940]. Define a *coding* $\theta : L_\alpha \to \alpha$ by $\theta(F_\beta) = \{\beta\}$ for each $\beta < \alpha$. It can be shown that θ is an *embedding*

$$\theta : \{L_\alpha\}\text{-rec}(\langle L_\alpha^*, \subseteq \rangle) \to \{L_\alpha\}\text{-rec}(\langle \alpha^*, \subseteq \rangle)$$

and that the relation $y \in (\theta \restriction \alpha)(x)$ is $\{L_\alpha\}$-r.e. over $\langle \alpha^*, \subseteq \rangle$. (This takes quite a bit of work, however.) It follows that the extension is inessential, hence conservative. Thus we have: *If* $\mathcal{R}$ *is a relation on* α, $\mathcal{R}$ *is* $\{L_\alpha\}$*-r.e. over* $\langle L_\alpha^*, \subseteq \rangle$ *iff* $\mathcal{R}$ *is* $\{L_\alpha\}$*-r.e. over* $\langle \alpha^*, \subseteq \rangle$. The connection between being $\{L_\alpha\}$-r.e. over $\langle L_\alpha^*, \subseteq \rangle$ and being Σ over L_α is again given by Example 2 in Section 6.

We need the fact that L_α is recursively listed. For this, see Barwise [1975] Theorem 3.7, p. 163 (or verify that the coding θ itself is essentially a recursive listing).

14. Indexed relational systems

In Chapter 7, we defined Indexed Relational Systems, and we investigated their properties. We now wish to apply the results of that work to admissible set recursion theories. Now, in Ch. 6, §15, we considered when recursion and ω-recursion theories gave rise to indexed relational systems. The key criteria were the existence of an effective pairing function, and having the unequals relation generated. Presumably some similar criteria could be developed for admissible sets, but the work would be lengthy and technical. So we leave that to others and simply restrict our considerations to two particular cases that are commonly considered, and cite the literature to establish that these give rise to indexed relational systems. We also consider the various other conditions discussed in Chapter 7.

I. Let $\mathbb{A}$ be a recursively listed admissible set. Let $\mathbb{A}^*$ be defined as in Section 6. Then $\langle \mathbb{A}^*, \subseteq \rangle$ is an augmented $\mathbb{A}$-structure. We claim

(1) The collection of relations $\mathbb{A}$-r.e. over $\langle \mathbb{A}^*, \subseteq \rangle$ constitutes an *indexed relational system*, taking basic to mean, simply, generated (this can be narrowed down to Δ_0 in fact).

(2) Taking $\mathcal{D}$*-finite* to mean *member of* $\mathbb{A}$, the special assumptions of Ch. 7, §7, apply. Thus the Rice–Shapiro Theorem holds.

(3) If $\mathbb{A}$ is also pure (no urelements) then $\mathbb{A}$ is a *Kleene–Mostowski system*, as defined in Ch. 7, §6. (In particular, this applies to admissible sets of the form L_α.)

We sketch the verification of our claims.

According to Example 2 in Section 6, $\mathbb{A}$-r.e. over $\langle \mathbb{A}^*, \subseteq \rangle$ is equivalent to being Σ on $\mathbb{A}$. In the literature, being Σ on $\mathbb{A}$ is the standard meaning of $\mathbb{A}$-r.e., something we rely on below.

Now, in Barwise [1975], Theorem 1.3, on p. 154, establishes that the $\mathbb{A}$-r.e. relations of *any* admissible set are parametrized, hence *universal relations* exist. One can extract from that argument a proof that the *iteration property* holds, taking the basic functions to be the $\mathbb{A}$-r.e. ones. Thus we in fact have an indexed relational system as defined in Ch. 7, §1.

Next we turn to the special assumptions in Ch. 7, §7. Let $\mathcal{D}$*-finite* mean *member of* $\mathbb{A}$. Take each $\mathcal{D}$-finite relation as a code for itself. Then all the special assumptions are obvious, except the $\mathcal{D}$-finite approximation condition. We argue for that as follows. Let $\mathcal{P}$ be generated, that is, $\mathbb{A}$-r.e. Since $\mathbb{A}$ is recursively listed, Theorem 4.1, from Barwise [1975], p. 164 applies, so $\mathcal{P}$ is the range of an $\mathbb{A}$-recursive function F whose domain is the collection of ordinals of $\mathbb{A}$. ($\mathbb{A}$-recursive means the graph of F is Σ on $\mathbb{A}$.) Now, for each ordinal $\alpha \in \mathbb{A}$, let $F''\alpha = \{F(\beta) \mid \beta < \alpha\}$. Let $\mathcal{C} = \{F''\alpha \mid \alpha \text{ an ordinal in } \mathbb{A}\}$. We leave it to the reader to check that $\mathcal{C}$ meets the $\mathcal{D}$-finite approximation conditions. Σ-replacement is needed to show that the members of $\mathcal{C}$ are $\mathcal{D}$-finite, as are unions of proper initial segments.

Finally, suppose in addition to being recursively listed, $\mathbb{A}$ is pure. That the generated relations of the indexed relational system, $\mathcal{R}_{\mathcal{A}}$, in question are the members of $\Sigma_1(\mathcal{R}_{\mathcal{A}})$ is immediate from the fact that $\mathbb{A}$-r.e. over $\langle \mathbb{A}^*, \subseteq \rangle$ is equivalent to Σ over $\mathbb{A}$. The problem is the existence of a strong pairing function, which must be 1–1 and *onto*. But, since $\mathbb{A}$ is recursively listed, there is an effective correspondence between $\mathbb{A}$ and Ord($\mathbb{A}$), by definition, and since $\mathbb{A}$ is pure, Ord($\mathbb{A}$) must be an admissible ordinal by Corollary 1.8 in Barwise [1975], p. 45. Then we can call on the existence of 1–1, onto effective pairing functions for admissible *ordinals*, from the next example, to create a strong pairing function for $\mathbb{A}$ itself. Thus $\mathbb{A}$ will be a Kleene–Mostowski system.

II. Let α be an admissible ordinal. Let α^* be the structure $\langle \alpha$, successor on $\alpha \rangle$. According to Example 4 in Section 6, the relations L_α-r.e. over $\langle \alpha^*, \subseteq \rangle$ are precisely the relations on α which are Σ on L_α. This is generally known as being *α-r.e.* (See, for instance, Simpson [1978], proposition 2.8, p. 360.) We claim

(1) The collection of α-r.e. relations constitutes an *indexed relational system*, taking basic to mean function with α-r.e. graph. Indeed, one can narrow this to mean primitive recursive, in the sense of Simpson [1978], p. 356.

(2) This indexed relational system is a *Kleene–Mostowski system.*

(3) Taking *$\mathcal{D}$-finite* to mean *subset of α which is a member of* L_α, the special assumptions of Ch. 7, §7, apply. Thus the Rice–Shapiro Theorem holds (note Simpson [1978], p. 371).

In brief, a verification may proceed as follows.

The existence of *universal relations* is asserted in Simpson [1978], Theorem 2.2, p. 358. He remarks that the proof is tedious but straightforward. The same can be said for the Iteration Property. But we do have an indexed relational system.

A strong pairing function for α is given in Simpson [1978], p. 356–7. It follows easily that we have a Kleene–Mostowski system.

A coding of the $\mathcal{D}$-finite sets is described in Simpson [1978], p. 360. That the special assumptions of Ch. 7, §7 hold, follows without much difficulty. Part of the argument is similar to that in the previous example, and uses the now-standard fact that if a relation is α-r.e., it is the range of an α-recursive function.

15. Indexed production systems

In Chapter 8, we defined Indexed Production Systems, and we investigated them under various extra assumptions. Now we sketch how that work applies to admissible set recursion theories discussed in the previous section.

I. Let $\mathbb{A}$ be a recursively listed admissible set, and consider the augmented $\mathbb{A}$-structure $\langle \mathbb{A}^*, \subseteq \rangle$.

(1) $\{\mathbb{A}\}$-rec($\langle \mathbb{A}^*, \subseteq \rangle$) is an *indexed production system*, where *allowable* means allowable in $\{\mathbb{A}\}$. In it the *output* and the *input place-fixing properties* hold, taking basic to mean, with $\mathbb{A}$-r.e. graph (this can be narrowed to Δ_0).

The verification of this must be left to the reader. It is relatively straightforward. But it is too long to include, and we cannot cite the literature, since these notions have not been investigated elsewhere.

(2) Take $\mathcal{D}$*-finite* to mean member of $\mathbb{A}$. Then we have a canonical coding, if we let each $\mathcal{D}$-finite set code itself.

Actually, the only part that needs serious checking is the existence of an operator F such that $F(\mathcal{P}) = \{c \mid D_c \subseteq \mathcal{P}\}$. Well, F is $[E^I_O]$ where E consists of

$$O\emptyset, \qquad \text{(ordinary axioms)}$$

$$Ox \rightarrow Iy \rightarrow x \cup \{y\} = z \rightarrow Oz,$$

$$O\square. \qquad \text{(infinitary axiom)}$$

(3) There is an effective pairing function, the usual Kuratowski one will do.

(4) Every $\{\mathbb{A}\}$ enumeration operator is montone and $\mathcal{D}$-compact ($\mathcal{D}$-compact since derivations must be $\mathbb{A}$-finite). Hence by Theorem 8.8.4, every operator can be put in Rogers' form.

(5) We note that Theorem 8.10.4, also applies, if $\mathbb{A}$ is pure.

(6) The Least-Fixed Point Theorem holds, since the hypotheses of either Theorem 8.11.2, or Theorem 8.15.2, may be applied. (In Theorem 8.11.2, one can take S to be Ord($\mathbb{A}$), and $<$ to be the usual ordering. In Theorem 8.15.2, one can take $<$ to be the transitive closure of $\in$.)

The the bulk of the material in Chapter 8, applies to $\{\mathbb{A}\}$-rec($\langle \mathbb{A}, \subseteq \rangle$).

II. Let α be an admissible ordinal, so that L_α is an admissible set. Let $\mathbb{F}(L_\alpha)$ be the family of all admissible sets related to $\langle L_\alpha, \in \rangle$, and let $\alpha^* = \langle \alpha$, successor on $\alpha \rangle$. We consider the augmented $\mathbb{F}(L_\alpha)$ structure $\langle \alpha^*, \subseteq \rangle$. This time the verification of our assertions must be left to those with some familiarity with α-recursion theory, though the effects should be clear.

(1) $\mathbb{F}(L_\alpha)$-rec($\langle \alpha^*, \subseteq \rangle$) is an *indexed production system*, where allowable means allowable in $\mathbb{F}(L_\alpha)$. In it the *output and the input place-fixing properties hold*, taking basic to mean α-r.e. (or even α-primitive recursive).

(2) Take $\mathcal{D}$*-finite* to mean: subset of α which is a member of L_α. Use the

$\mathcal{D}$-finite coding described in Simpson [1978], p. 360. This gives us a canonical coding.

(3) An effective pairing function was given in the last section.

(4) Every $\mathbb{F}(L_\alpha)$ enumeration operator is monotone, and $\mathcal{D}$-compact. Hence every operator can be put in Rogers' form.

(5) Theorem 8.10.4, applies.

(6) The Least-Fixed Point Theorem holds. (In Theorem 8.11.2, take S to be α, and $<$ to be $\in$. In Theorem 8.15.2, take $<$ to be $\in$.)

Thus again, the material in Chapter 8, applies.

16. Remarks on α-recursion theory

For this section, let α be an admissible ordinal, and let $\mathbb{F}(L_\alpha)$ be the family of admissible sets related to L_α. We wish to discuss how our $\mathbb{F}(L_\alpha)$-rec($\langle \alpha^*, \subseteq \rangle$) relates to α-recursion theory as it is commonly developed.

First a few preliminary results. Recall the definition of semi-hyperregular from Ch. 8, §13.

LEMMA 16.1. *In* $\mathbb{F}(L_\alpha)$-rec($\langle \alpha^*, \subseteq \rangle$), *allowable = semi-hyperregular.*

PROOF. Everything semi-hyperregular must be allowable, by definition.

Suppose $\mathcal{P}$ is allowable; let Φ be an $\mathbb{F}(L_\alpha)$ enummeration operator; we show $\Phi(\mathcal{P})$ satisfies weak replacement, which means $\mathcal{P}$ is semi-hyperregular.

$\mathcal{P}$ is allowable, hence so is $\Phi(\mathcal{P})$ by theorem 8.1. Say $\Phi(\mathcal{P})$ is Σ on $\mathbb{A} \in \mathbb{F}(L_\alpha)$. But $\mathbb{A}$ is admissible, so Σ-collection applies (item (4) in Section 2). Since $\mathbb{A}$ and L_α have the same notion of $\mathcal{D}$-finiteness, it is immediate that $\Phi(\mathcal{P})$ satisfies weak replacement.

COROLLARY 16.2. *In* $\mathbb{F}(L_\alpha)$-rec($\langle \alpha^*, \subseteq \rangle$) *every* $\mathbb{F}$-*enumeration operator is* $\mathcal{D}$-*metacompact.*

PROOF. By the above and Theorem 8.14.5. One also needs that the $\mathcal{D}$-finite union of $\mathcal{D}$-finite sets is $\mathcal{D}$-finite. This is not hard.

Then Theorem 8.12.1, immediately gives us the following (recall $\mathcal{Q} \leqslant \mathcal{P}$ means $\mathcal{Q} = \Phi(\mathcal{P})$ for some operator Φ).

FACT. In $\mathbb{F}(L_\alpha)$-rec($\langle \alpha^*, \subseteq \rangle$), if $\mathcal{P}$ is an allowable set, $\mathcal{Q} \leqslant \mathcal{P}$ if and only if there is some α-r.e. relation R such that

$$D_x \subseteq \mathcal{Q} \Leftrightarrow (\exists y)[R(y, x) \wedge D_y \subseteq \mathcal{P}].$$

A comparison of this with DiPaola [1978] shows that, *for allowable sets*, our notion of $\leqslant$ is equivalent to his notion of α-enumeration reducibility.

We leave to the reader the demonstration of the following, closely related, results connecting our terms from Ch. 2, §4, with those common in the field (see Simpson [1974], p. 168).

FACTS. In $\mathbb{F}(L_\alpha)$-rec($\langle\alpha^*, \subseteq\rangle$), *if* $\mathcal{R} \times \bar{\mathcal{R}}$ *is allowable*, then

(1) $\mathcal{P}$ is α-recursive in $\mathcal{R}$ iff $\mathcal{P}$ is bi-generated in $\mathcal{R}$.

(2) f is weakly α-recursive in $\mathcal{R}$ iff f is generated in $\mathcal{R}$.

Thus for allowable sets, our notions coincide with those that are standard. However, generally, there will be sets that are not allowable.

PROPOSITION 16.3. *If* α *is not a cardinal,* $\mathbb{F}(L_\alpha)$-rec($\langle\alpha^*, \subseteq\rangle$) *is a partial production system.*

PROOF. Lemma 16.1 and Proposition 8.13.6.

Since, generally, there are relations that are not allowable, it is pertinent to try and relate those that are to concepts that are standard in the literature.

See Simpson [1974], p. 169, for definitions of the terms *α-regular* and *α-hyperregular.*

THEOREM 16.4. *In* $\mathbb{F}(L_\alpha)$-rec($\langle\alpha^*, \subseteq\rangle$), $\mathcal{P} \times \bar{\mathcal{P}}$ *is allowable in our sense if and only if* $\mathcal{P}$ *is a* α*-regular and* α*-hyperregular.*

PROOF. (1) Suppose $\mathcal{P} \times \bar{\mathcal{P}}$ is allowable. By Lemma 16.1, then, it is semi-hyperregular. Then by Lemma 8.13.3, for a partial function f with $f \leqslant \mathcal{P} \times \bar{\mathcal{P}}$, if $K \subseteq \operatorname{dom} f$ and K is $\mathcal{D}$-finite, then $f''K$ is also $\mathcal{D}$-finite. By the Facts above, $f \leqslant \mathcal{P} \times \bar{\mathcal{P}}$ is equivalent to f being α-recursive in $\mathcal{P}$. Then it follows from a result of Sacks (see Theorem 1.6, in Simpson [1974], p. 170) that $\mathcal{P}$ is α-regular and α-hyperregular.

(2) Suppose $\mathcal{P}$ is α-regular and α-hyperregular. Using the result contained in the footnote on p. 170 in Simpson [1974], $\langle L_\alpha, \in, \mathcal{P}\rangle$ must be admissible. It is, of course, related to L_α, hence it is in the family $\mathbb{F}(L_\alpha)$. In it, $\mathcal{P} \times \bar{\mathcal{P}}$ is Σ, hence $\mathcal{P} \times \bar{\mathcal{P}}$ is allowable in $\mathbb{F}(L_\alpha)$.

Thus we see that our notions agree with standard ones essentially on the α-regular, α-hyperregular sets, which is apparently a severe restriction. After all, the customary notion of α-recursive-in makes sense for any sets, the only question is whether or not it holds. But by imposing the restriction to α-regular, α-hyperregular, we gain the advantage of having an indexed production system to work with. It is interesting that many of of the results of α-degree theory turn out to hold under the restriction to the α-regular, α-hyperregular sets, and thus are quite meaningful in our version of α-recursion theory. As an example, consider the generalization of the Fridberg–Muchnik Theorem (Simpson [1974], p. 181). We suspect that the whole point of such restrictions in the literature is precisely to secure the advantages of working in a full production system, though the effects appear in some disguised form.

CHAPTER TEN

EFFECTIVE OPERATORS OF HIGHER TYPES

1. Introduction

We suggest, at this point, that you go back and re-read the informal description of enumeration operators in Ch. 1, §1 and §6. We pick up those anthropomorphic ideas one last time.

Recall, you had a set of "work boxes" and instructions for filling them. One of your boxes was labeled "input" and one was labeled "output". Somewhere there was a demon who filled your input box, let you work, and observed what turned up in your output box as you followed your instructions. That demon had, in you following your instructions, an enumeration operator. Let us say that, as such, you are a member of the office staff at level 1 in the corporate hierarchy.

Now, suppose you are promoted. You get the demon's job, and he (it?) has been kicked upstairs. You are now at level 2, and you oversee the work of several level 1 employees. They each have their instructions, and you have yours. You, like them, still have "work boxes", and input and output boxes. But you now have greater responsibility and can do things you could not do before. You now put things in the level 1 employee's input boxes as they work in their separate offices, and you observe their output boxes. In effect, you (following your instructions) run the interoffice mail system.

But the demon who was your boss is still around, and is still your boss. He still puts things in your input box, and he still watches your output box. But in addition to the items he puts in your input box, the things you manipulate, he also supplies you with outside "consulting firms" over which you have no control. Sometimes your instructions call on you to take material from one of your "work boxes" and send it to one of the consulting firms the demon has supplied you with. If the firm sends you back some material, your instructions tell you what to do with it, what box to put it in.

These outside consulting firms are rather like the offices you oversee: you put material in, material comes back. But a little thought shows that the offices in your firm that you oversee are not really necessary. You still have your work boxes, so their functions, their instructions, could be incorporated into your own. Middle management should be able to take over if called on. What is really new here is these outside consulting firms. You don't know how they operate, all you can do is send them material and see what comes back.

What the demon now has in you, as a level 2 employee, is an operator that can be supplied both with sets (by putting them in your input box) and with type 1 operators (by giving you consulting firms). And you return a set (in your output box). In short, we say he has an *effective type 2 operator.*

This sort of description can be continued through another promotion, and yet another, up through the type levels of our mythical corporate hierarchy. But things rapidly get quite complicated, so instead, let us begin the conversion of our description into a more formal approach.

In Ch. 1, §10 we proved that the intersection of two enumeration operators is another enumeration operator (Proposition 1.10.3). Our proof went as follows. Let φ be the set of axioms:

$$Ix \to Ax, \qquad Ix \to Cx,$$

$$By \to Dy \to Oy.$$

Then, if $\Phi = [E^A_B]$ and $\Psi = [F^C_D]$ are enumeration operators, the combined set of axioms, φ, E and F, call it H, determines an enumeration operator $[H^I_O]$ which, in fact, is $\Phi \cap \Psi$.

This is quite suggestive. We might think of the set φ of axioms as defining a type 2 operator; it needs supplementation with Φ and Ψ (as outside consulting firms) before anything can be done.

In the present chapter, using elementary formal systems, we develop this idea further, and provide a natural hierarchy of effective higher type operators for all type levels $n \geq 1$, with enumeration operators being the special case of type 1. We will see that several of the proofs from Chapter 1, like the one above, actually establish the existence of effective type 2 operators, and that similar ideas extend to higher type levels yet.

2. Operators with several inputs

Before we get into a discussion of higher type operators, it is necessary to fill a minor gap in the development of Chapter 1. Enumeration Operators, as we defined them, take single inputs, that is, $\Phi(\mathscr{P})$ makes sense but

$\Phi(\mathcal{P}, \mathcal{R})$ does not. It is time now to change this situation, and there are two natural ways it might be done. We illustrate both by examples, and show the two approaches are equivalent for enumeration operators. For this section, we work over a fixed structure, $\mathfrak{A}$.

I. First, we may use extra input symbols in elementary formal systems. For example, let I_1, I_2 and I_3 be, respectively, a one-place, a two-place and a three-place predicate symbol. Let E be a set of elementary formal system axioms over $\mathfrak{A}$, with none of I_1, I_2 or I_3 in the conclusion of any axiom. An obvious meaning can be assigned to $[E_O^{I_1,I_2,I_3}]$, in a way paralleling the definition on Ch. 1, §7. Thus, if $\mathcal{P} \in [\mathcal{A}]^1$, $\mathcal{Q} \in [\mathcal{A}]^2$ and $\mathcal{R} \in [\mathcal{A}]^3$ then $[E_O^{I_1,I_2,I_3}](\mathcal{P}, \mathcal{Q}, \mathcal{R})$ is the relation represented by O using the axioms E in which I_1 represents $\mathcal{P}$, I_2 represents $\mathcal{Q}$ and I_3 represents $\mathcal{R}$.

II. A second approach is to keep the single-input style of elementary formal system, but tag different inputs differently, after first "filling them out" to be of equal lengths. For example, let $[E_O^I]$ be of order $\langle 4, \boldsymbol{n} \rangle$ and let $\mathcal{P} \in [\mathcal{A}]^1$, $\mathcal{Q} \in [\mathcal{A}]^2$ and $\mathcal{R} \in [\mathcal{A}]^3$ as above. We might agree that $[E_O^I](\mathcal{P}, \mathcal{Q}, \mathcal{R})$ is to mean

$$[E_O^I](\mathcal{P}' \cup \mathcal{Q}' \cup \mathcal{R}')$$

where

$$\mathcal{P}' = \{1\} \times \mathcal{P} \times \{\langle 0, 0 \rangle\},$$

$$\mathcal{Q}' = \{2\} \times \mathcal{Q} \times \{0\}, \qquad \mathcal{R}' = \{3\} \times \mathcal{R}.$$

Thus, first components are "tags", and there may be "padding" on the right, to make everything a 4-tuple. Of course, this approach supposes that $\mathcal{A}$ has enough "tags". We used natural numbers above for convenience, not for necessity.

That the two approaches are equivalent is easy to see.

Suppose $\Phi(\mathcal{P}, \mathcal{Q}, \mathcal{R})$ is defined as an enumeration operator using approach I; thus

$$\Phi(\mathcal{P}, \mathcal{Q}, \mathcal{R}) = [E_O^{I_1,I_2,I_3}](\mathcal{P}, \mathcal{Q}, \mathcal{R}).$$

Let E' be the following set of axioms, where I is a new predicate symbol:

those of E,

$$I1, x, 0, 0 \to I_1 x,$$

$$I2, x, y, 0 \to I_2 x, y,$$

$$I3, x, y, z \to I_3 x, y, z.$$

Clearly $\Phi(\mathcal{P}, \mathcal{Q}, \mathcal{R}) = [E_O^{\prime I}](\mathcal{P}, \mathcal{Q}, \mathcal{R})$ in sense II.

Suppose $\Phi(\mathcal{P}, \mathcal{Q}, \mathcal{R})$ is defined using approach II; thus

$$\Phi(\mathcal{P}, \mathcal{Q}, \mathcal{R}) = [E_O^I](\mathcal{P}' \cup \mathcal{Q}' \cup \mathcal{R}').$$

Let E' now be the following set of axioms, where I_1, I_2 and I_3 are new to E:

those of E,

$I_1x \to I1, x, 0, 0,$

$I_2x, y \to I2, x, y, 0,$

$I_3x, y, z \to I3, x, y, z.$

Clearly $\Phi(\mathcal{P}, \mathcal{Q}, \mathcal{R}) = [E_O^{\prime I_1, I_2, I_3}](\mathcal{P}, \mathcal{Q}, \mathcal{R})$ in sense I.

Sense I is most convenient in actually giving elementary formal systems. But sense II allows all the theory developed for single-input operators to be applied more generally. We will use the two approaches interchangeably, often not specifying which, if it does not matter.

3. Effective type 2 operators

Ultimately we want a notion of effective operator of type n for every n. But as one goes up the type hierarchy, notation becomes something fierce. Consequently we have compromised our integrity by treating things in detail only through the type 3 level, but this is quite representative, and we feel nothing essential is lost. Also, we think it easiest to introduce things through type 2 now, and investigate that for a while, and then introduce the type 3 level later.

Let $\mathfrak{A}$ be a structure, fixed for this section. By a *type* 0 *object* over $\mathfrak{A}$ we mean a set or relation on $\mathcal{A}$. By an *effective type* 0 *object* we mean an r.e. set or relation. The *type* 1 *operators, or objects* over $\mathfrak{A}$ are the monotone operators on $\mathcal{A}$ (allowing multiple inputs). By an *effective type* 1 *operator* over $\mathfrak{A}$ we mean an enumeration operator as defined in Chapter 1, and modified in the previous section to take multiple inputs. Note that type 1 operators map type 0 objects to type 0 objects, and effective type 1 operators map effective type 0 objects to effective type 0 objects.

The intention is that a type 2 operator should map objects of types 1 and 0 to type 0 objects. In order to say this better we need some notation.

At the type 0 level the predicate letters of elementary formal systems serve as variables. We need variables for type 1 level. Suppose $I_1, I_2, \ldots, I_k$

and O are distinct predicate letters, where I_i is n_i-place, and O is m-place. Then we use

$$\begin{bmatrix} I_1, I_2, \ldots, I_k \\ \\ O \end{bmatrix}$$

as a variable, ranging over all multiple input type 1 operators that take as input k relations which are, respectively, n_1-ary, n_2-ary, $\ldots$, n_k-ary, and give as output m-ary relations. We say the variable is of *order* $\langle n_1, n_2, \ldots, n_k; m\rangle$. If $[{}^I_O]$ is of order $\langle n; m\rangle$, we will often write the order as $\langle n, m\rangle$, to agree with the style of earlier chapters.

Let $[E{}^I_O]$ be an enumeration operator. Then E is a set of elementary formal system axioms (in which the symbol I does not occur in the conclusion of any axiom). In using the axioms E it is clear that we think of E as an *incomplete* elementary formal system, in that we need supplementation by a "black box" saying what I represents. Thus $\boldsymbol{y} \in [E{}^I_O](\mathcal{P})$ means there is a derivation of $O\boldsymbol{y}$ using the axioms E and the additional rule: if $\boldsymbol{x} \in \mathcal{P}$ (which our "black box" tells us) then $I\boldsymbol{x}$ may be used as a line in the derivation.

We will think of effective type 2 operators similarly, as being given by incomplete elementary formal systems, but now the supplementation necessary consists of "black boxes" representing type 1 as well as type 0 objects. Before giving a proper definition, we present some examples.

EXAMPLE 1. Let I, O, J and K be distinct predicate symbols, with I and J n-place and O and K k-place. Let φ be the following set of elementary formal system axioms:

$$J\boldsymbol{x} \to I\boldsymbol{x}, \qquad O\boldsymbol{y} \to K\boldsymbol{y}.$$

Then, we use the notation

$$\begin{bmatrix} \begin{bmatrix} I \\ O \end{bmatrix}, & J \\ \varphi & \\ K & \end{bmatrix}$$

to denote a certain effective type 2 operator. First, the inputs are objects that can be substituted for $[{}^I_O]$ and J, and thus are type 1 operators of order $\langle n; k\rangle$ and n-ary relations. Second, outputs are k-ary relations. We say the *order* of this effective type 2 operator is $\langle\langle n, k\rangle, n; k\rangle$.

Finally we describe the intended behavior of this operator. Suppose Φ is a type 1 operator of order $\langle n, k\rangle$, and $\mathscr{P}$ is a k-ary relation. Then we intend that

$$\mathbf{y} \in \left[\begin{matrix} \begin{bmatrix} I \\ O \end{bmatrix}, & J \\ \varphi & \\ K & \end{matrix}\right](\Phi, \mathscr{P})$$

should mean the following. There is a derivation of $K\mathbf{y}$ over $\mathfrak{A}$, from the axioms φ, using the usual elementary formal system rules, together with:

If $\mathbf{x} \in \mathscr{P}$ then $J\mathbf{x}$ may be used as a line in the derivation.

At any point in the derivation, we may form the set $\mathscr{R}$ consisting of all $\mathbf{x}$ such that $I\mathbf{x}$ has been derived, give $\mathscr{R}$ to Φ, and if z comes out [that is, if $z \in \Phi(\mathscr{R})$] we may use Oz as a line in the derivation.

There are two points to be noted.

First, it would be inconvenient and somewhat unnatural if the use of Φ late in the derivation from φ gave us less output than it would have if used earlier. But this cannot happen since Φ, being a type 1 operator, must be monotone.

Second, if Φ is not compact, use of Φ in a derivation may not produce anything, unless we can arrange for Φ to be given infinite input. Since we definitely will want to consider non-compact Φ, we commit ourselves to allowing infinite (well-ordered) elementary formal system derivations. That this suggests the ω-rule is not coincidental.

We leave it to the reader to convince himself that the effective type 2 operator of this example has the following behavior:

$$\left[\begin{matrix} \begin{bmatrix} I \\ O \end{bmatrix}, & J \\ \varphi & \\ K & \end{matrix}\right](\Phi, \mathscr{P}) = \Phi(\mathscr{P}).$$

The above example makes no special use of any peculiarities of the structure $\mathfrak{A}$. Here is one that does.

EXAMPLE 2. This time we work over the structure of arithmetic, $\mathfrak{S}(\mathbb{N})$. Let I and J be n-place and O and K be 1-place. Let φ be the set of axioms:

axioms for multiplication,

$J\mathbf{x} \to I\mathbf{x}$,

$Oy \to y = 2z \to Ky$.

Consider the effective type 2 operator:

$$\varphi\begin{bmatrix} \begin{bmatrix} I \\ O \end{bmatrix}, & J \\ & \\ K & \end{bmatrix}.$$

It is easy to check that

$$\varphi\begin{bmatrix} \begin{bmatrix} I \\ O \end{bmatrix}, & J \\ & \\ K & \end{bmatrix}(\Phi, \mathscr{P}) = \{y \mid y \text{ is even and } y \in \Phi(\mathscr{P})\}.$$

There is still a serious deficiency in the above machinery, however. Return to Example 1, and suppose now that $n = k$, and we want an effective type 2 operator which, when supplied with Φ and $\mathscr{P}$ as inputs, gives out $\Phi(\Phi(\mathscr{P}))$. We would need to apply Φ to two quite different inputs, but as things stand, we have only a single predicate symbol I available to represent the input given to Φ, so in an elementary formal system derivation we could not keep the two applications of Φ separate. Our way out of this, for this particular example, is to associate with the *n-place* symbols I and O, two other predicate symbols I' and O' that are $1 + n$*-place*. And we suppose our rule for type 1 input, given above, is modified to the following:

Let a be a fixed member of $\mathscr{A}$. At any point in the derivation we may form the set $\mathscr{R}_a$ consisting of all $\boldsymbol{x}$ such that $I'a, \boldsymbol{x}$ has been derived, and if $\boldsymbol{y} \in \Phi(\mathscr{R}_a)$, we may use $O'a, \boldsymbol{y}$ as a line in the derivation.

EXAMPLE 3. Let I, O, J and K all be n-place, and let I' and O' be $1 + n$-place. Let Ψ be the set of axioms:

$$J\boldsymbol{x} \rightarrow I'1, \boldsymbol{x},$$

$$O'1, \boldsymbol{x} \rightarrow I'2, \boldsymbol{x},$$

$$O'2, \boldsymbol{x} \rightarrow K\boldsymbol{x}.$$

[Here 1 and 2 are any two distinct members of $\mathscr{A}$.]

Now consider the effective type 2 operator

$$\varphi\begin{bmatrix} \begin{bmatrix} I \\ O \end{bmatrix}, & J \\ & \\ K & \end{bmatrix}$$

whose behavior is characterized as before, but with the modified rule for type 1 inputs. We leave it to the reader to verify that, now,

$$\varphi\begin{bmatrix}\begin{bmatrix}I\\O\end{bmatrix}, & J\\ & \\ K & \end{bmatrix}(\Phi, \mathcal{P}) = \Phi(\Phi(\mathcal{P})).$$

A few more remarks, before we give a proper definition.

First, in Example 3, I was n-place and I' was $1+n$-place. It is just as natural to allow I' to be $q+n$-place, for some fixed $q \geqslant 0$. This covers Examples 1 and 3, and provides great freedom generally.

Second, it is not really necessary to have two symbols I and I'; I can serve very well for both. We can tell the two uses apart with no trouble.

Now we give our official definition of an *effective type* 2 *operator*. To keep notation readable, we assume all type 1 inputs are themselves single-input operators. Modifications to cover multiple-input type 1 operators are obvious. Let

$$\begin{bmatrix}I_1\\O_1\end{bmatrix}, \begin{bmatrix}I_2\\O_2\end{bmatrix}, \dots, \begin{bmatrix}I_n\\O_n\end{bmatrix}, \qquad J_1, J_2, \dots, J_m, K$$

be a list of type 1 and type 0 variables. We say the list is *independent* if $I_1, O_1, \dots, I_n, O_n, J_1, \dots, J_m, K$ are all distinct. Let us suppose now that we have an independent list. And let us suppose that I_p is i_p-place, O_p is o_p-place, J_p is j_p-place and K is k-place.

Let φ be a set of elementary formal system axioms over $\mathfrak{A}$, meeting the two conditions:

$J_1, J_2, \dots, J_m$ never occur in the conclusion of any axiom.

If I_p and O_p occur in the axioms, they do so as $q_p + i_p$- and $q_p + o_p$-place predicate symbols, for some single $q_p \geqslant 0$.

Then we say the symbol

$$\varphi\begin{bmatrix}\begin{bmatrix}I_1\\O_1\end{bmatrix}, \dots, \begin{bmatrix}I_n\\O_n\end{bmatrix}, J_1, \dots, J_m\\ \\ K\end{bmatrix}$$

denotes an effective type two operator. (We will generally follow the custom of listing the type 1 inputs before the type 0 inputs.) We say the *order* of this operator is $\langle\langle i_1, o_1\rangle, \dots, \langle i_n, o_n\rangle, j_1, \dots, j_m; k\rangle$.

The behavior of the operator denoted above is specified as follows. Let $\Phi_1, \dots, \Phi_n$ be a list of type 1 operators, with each Φ_p of the order corresponding to type 1 variable $\begin{bmatrix} I_p \\ O_p \end{bmatrix}$, namely $\langle i_p, o_p \rangle$. Let $\mathcal{P}_1, \ldots, \mathcal{P}_m$ be a list of type 0 objects, with $\mathcal{P}_p$ being j_p-ary, so that it corresponds to the type 0 variable J_p. Then we say

$$y \in \left[\begin{matrix} & \begin{bmatrix} I_1 \\ O_1 \end{bmatrix}, \ldots, \begin{bmatrix} I_n \\ O_n \end{bmatrix}, J_1, \ldots, J_m \\ \varphi & \\ & K \end{matrix}\right] (\Phi_1, \ldots, \Phi_n, \mathcal{P}_1, \ldots, \mathcal{P}_m),$$

provided Ky is the last line of a well-ordered (possibly infinite) sequence of formulas constructed according to the usual rules for an elementary formal system derivation from φ over $\mathfrak{A}$, together with the extra rules:

Rule 0: If $x \in \mathcal{P}_p$ then $J_p x$ may be used as a line in the derivation.

Rule 1: Suppose I_p and O_p are used in the axioms φ as $q_p + i_p$- and $q_p + o_p$-place predicate symbols respectively. Let a be a fixed sequence of members of $\mathcal{A}$, of length q_p. Then $O_p a, y$ may be used as a line in the derivation provided $y \in \Phi_p(\mathcal{R}_a)$, where $\mathcal{R}_a$ is the set of x such that $I_p a, x$ occurs in the derivation before that line.

4. Properties of effective type 2 operators

By our definition, effective operators of types 1 and 2 have objects of type 0 as outputs. We now introduce a convention that will allow us to think of them as having objects of other types as outputs as well. Say

$$\left[\begin{matrix} & \begin{bmatrix} I \\ O \end{bmatrix}, J \\ \varphi & \\ & K \end{matrix}\right]$$

is an effective type 2 operator, where I is i-place, O is o-place, J is j-place and K is k-place. For convenience let us temporarily write F for this operator. Then F, when supplied with a type 1 operator Φ of order $\langle i, o \rangle$, and a type 0 object $\mathcal{P}$ which is j-ary, gives as output the type 0 object $F(\Phi, \mathcal{P})$, a k-ary relation.

Now, suppose we only supply Φ as input. What we get is incomplete, $F(\Phi, \text{-})$. When supplied with a type 0 input $\mathcal{P} \in [\mathcal{A}]^j$, it gives back a type 0 output $F(\Phi, \mathcal{P}) \in [\mathcal{A}]^k$. But this is a description of a map from $[\mathcal{A}]^j$ to $[\mathcal{A}]^k$. It is easy to check that it is monotone. Thus we have described a type 1 operator, of order $\langle j, k \rangle$. We will denote it by $(\lambda x)F(\Phi, x)$. So we may also think of F as giving us a map from type 1 to type 1: when Φ is given as input, $(\lambda x)F(\Phi, x)$ is produced.

When we are thinking of F as mapping to type 0 [given Φ and $\mathcal{P}$ we get $F(\Phi, \mathcal{P})$] we will say the order of F is $\langle\langle i, o\rangle, j; k\rangle$ as in Section 3. When we are thinking of F as mapping to type 1 [given Φ we get $(\lambda x)F(\Phi, x)$] we will say the order is $\langle\langle i, o\rangle; \langle j, k\rangle\rangle$.

As another example, suppose $[E_K^{I,J}]$ is a type 1 operator (taking 2 inputs) where I is i-place, J is j-place and K is k-place. We may think of this as being of order $\langle i, j; k\rangle$. But we may also think of it as giving a map of order $\langle i; \langle j, k\rangle\rangle$: when $\mathcal{P} \in [\mathcal{A}]^i$ is supplied, we get the type 1 operator $(\lambda x)[E_K^{I,J}](\mathcal{P}, x)$.

We believe these examples are sufficiently representative, and we do not give a proper definition, which is rather involved. But, from now on, we will feel free to consider operators whose output is not necessarily a type 0 object. We now turn to some more specific examples.

EXAMPLE 1. Let I and J be n-place and O and K be k-place. Consider the set φ of elementary formal system axioms, over $\mathfrak{A}$:

$$\boldsymbol{Jx \to Ix}, \qquad \boldsymbol{Oy \to Ky}.$$

This is Example 1 of Section 3, and thus it is an effective operator of order $\langle\langle n, k\rangle, n; k\rangle$ such that

$$\begin{bmatrix} & \begin{bmatrix} I \\ O \end{bmatrix}, & J \\ \varphi & & \\ & K & \end{bmatrix}(\Phi, \mathcal{P}) = \Phi(\mathcal{P}).$$

But we may also think of φ as defining an effective operator of order $\langle\langle n, k\rangle; \langle n, k\rangle\rangle$, and it should be clear that

$$(\lambda x)\begin{bmatrix} & \begin{bmatrix} I \\ O \end{bmatrix}, & J \\ \varphi & & \\ & K & \end{bmatrix}(\Phi, x) = \Phi$$

for all type 1 operators Φ of order $\langle n, k\rangle$. Looked at this way, we have the *identity operator* on type 1 objects of order $\langle n, k\rangle$, which is thus effective.

EXAMPLE 2. In Chapter 1 we showed the collection of enumeration operators over $\mathfrak{A}$ was closed under $\cap$, $\cup$, $\times$ and composition. Actually, the proofs we gave establish the stronger fact that these operations are given by effective type 2 operators. We show this for $\cap$ and leave the rest to the reader. Please compare the following with the proof of Proposition 1.10.3.

Let A, C and I be n-place, and B, D and O be m-place. Let φ be the following set of axioms; in which all predicate letters occur with the same arity as in the type 2 variables:

$$\boldsymbol{Ix \to Ax}, \qquad \boldsymbol{Ix \to Cx},$$

$$\boldsymbol{By \to Dy \to Oy}.$$

Then

$$\left[\begin{matrix} & \begin{bmatrix} A \\ B \end{bmatrix}, \begin{bmatrix} C \\ D \end{bmatrix}, \quad I \\ \varphi & \\ & O \end{matrix}\right]$$

is an effective type 2 operator, and it is easy to see that for *any* type 1 operators Φ and Ψ, of order $\langle n, m\rangle$, effective or not,

$$(\lambda x)\left[\begin{matrix} & \begin{bmatrix} A \\ B \end{bmatrix}, \begin{bmatrix} C \\ D \end{bmatrix}, \quad I \\ \varphi & \\ & O \end{matrix}\right](\Phi, \Psi, x) = \Phi \cap \Psi.$$

REMARK. An effective type 2 operator, given effective input, yields effective output. Thus, if

$$\left[\begin{matrix} & \begin{bmatrix} I \\ O \end{bmatrix}, \quad J \\ \varphi & \\ & K \end{matrix}\right]$$

is an effective type 2 operator, Φ is an enumeration operator, and $\mathcal{P}$ is recursively enumerable, then

$$\left[\begin{matrix} & \begin{bmatrix} I \\ O \end{bmatrix}, \quad J \\ \varphi & \\ & K \end{matrix}\right](\Phi, \mathcal{P})$$

is recursively enumerable. Similarly,

$$(\lambda x)\left[\begin{matrix} & \begin{bmatrix} I \\ O \end{bmatrix}, \quad J \\ \varphi & \\ & K \end{matrix}\right](\Phi, x)$$

is an enumeration operator. In each case, appropriate axioms for the output can be obtained by simply combining axioms for the inputs with the axioms of φ.

This remark, together with Example 2, provides a re-proof of Proposition 1.10.3 (1).

EXAMPLE 3. In Ch. 1, §12 we proved the Least Fixed Point Theorem for enumeration operators. The proof took the following form. Let $[E^I_O]$ be an enumeration operator of order $\langle n, n\rangle$. If we let E' be the axioms of E together with the axiom $O\mathbf{x} \to I\mathbf{x}$, then using E' the symbol O will represent the least fixed point of $[E^I_O]$, which is thus r.e. As a matter of fact, the proof of Theorem 1.12.1 actually establishes something stronger.

Let φ be the following set of axioms; where I, O and K are n-place:

$$O\mathbf{x} \to I\mathbf{x}, \qquad O\mathbf{x} \to K\mathbf{x}.$$

Then the operator

$$\begin{bmatrix} & \begin{bmatrix} I \\ O \end{bmatrix} \\ \varphi & \\ & K \end{bmatrix}$$

of order $\langle\langle n, n\rangle; n\rangle$ is a "least fixed point" operator, in that, for *any* type 1 object Φ of order $\langle n, n\rangle$,

$$\begin{bmatrix} & \begin{bmatrix} I \\ O \end{bmatrix} \\ \varphi & \\ & K \end{bmatrix}(\Phi)$$

is the least fixed point of Φ. [Since type 1 operators are monotone, Φ must have a least fixed point, as the first 4 parts of the proof of Theorem 1.12.1 shows.] This, together with the above remarks on effective output, given effective input, provides an alternative approach to the proof of Theorem 1.12.1.

EXAMPLE 4. Let D be the operator of order $\langle 1, 1\rangle$ on $\mathscr{A}$ defined by

$$D(\mathscr{P}) = \begin{cases} \{0\} & \text{if } \mathscr{P} = \mathscr{A}, \\ \emptyset & \text{if } \mathscr{P} \neq \mathscr{A}. \end{cases}$$

D is trivially monotone, but is not compact (provided $\mathscr{A}$ is infinite). Now,

ω-elementary formal systems and type 2 elementary formal systems involving D can be intertranslated to establish:

A relation $\mathcal{R}$ is ω-r.e. if and only if there is some effective type 2 operator F such that $\mathcal{R} = F(D)$.

More generally, if Φ is an ω-enumeration operator, then there is an effective type 2 operator F such that, for all $\mathcal{P}$, $\Phi(\mathcal{P}) = F(D, \mathcal{P})$ and conversely.

We leave a general proof of this to the reader, and merely indicate how one can translate the particular ω-enumeration operator which was defined in Ch. 8, §9 as an example of a non-compact operator. Recall: we work over the structure $\mathfrak{S}(\mathbb{N})$, and Φ is the ω-enumeration operator of order $\langle 1, 1\rangle$ with axioms:

$$\text{axioms for } <,$$

$$Iy \to x < y \to Gx,$$

$$G\forall \to O5.$$

Now, let φ be the set of axioms:

$$\text{axioms for } <,$$

$$Iy \to x < y \to Gx,$$

$$H0 \to O5,$$

and consider the effective type 2 operator

$$\begin{bmatrix} & \begin{bmatrix} G \\ H \end{bmatrix}, & I \\ \varphi & & \\ & O & \end{bmatrix}$$

It is easy to see that, for $\mathcal{P} \in [\mathbb{N}]^1$,

$$\begin{bmatrix} & \begin{bmatrix} G \\ H \end{bmatrix}, & I \\ \varphi & & \\ & O & \end{bmatrix}(D, \mathcal{P}) = \Phi(\mathcal{P})$$

This, in effect, makes ω-recursion theory a branch of the recursion theory of higher type operators. It is the analog, for our development, of a famous result of Kleene [1959] relating the hyperarithmetic functions with those recursive in the type 2 existential quantifier. In our case the switch from existential to (essentially) universal quantifier arises from the fact that we consider (essentially) enumeration reducibility as basic, rather than relative recursiveness.

5. Effective type 3 operators

Extending the pattern of earlier sections, we now define the effective type 3 operators. An effective type 3 operator will be specified by a set of elementary formal system axioms φ, and will take as inputs, type 2 operators. If F is a type 2 operator, how can we make use of F in a derivation from φ, since F needs to be given type 1 operators to work? The idea is, during the course of a derivation from φ we will be allowed to form "subsidiary operators". Thus, let I and O be predicate letters of φ (with I not occurring in the conclusion of any axiom). We can, temporarily, think of φ as instructions for turning I's into O's; but this is the description of a type 1 operator. As such, it can be given to F as input, and the output can be suitably entered into a "main" derivation from φ. We have yet to introduce notation that will allow us to associate F with the variables I and O, and that will tell us how to label the outputs of F when entered as lines in the derivation from φ. That will be dealt with below.

Again, the structure $\mathfrak{A}$ is fixed for this section, and all work is over it.

We begin by defining the type 2 variables. Let $t_1, t_2, \ldots, t_k$ be a list of type 1 and type 0 variables, and let O be a type 0 variable. We say the list $t_1, t_2, \ldots, t_k, O$ is *independent* if no two of them share a predicate letter. If the list is independent, we say

$$\begin{bmatrix} t_1, t_2, \ldots, t_k \\ \\ O \end{bmatrix}$$

is a *type* 2 *variable*. For example,

$$\begin{bmatrix} \begin{bmatrix} I \\ O \end{bmatrix}, \begin{bmatrix} J, K \\ L \end{bmatrix}, M \\ N \end{bmatrix}$$

is a type 2 variable.

Further, suppose that the order of t_i is s_i, and that O is 0-place. Then we say the *order* of

$$\begin{bmatrix} t_1, t_2, \ldots, t_k \\ \\ O \end{bmatrix}$$

is $\langle s_1, s_2, \ldots, s_k\,; 0\rangle$.

Next we say what can be "substituted" for a type 2 variable, that is, we say what the *type* 2 *operators* are (in Section 3 we only defined the *effective* type 2 operators). The only restriction we want to make is that they should be *monotone*, as we now define it.

On type 0 objects a partial ordering, $\subseteq$ (subset) already exists. We now define a similar notion on type 1 objects. Let Φ and Ψ be two type 1 operators of the same order. Suppose, for all (appropriate) inputs $\mathcal{P}_1, \dots, \mathcal{P}_k$ we have $\Phi(\mathcal{P}_1, \dots, \mathcal{P}_j) \subseteq \Psi(\mathcal{P}_1, \dots, \mathcal{P}_j)$. Then we write $\Phi \subseteq \Psi$.

Now, let

$$\begin{bmatrix} t_1, \dots, t_k \\ \\ O \end{bmatrix}$$

be a type 2 variable. Let $\sigma(t_i)$ be the set of objects of the type and order corresponding to the variable t_i, and similarly for $\sigma(O)$. Consider a map

$$F : \sigma(t_1) \times \cdots \times \sigma(t_k) \to \sigma(O).$$

We say F is *monotone* if, for all $x_i, y_i \in \sigma(t_i)$ we have

$$x_1 \subseteq y_1 \wedge \cdots \wedge x_k \subseteq y_k \;\Rightarrow\; F(x_1, \dots, x_k) \subseteq F(y_1, \dots, y_k).$$

If F is monotone we call it a *type* 2 *operator* or a *type* 2 *object* whose order is the order of

$$\begin{bmatrix} t_1, \dots, t_k \\ \\ O \end{bmatrix}.$$

It should be clear that the effective type 2 operators as defined in Section 3, are monotone in the above sense, hence are among the type 2 objects as just defined.

Next we say how we will denote effective type 3 operators.

Let $\boldsymbol{t}$ be a list of type 2 variables, let $\boldsymbol{o}$ be a list of type 1 variables (possibly empty) and let $\boldsymbol{z}$ be a list of type 0 variables (possibly empty). Also let O be a type 0 variable. Suppose the combined list, $\boldsymbol{t}, \boldsymbol{o}, \boldsymbol{z}, O$ is independent. Further, let φ be a set of elementary formal system axioms over $\mathfrak{A}$. Then we say

$$\begin{bmatrix} & \boldsymbol{t}, \boldsymbol{o}, \boldsymbol{z} \\ \varphi & \\ & O \end{bmatrix}$$

denotes an effective type 3 *operator* provided the following syntactic conditions are met.

First, the type 0 and 1 variables in $\boldsymbol{z}$ and $\boldsymbol{o}$ must meet the conditions given in Section 3, under the definition of effective type 2 operator.

Then there are conditions on the type 2 variables in $\boldsymbol{t}$. Say one of them is

$$t = \begin{bmatrix} \begin{bmatrix} I_1, \ldots, I_i \\ O_1 \end{bmatrix}, \ldots, \begin{bmatrix} I_n, \ldots, I_m \\ O_k \end{bmatrix}, J_1, \ldots, J_j \\ \\ K \end{bmatrix}$$

and let us say that, for each predicate symbol S in t, it occurs there as a $p(S)$-place symbol.

The first condition on the type 2 variable t is: none of $I_1, \ldots, I_i, \ldots, I_n, \ldots, I_m$ may occur in the conclusion of any axiom in φ. (The reason is, later on we will want to form "subsidiary type 1 operators" whose inputs will be represented by $I_1, \ldots, I_m$ and so each I_i must satisfy restrictions appropriate to a type 1 input variable.)

The second condition on t is: there is an integer $q \geqslant 0$ such that, for any predicate symbol S in t, any occurrence of S *in the axioms* is as a $q + p(S)$-place predicate symbol.

This ends the list of conditions necessary for us to say

$$\begin{bmatrix} & \boldsymbol{t}, \boldsymbol{o}, \boldsymbol{z} \\ \varphi & \\ & O \end{bmatrix}$$

denotes an effective type 3 operator.

Next we must say how the effective type 3 operator denoted by the above behaves. To keep things relatively simple, we give in detail the rules for when there is only one type 2 variable t in $\boldsymbol{t}$, and just outline what to do if there are more. And as a further simplification, we only discuss the case where t has the form

$$\begin{bmatrix} \begin{bmatrix} I \\ J \end{bmatrix} \\ K \end{bmatrix}.$$

More complicated, multiple-input variables are treated analogously.

Thus let us suppose

$$\begin{bmatrix} & \begin{bmatrix} \begin{bmatrix} I \\ J \end{bmatrix} \\ K \end{bmatrix}, \boldsymbol{o}, \boldsymbol{z} \\ \varphi & \\ & O \end{bmatrix}$$

denotes an effective type 3 operator, where $\boldsymbol{o}$ is a list of type 1 variables and $\boldsymbol{z}$ is a list of type 0 variables. Let F be a type 2 operator of the same order as the type 2 variable

$$\begin{bmatrix} \begin{bmatrix} I \\ J \end{bmatrix} \\ K \end{bmatrix}$$

Let $\boldsymbol{\Phi}$ be a sequence of type 1 operators, in which the ith term is of the order corresponding to the ith variable in the list $\boldsymbol{o}$. Similarly let $\mathscr{P}$ be a list of type 0 objects, matched up with the list $\boldsymbol{z}$. Then

$$\begin{bmatrix} \begin{bmatrix} \begin{bmatrix} I \\ J \end{bmatrix} \\ K \end{bmatrix}, \boldsymbol{o}, \boldsymbol{z} \\ \varphi \\ O \end{bmatrix} (F, \boldsymbol{\Phi}, \mathscr{P})$$

is to be a certain relation; we proceed to say what.

We intend that

$$\boldsymbol{v} \in \begin{bmatrix} \begin{bmatrix} \begin{bmatrix} I \\ J \end{bmatrix} \\ K \end{bmatrix}, \boldsymbol{o}, \boldsymbol{z} \\ \varphi \\ O \end{bmatrix} (F, \boldsymbol{\Phi}, \mathscr{P})$$

is to mean: $O\boldsymbol{v}$ is the last line of what we call a *main derivation* from φ over $\mathfrak{A}$. In that derivation, we follow all the usual elementary formal system rules, from Ch. 1, §3, and rules 0 and 1 from Section 3 of this chapter, for making use of the type 0 and type 1 inputs $\mathscr{P}$ and $\boldsymbol{\Phi}$. And we are allowed an additional rule, rule 2, for making use of the type 2 input F. Before we can state that rule, however, we must say what *subsidiary operators* are, and how they behave.

Recall $p(S)$ is the a-rity of the predicate symbol S in its occurrence in an input variable. By our assumptions, there is an integer q such that, in the elementary formal system axioms of φ, I occurs as a $q+p(I)$-place predicate symbol, J occurs as a $q+p(J)$-place predicate symbol, and K occurs as a $q+p(K)$-place predicate symbol.

Subsidiary operators. Suppose the main derivation has been completed through κ lines. Let $\boldsymbol{a}$ be a list of constants, from the domain $\mathscr{A}$, of length q. Then we may *create, after κ lines, the subsidiary operator*

$$(\lambda \boldsymbol{a}, x)\left[\begin{array}{ccc} & \boldsymbol{o}, z, I & \\ \varphi & & \\ & J & \end{array}\right](\boldsymbol{\Phi}, \mathcal{P}, x)$$

with parameters $\boldsymbol{a}$. It is to be a type 1 operator of the same order as the variable $[{}^{I}_{J}]$. Now, we must say how this subsidiary operator behaves. That is, if $\mathcal{R}$ is a $p(I)$-place relation, we must say when

$$\boldsymbol{w} \in \left((\lambda \boldsymbol{a}, x)\left[\begin{array}{ccc} & \boldsymbol{o}, z, I & \\ \varphi & & \\ & J & \end{array}\right](\boldsymbol{\Phi}, \mathcal{P}, x)\right)(\mathcal{R}).$$

Well, this is to happen if $J\boldsymbol{a}, \boldsymbol{w}$ is the last line of a *subsidiary derivation.* A subsidiary derivation is: a continuation of the main derivation from the first κ lines; which follows the usual elementary formal system rules, and rules 0 and 1 for using $\mathcal{P}$ and $\boldsymbol{\Phi}$; and which allows the extra rule that, if $\boldsymbol{u} \in \mathcal{R}$ then $I\boldsymbol{a}, \boldsymbol{u}$ may be used as a line in the subsidiary derivation.

Now we have said what subsidiary operators are, and we are finally in a position to state rule 2, for making use of F in the main derivation.

RULE 2. Suppose the main derivation has been completed through κ lines. Then we may create, after κ lines, the subsidiary operator

$$(\lambda \boldsymbol{a}, x)\left[\begin{array}{ccc} & \boldsymbol{o}, z, I & \\ \varphi & & \\ & J & \end{array}\right](\boldsymbol{\Phi}, \mathcal{P}, x)$$

with parameters $\boldsymbol{a}$; denote it by $[\Psi^{I}_{\boldsymbol{a}J}]$ for convenience. $[\Psi^{I}_{\boldsymbol{a}J}]$ is of the order of $[{}^{I}_{J}]$, hence is appropriate as input for F. If $\boldsymbol{y} \in F([\Psi^{I}_{\boldsymbol{a}J}])$ then $K\boldsymbol{a}, \boldsymbol{y}$ may be entered as line γ for any $\gamma > \kappa$, in the main derivation.

REMARKS. (1) The behavior of a subsidiary operator is thus determined by running subsidiary derivations; but each different $p(I)$-ary input $\mathcal{R}$ requires a different subsidiary derivation. All of them together give the behavior of the subsidiary operator $[\Psi^{I}_{\boldsymbol{a}J}]$. But once rule 2 is used, those subsidiary derivations are "erased" and we return to the main derivation. Thus we might picture things as follows.

main derivation,
consisting of κ lines

here $[\Psi^{I}_{\boldsymbol{a}J}]$ is created

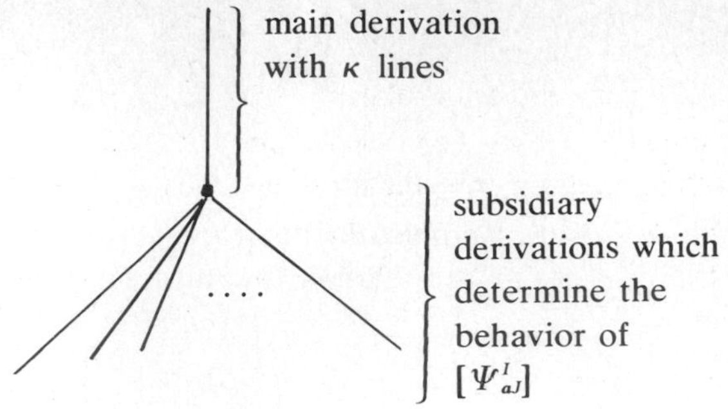

Then, if $\boldsymbol{y} \in F([\Psi^I_{aJ}])$, we might continue

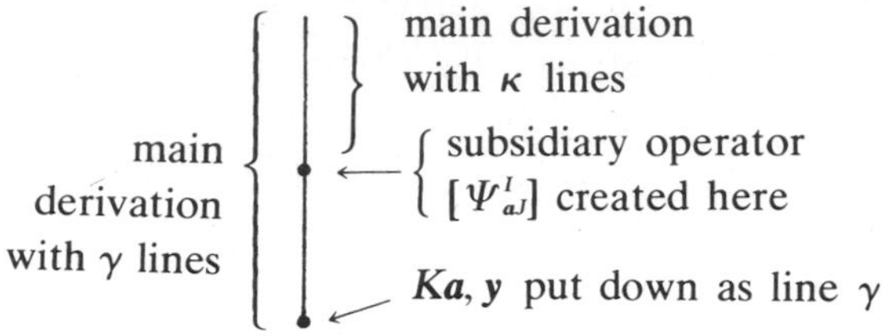

(2) A subsidiary operator, involving the parameters $\boldsymbol{a}$, can be created at more than one point in the main derivation. Let us, for the moment, use

$$[\Psi^I_{aJ}]_\kappa \quad \text{and} \quad [\Psi^I_{aJ}]_\delta$$

for two such creations, one after κ lines and one after δ lines, with $\delta > \kappa$. Since later in the main derivation we have more to work with, so to speak, it is easy to see that, as type 1 operators

$$[\Psi^I_{aJ}]_\kappa \subseteq [\Psi^I_{aJ}]_\delta .$$

But our type 2 input, F, must be *monotone*, so

$$F([\Psi^I_{aJ}]_\kappa) \subseteq F([\Psi^I_{aJ}]_\delta).$$

Thus, by delaying the application of rule 2, nothing is lost.

(3) Next, what happens if there are more than one type 2 inputs. Say we have

$$\left[\begin{array}{l} \left[\begin{array}{c} \left[\begin{array}{c} I \\ J \end{array}\right] \\ K \end{array}\right], \left[\begin{array}{c} \left[\begin{array}{c} L \\ M \end{array}\right] \\ N \end{array}\right], \boldsymbol{o}, \boldsymbol{z} \\ \varphi \\ \quad O \end{array}\right] (F, G, \boldsymbol{\Phi}, \boldsymbol{\mathcal{P}}).$$

We outline the procedure without giving a full set of rules.

As above, we run a main derivation. At various points we may create subsidiary operators, for use as input to F. But those subsidiary operators will still contain a type 2 input, G. So, in running subsidiary derivations, we allow the further creation of sub-subsidiary operators, to be used as input for G in the subsidiary derivations, and whose behavior is determined by running sub-subsidiary derivations in the obvious way. We leave an exact formulation of the rules to any reader who desires it (its rather messy), and we consider some examples in the next section, which we think are sufficiently representative.

6. Examples of effective type 3 operators

We give several examples of effective type 3 operators. All our examples are over the structure $\mathfrak{A}$.

EXAMPLE 1. Let I and J be n-place, O and K be k-place, and L and M be m-place, in both the variables and in the elementary formal system axioms below. Consider the following set φ of axioms:

$$Jx \to Ix, \qquad Oy \to Ky, \qquad Lz \to Mz.$$

Then

$$\left[\begin{array}{l} \left[\begin{array}{c} \left[\begin{array}{c} J \\ K \end{array}\right] \\ L \end{array}\right], \left[\begin{array}{c} I \\ O \end{array}\right] \\ \varphi \\ \quad M \end{array}\right]$$

denotes an effective type 3 operator. We consider its behavior by looking at

$$\left[\begin{array}{l} \left[\begin{array}{c} \left[\begin{array}{c} J \\ K \end{array}\right] \\ L \end{array}\right], \left[\begin{array}{c} I \\ O \end{array}\right] \\ \varphi \\ \quad M \end{array}\right] (F, \Phi) \qquad (*)$$

where F is of type 2, Φ is of type 1, and each are of the appropriate orders. We see what formulas of the form Mz have a main derivation, using (*).

Right at the start of our main derivation we may create the subsidiary operator (with no parameters)

$$(\lambda x)\begin{bmatrix} & \begin{bmatrix} I \\ O \end{bmatrix}, & J \\ \varphi & & \\ & K & \end{bmatrix}(\Phi, x)$$

which we denote $[\varphi{}^{J}_{K}]$ for short. Now, by Example 1 of Section 4, this is simply the operator Φ.

Then by rule 2, we may use Lz as line 1 in the main derivation provided $z \in F([\varphi{}^{J}_{K}])$, that is, provided $z \in F(\Phi)$. But $Lz \to Mz$ is one of the axioms in φ. It follows that there is a main derivation ending with Mz for every $z \in F(\Phi)$. Thus

$$F(\Phi) \subseteq \begin{bmatrix} & \begin{bmatrix} \begin{bmatrix} J \\ K \end{bmatrix} \\ L \end{bmatrix}, & \begin{bmatrix} I \\ O \end{bmatrix} \\ \varphi & & \\ & M & \end{bmatrix}(F, \Phi)$$

On the other hand, suppose there is a main derivation, using (*), which ends with Mz. The only way Mz can occur as a line in this derivation is if Lz occurs earlier. And the only way Lz can occur is by use of rule 2, which means we must have

$$z \in F([\varphi{}^{J}_{K}])$$

where $[\varphi{}^{J}_{K}]$ is the subsidiary operator created after line κ, say. But it should be clear that no matter where in the main derivation we create $[\varphi{}^{J}_{K}]$, it will still be the operator Φ. Thus the only way Mz could be derived is if $z \in F(\Phi)$. Thus

$$\begin{bmatrix} & \begin{bmatrix} \begin{bmatrix} J \\ K \end{bmatrix} \\ L \end{bmatrix}, & \begin{bmatrix} I \\ O \end{bmatrix} \\ \varphi & & \\ & M & \end{bmatrix}(F, \Phi) \subseteq F(\Phi).$$

In summary, then,

$$\begin{bmatrix} & \begin{bmatrix} \begin{bmatrix} J \\ K \end{bmatrix} \\ L \end{bmatrix}, & \begin{bmatrix} I \\ O \end{bmatrix} \\ \varphi & & \\ & M & \end{bmatrix}(F, \Phi) = F(\Phi)$$

and thus "point evaluation" is an effective type 3 operator.

EXAMPLE 2. Use the same set-up as in Example 1. Let x^1 be a type 1 place holder; then under the obvious interpretation of notation,

$$(\lambda x^1)\begin{bmatrix} \begin{bmatrix}\begin{bmatrix} J\\ K\end{bmatrix}\\ L\end{bmatrix}, \begin{bmatrix} I\\ O\end{bmatrix} \\ \varphi \\ M\end{bmatrix}(F, x^1) = F$$

and thus there are effective type 3 identity operators.

EXAMPLE 3. Let I, J, K, L, M, N, O, P and Q be predicate symbols as follows:

symbol	a-rity in variables	a-rity in axioms
I	n	n
J	m	m
K	k	k
L	n	n
M	m	m
N	k	k
O	n	$1+n$
P	m	$1+m$
Q	k	k

and let φ be the following set of elementary formal system axioms (we assume 1 and 2 are two distinct members of $\mathscr{A}$):

$$I\mathbf{x} \rightarrow O1, \mathbf{x}, \qquad L\mathbf{x} \rightarrow O2, \mathbf{x},$$

$$P1, \mathbf{y} \rightarrow J\mathbf{y}, \qquad P2, \mathbf{y} \rightarrow M\mathbf{y},$$

$$K\mathbf{z} \rightarrow N\mathbf{z} \rightarrow Q\mathbf{z}.$$

Then

$$\begin{bmatrix} \begin{bmatrix}\begin{bmatrix} I\\ J\end{bmatrix}\\ K\end{bmatrix}, \begin{bmatrix}\begin{bmatrix} L\\ M\end{bmatrix}\\ N\end{bmatrix}, \begin{bmatrix} O\\ P\end{bmatrix} \\ \varphi \\ Q\end{bmatrix}$$

denotes an effective type 3 operator. By techniques similar to those of Example 1 it can be shown that, for type 2 operators F and G, and a type 1 operator Φ, of appropriate orders, we have:

$$\begin{bmatrix} \begin{bmatrix} \begin{bmatrix} I \\ J \end{bmatrix} \\ K \end{bmatrix}, \begin{bmatrix} \begin{bmatrix} L \\ M \end{bmatrix} \\ N \end{bmatrix}, \begin{bmatrix} O \\ P \end{bmatrix} \\ \varphi \\ Q \end{bmatrix} (F, G, \Phi) = F(\Phi) \cap G(\Phi).$$

This is the first example we have presented with two type 2 inputs. Still, a careful check will show that, in this case, sub-subsidiary derivations are not really needed.

Again, using x^1 as a type 1 place holder, under the obvious interpretation of the notation, we have

$$(\lambda x^1) \begin{bmatrix} \begin{bmatrix} \begin{bmatrix} I \\ J \end{bmatrix} \\ K \end{bmatrix}, \begin{bmatrix} \begin{bmatrix} L \\ M \end{bmatrix} \\ N \end{bmatrix}, \begin{bmatrix} O \\ P \end{bmatrix} \\ \varphi \\ Q \end{bmatrix} (F, G, x^1) = F \cap G.$$

Similarly one can show there are effective type 3 operators for $\cup$ and $\times$.

EXAMPLE 4. Let I, N be n-place, L, P be p-place, Q, M be m-place, O, J be j-place, and K, R be r-place, both in variables and in axioms. Let φ be the set of axioms:

$$Ix \rightarrow Nx, \qquad Oy \rightarrow Jy,$$

$$Lz \rightarrow Pz, \qquad Qw \rightarrow Mw, \qquad Kt \rightarrow Rt.$$

Then

$$\begin{bmatrix} \begin{bmatrix} \begin{bmatrix} I \\ J \end{bmatrix} \\ K \end{bmatrix}, \begin{bmatrix} \begin{bmatrix} L \\ M \end{bmatrix}, & N \\ O & \end{bmatrix}, \begin{bmatrix} P \\ Q \end{bmatrix} \\ \varphi \\ R \end{bmatrix}$$

denotes an effective type 3 operator.

Now, let F be a type 2 operator of order $\langle\langle n, j\rangle; r\rangle$, let G be a type 2 operator of order $\langle\langle p, m\rangle; \langle n, j\rangle\rangle$, and let Φ be a type 1 operator of order $\langle p, m\rangle$. We claim

$$\begin{bmatrix} \begin{bmatrix}\begin{bmatrix} I \\ J\end{bmatrix} \\ K\end{bmatrix}, \begin{bmatrix}\begin{bmatrix} L \\ M\end{bmatrix}, \; N \\ O\end{bmatrix}, \begin{bmatrix} P \\ Q\end{bmatrix} \\ \varphi \\ R \end{bmatrix}(F, G, \Phi) = F(G(\Phi)) = F((\lambda x)G(\Phi, x)).$$

We sketch a verification of this claim. This time, sub-subsidiary derivations play a role.

Begin a main derivation for the above. At the very start we may create the subsidiary operator (with no parameters)

$$(\lambda x)\begin{bmatrix} \begin{bmatrix}\begin{bmatrix} L \\ M\end{bmatrix}, \; N \\ O\end{bmatrix}, \begin{bmatrix} P \\ Q\end{bmatrix}, \; I \\ \varphi \\ J \end{bmatrix}(G, \Phi, x).$$

Let us denote this by $[\varphi\,^{I}_{J}]$ for short. The intention is that we will use this as input for F, by rule 2, to get line 1 of the main derivation.

First the behavior of $[\varphi\,^{I}_{J}]$ must be determined, so we must run subsidiary derivations. Let $\mathscr{P} \subseteq \mathscr{A}^n$; we run a subsidiary derivation for

$$\begin{bmatrix} \begin{bmatrix}\begin{bmatrix} L \\ M\end{bmatrix}, \; N \\ O\end{bmatrix}, \begin{bmatrix} P \\ Q\end{bmatrix}, \; I \\ \varphi \\ J \end{bmatrix}(G, \Phi, \mathscr{P}).$$

Now, in this subsidiary derivation we may enter $I\boldsymbol{x}$ for any $\boldsymbol{x} \in \mathscr{P}$. One of our axioms is $I\boldsymbol{x} \rightarrow N\boldsymbol{x}$, so we may enter $N\boldsymbol{x}$ for each $\boldsymbol{x} \in \mathscr{P}$. Let us say this has been done, and so, say, κ lines of the subsidiary derivation have been completed.

Next, let us further create, after these κ lines, the sub-subsidiary operator

$$(\lambda y)\begin{bmatrix} \begin{bmatrix} P \\ Q\end{bmatrix}, \; I, \; L \\ \varphi \\ M \end{bmatrix}(\Phi, \mathscr{P}, y).$$

Let us call it $[\varphi\,^{L}_{M}]$ for short. The intention is to use it as input for G in a rule 2 application in the subsidiary derivation.

So now we must determine the behavior of $[\varphi\,^{L}_{M}]$. We must run sub-subsidiary derivations. Let $\mathscr{Q} \subseteq \mathscr{A}^p$; we run a sub-subsidiary derivation for

$$\begin{bmatrix} & \begin{bmatrix} P \\ Q \end{bmatrix}, \; I, \; L \\ \varphi & \\ & M \end{bmatrix} (\Phi, \mathscr{P}, \mathscr{Q}).$$

In such a sub-subsidiary derivation, we may enter Lz for exactly the $z \in \mathscr{Q}$. One of the axioms is $Lz \to Pz$, so we may derive Pz for exactly the $z \in \mathscr{Q}$. Then by rule 1, applied in our sub-subsidiary derivation, we may enter Qw for precisely those $w \in \Phi(\mathscr{Q})$. Again, one of the axioms is $Qw \to Mw$, so, in a sub-subsidiary derivation, we can get Mw for exactly those $w \in \Phi(\mathscr{Q})$. Thus we have

$$\begin{bmatrix} & \begin{bmatrix} P \\ Q \end{bmatrix}, \; I, \; L \\ \varphi & \\ & M \end{bmatrix} (\Phi, \mathscr{P}, \mathscr{Q}) = \Phi(\mathscr{Q}).$$

It follows that the sub-subsidiary operator $[\varphi^L_M]$, created after κ lines in the subsidiary derivation, is simply Φ itself.

We now return to the subsidiary derivation. After κ lines we have derived Nx for each $x \in \mathscr{P}$, and we have created the sub-subsidiary operator $[\varphi^L_M] = \Phi$. Then an application of rule 2 in the subsidiary derivation allows us to enter, as the next line, Oy for any $y \in G(\Phi, \mathscr{P})$. One of our axioms is $Oy \to Jy$, thus there is a subsidiary derivation of Jy for each $y \in G(\Phi, \mathscr{P})$. This means we have worked out the behavior of our subsidiary operator $[\varphi^I_J]$ far enough to conclude

$$G(\Phi, \mathscr{P}) \subseteq \begin{bmatrix} & \begin{bmatrix} \begin{bmatrix} L \\ M \end{bmatrix}, \; N \\ \quad O \end{bmatrix}, \begin{bmatrix} P \\ Q \end{bmatrix}, \; I \\ \varphi & \\ & J \end{bmatrix} (G, \Phi, \mathscr{P})$$

for any $\mathscr{P} \subseteq \mathscr{A}^n$. Thus

$$(\lambda x) G(\Phi, x) \subseteq [\varphi^I_J]. \qquad (*)$$

This is far enough to carry the investigation of the subsidiary operator $[\varphi^I_J]$ for the moment. Now let us return to the main derivation. Recall, we created $[\varphi^I_J]$ at the very start. Then rule 2 says we may enter, as line 1 in the main derivation, Kt for any $t \in F([\varphi^I_J])$. But F is a type 2 operator, hence is monotone. We have (*), hence $F((\lambda x)G(\Phi, x)) \subseteq F([\varphi\,{}^{I}_{J}])$. Thus we may enter, as line 1 in the main derivation, $\boldsymbol{Kt}$ for any $\boldsymbol{t} \in F((\lambda x)G(\Phi, x))$.

Further, one of our axioms is $\boldsymbol{Kt} \rightarrow \boldsymbol{Rt}$. Thus there is a main derivation of $\boldsymbol{Rt}$ for each $\boldsymbol{t} \in F((\lambda x)G(\Phi, x))$. We have thus shown

$$F((\lambda x)G(\Phi, x)) \subseteq \begin{bmatrix} \begin{bmatrix} \begin{bmatrix} I \\ J \end{bmatrix} \\ K \end{bmatrix}, \begin{bmatrix} \begin{bmatrix} L \\ M \end{bmatrix}, N \\ O \end{bmatrix}, \begin{bmatrix} P \\ Q \end{bmatrix} \\ \varphi \\ R \end{bmatrix} (F, G, \Phi).$$

We leave it to the reader to show this is actually equality.

REMARK. This example shows that composition, for certain orders of type 2 operators, is given by an effective type 3 operator. Other arrangements of orders may be handled similarly. In short, type 2 composition is effective at the type 3 level.

EXAMPLE 5. Let I, J, K, L, M, N, P and Q be predicate symbols as follows:

symbol	a-rity in variables	a-rity in axioms
I	n	$1+n$
J	m	$1+m$
K	k	$1+k$
L	k	$1+k$
M	n	$1+n$
N	m	$1+m$
P	k	k
Q	k	k

and let φ be the following set of elementary formal system axioms (again we assume 1 and 2 are two distinct members of $\mathscr{A}$).

$$I1, \boldsymbol{x} \rightarrow M1, \boldsymbol{x}, \qquad N1, \boldsymbol{y} \rightarrow J1, \boldsymbol{y},$$

$$P\boldsymbol{z} \rightarrow K1, \boldsymbol{z}, \qquad I2, \boldsymbol{x} \rightarrow M2, \boldsymbol{x},$$

$$N2, \boldsymbol{y} \rightarrow J2, \boldsymbol{y}, \qquad L1, \boldsymbol{z} \rightarrow K2, \boldsymbol{z}, \qquad L2, \boldsymbol{z} \rightarrow Q\boldsymbol{z}.$$

Then

$$\begin{bmatrix} \begin{bmatrix} \begin{bmatrix} I \\ J \end{bmatrix}, \ K \\ L \end{bmatrix}, \begin{bmatrix} M \\ N \end{bmatrix}, \ P \\ \varphi \\ Q \end{bmatrix}$$

denotes an effective type 3 operator.

Let F be a type 2 operator, Φ be a type 1 operator, and $\mathscr{P}$ be a type 0 object, all of appropriate orders. Then

$$\begin{bmatrix} \begin{bmatrix} \begin{bmatrix} I \\ J \end{bmatrix}, \ K \\ L \end{bmatrix}, \begin{bmatrix} M \\ N \end{bmatrix}, \ P \\ \varphi \\ Q \end{bmatrix} (F, \Phi, \mathscr{P}) = F(\Phi, F(\Phi, \mathscr{P})).$$

We leave this to the reader.

EXAMPLE 6. Again let I, J, K, L, M, N, P, Q be predicate symbols, now such that

symbol	a-rity in variables	a-rity in axioms
I	n	$1+n$
J	m	$1+m$
K	n	$1+n$
L	m	$1+m$
M	n	$1+n$
N	m	$1+m$
P	n	n
Q	m	m

This time let φ be the following set of axioms:

$$I1, \mathbf{x} \to M1, \mathbf{x}, \qquad N1, \mathbf{y} \to J1, \mathbf{y},$$

$$I2, \mathbf{x} \to K1, \mathbf{x}, \qquad L1, \mathbf{y} \to J2, \mathbf{y},$$

$$P\mathbf{x} \to K2, \mathbf{x}, \qquad L2, \mathbf{y} \to Q\mathbf{y}.$$

Then

$$\left[\begin{matrix} \left[\begin{matrix}\begin{bmatrix} I \\ J\end{bmatrix}, & K \\ & L\end{matrix}\right], \begin{bmatrix} M\\ N\end{bmatrix}, P \\ \varphi \\ \quad Q \end{matrix}\right]$$

denotes an effective type 3 operator.

Let F be type 2, Φ be type 1 and $\mathscr{P}$ be type 0, all of appropriate orders. Then

$$\left[\begin{matrix} \left[\begin{matrix}\begin{bmatrix} I \\ J\end{bmatrix}, & K \\ & L\end{matrix}\right], \begin{bmatrix} M\\ N\end{bmatrix}, P \\ \varphi \\ \quad Q \end{matrix}\right](F, \Phi, \mathscr{P}) = F((\lambda x)F(\Phi, x), \mathscr{P}).$$

7. The least fixed point theorem, again

Let $\mathfrak{A}$ be a structure, fixed for this section. All work is with respect to it.

Let F be a type 2 operator of order $\langle\langle n, m\rangle : \langle n, m\rangle\rangle$. We will show that F has a least fixed point, Φ, and there is an effective type 3 operator that maps such F to their least fixed points. If one supplies an effective type 3 operator with effective input, one gets effective output (combine the axioms for the input for those for the type 3 operator). So a consequence of these results is: an *effective* type 2 operator F of order $\langle\langle n, m\rangle ; \langle n, m\rangle\rangle$ has an *effective* least fixed point Φ.

In Ch. 1, §12 and in Example 3 (Section 4 of the present chapter) we showed similar results one type level lower down. And similar results can also be obtained at higher type levels (once these are defined) with proofs much like those of the present section. The work here must serve as representative of what goes on from this type level up.

THEOREM 7.1. *Let F be a type 2 operator of order $\langle\langle n, m\rangle, n ; m\rangle$. There is a type 1 operator Φ of order $\langle n, m\rangle$ such that*

(1) $(\lambda x)F(\Phi, x) = \Phi$,

(2) *If* $(\lambda x)F(\Psi, x) = \Psi$ *then* $\Phi \subseteq \Psi$.

PROOF. Define, for each ordinal α, a map Φ_F^α of order $\langle n, m\rangle$ as follows.

$$\Phi_F^0 = (\lambda x)\emptyset, \qquad \Phi_F^{\alpha+1} = (\lambda x)F(\Phi_F^\alpha, x)$$

for limit ordinals λ,

$$\Phi_F^\lambda = (\lambda x) \bigcup_{\alpha<\lambda} \Phi_F^\alpha(x).$$

F is a type 2 operator, hence monotone. Then an easy induction on α shows that each Φ_F^α is monotone. Hence each Φ_F^α is a type 1 operator, of order $\langle n, m\rangle$.

Similarly, by using the monotonicity of F, an induction on α shows $\Phi_F^\alpha \subseteq \Phi_F^{\alpha+1}$. It follows that

$$\alpha \leq \beta \Rightarrow \Phi_F^\alpha \subseteq \Phi_F^\beta.$$

That is, the Φ_F^α sequence is increasing in α.

Since, for each α and each $\mathscr{P} \subseteq \mathscr{A}^n$ we have $\Phi_F^\alpha(\mathscr{P}) \subseteq \mathscr{A}^m$, cardinality considerations say that from some point α_0 on, the Φ_F^α sequence must stop growing, that is, $\Phi_F^{\alpha_0} = \Phi_F^{\alpha_0+1} = \Phi_F^{\alpha_0+2} = \cdots$.

Then $\Phi_F^{\alpha_0}$ is a fixed point for F, since

$$\Phi_F^{\alpha_0} = \Phi_F^{\alpha_0+1} = (\lambda x)F(\Phi_F^{\alpha_0}, x).$$

Finally, suppose $(\lambda x)F(\Psi, x) = \Psi$. Then an induction on α gives, for each α, $\Phi_F^\alpha \subseteq \Psi$, hence in particular, $\Phi_F^{\alpha_0} \subseteq \Psi$. Thus Φ^{α_0} is the *least* fixed point for F.

THEOREM 7.2. *There is an* effective *type* 3 *operator* $[\varphi]$ *such that, for each type* 2 *operator* F *of order* $\langle\langle n, m\rangle, n; m\rangle$, $(\lambda x)[\varphi](F, x)$ *is the least fixed point of* F, *whose existence is guaranteed by the previous theorem.*

PROOF. Let I, K and M be n-place predicate symbols, and J, L and N be m-place, in their occurrences in variables below. In the axioms, I, J, K and L occur with one extra place. Consider the effective type 3 operator denoted by

$$\left[\begin{array}{l} \left[\begin{array}{c}\left[\begin{array}{c} I \\ J\end{array}\right], K \\ L\end{array}\right], M \\ \varphi \\ \quad N \end{array}\right]$$

where φ consists of the following elementary formal system axioms:

$$I1, \mathbf{x} \to K1, \mathbf{x}, \qquad L1, \mathbf{y} \to J1, \mathbf{y},$$

$$I2, \mathbf{y} \to K1, \mathbf{x}, \qquad L1, \mathbf{y} \to J2, \mathbf{y},$$

$$M\mathbf{x} \to K2, \mathbf{x}, \qquad L2, \mathbf{y} \to N\mathbf{y}.$$

We generally write $[\varphi]$ for this operator.

We will show that, for any type 2 operator F of order $\langle\langle n, m\rangle, n; m\rangle$, $(\lambda x)[\varphi](F, x)$ is the least fixed point of F, $\Phi_F^{\alpha_0}$ in the notation of the proof of Theorem 7.1. (That notation will be used throughout the present proof.) We will show this by showing, for any F and any $\mathscr{P} \subseteq \mathscr{A}^n$,

$$[\varphi](F, \mathscr{P}) = \Phi_F^{\alpha_0}(\mathscr{P}).$$

Now let F, of order $\langle\langle n, m\rangle, n; m\rangle$ and $\mathscr{P} \subseteq \mathscr{A}^n$ be fixed for the rest of this proof. We want to show there is an elementary formal system derivation, from $[\varphi](F, \mathscr{P})$ of $N\mathbf{y}$ if and only if $\mathbf{y} \in \Phi_F^{\alpha_0}(\mathscr{P})$. In fact, we describe a sort of "maximal" elementary formal system derivation procedure; if $N\mathbf{y}$ has any elementary formal system derivation from $[\varphi](F, \mathscr{P})$, we can produce a derivation of $N\mathbf{y}$ by following our "maximal" procedure, and thus it is all we need to consider.

To begin: at the every start, create the subsidiary operator

$$(\lambda 1, x)[\varphi_J^{M,I}]_0(\mathscr{P}, x).$$

The subscript 0 is to distinguish this from later creations of the "same" subsidiary operator. We will continue the use of ordinal subscripts for this purpose. To determine the behavior of this subsidiary operator, we run subsidiary derivations. But, the only way we could derive $J1, \mathbf{y}$ in a subsidiary derivation is by using the axiom $L1, \mathbf{y} \to J1, \mathbf{y}$. To do this we need $L1, \mathbf{y}$ which can only be introduced by an application of rule 2, and there are no previously created subsidiary operators to use as input for F. In short, $J1, \mathbf{y}$ has no subsidiary derivation for any $\mathbf{y}$. The subsidiary operator just created is simply $(\lambda x)\emptyset$, or in the notation of the previous proof,

$$(\lambda 1, x)[\varphi_J^{M,I}]_0(\mathscr{P}, x) = \Phi_F^0.$$

Thus our "maximal" derivation has begun.

Now suppose our "maximal" derivation has been carried out to the point that the above subsidiary operator has been created α times, and suppose that for each $\beta \leqslant \alpha$ we have determined, about the βth creation, that

$$(\lambda 1, x)[\varphi_J^{M,I}]_\beta(\mathscr{P}, x) = \Phi_F^\beta.$$

We say how to continue the derivation.

After the steps of the derivation just described, create the subsidiary operator in question yet another time; denote it

$$(\lambda 1, x)[\varphi_J^{M,I}]_{\alpha+1}(\mathcal{P}, x).$$

We will show that, in fact, this subsidiary operator is equal to $\Phi_F^{\alpha+1}$, and thus the pattern we have followed through α creations continues another time.

To investigate the behavior of this subsidiary operator, we run subsidiary derivations. Let $\mathcal{Q} \subseteq \mathcal{A}^n$, and run them for

$$((\lambda 1, x)[\varphi_J^{M,I}]_{\alpha+1}(\mathcal{P}, x))(\mathcal{Q}).$$

In any subsidiary derivation we may put down $I1, \boldsymbol{x}$ for any $\boldsymbol{x} \in \mathcal{Q}$. But $I1, \boldsymbol{x} \to K1, \boldsymbol{x}$ is an axiom (the only one with $K1$, in the conclusion) so we may derive $K1, \boldsymbol{x}$ for exactly those $\boldsymbol{x} \in \mathcal{Q}$.

Rule 2 may be applied. It says we may put down $L1, \boldsymbol{y}$ for any

$$\boldsymbol{y} \in F((\lambda 1, x)[\varphi_J^{M,I}]_\beta(\mathcal{P}, x), S)$$

where: $S = \{\boldsymbol{x} \mid K1, \boldsymbol{x} \text{ has been derived}\}$ and $\beta \leqslant \alpha$ so the operator is a previously created subsidiary one. Our assumptions about the earlier parts of the derivation, and the remarks about K, say we can put down $L1, \boldsymbol{y}$ for any

$$\boldsymbol{y} \in F(\Phi_F^\beta, \mathcal{Q}) \quad \text{for } \beta \leqslant \alpha.$$

Now $\beta \leqslant \alpha \Rightarrow \Phi_F^\beta \subseteq \Phi_F^\alpha$, and F is monotone, so it follows that we may derive $L1, \boldsymbol{y}$ for exactly those $\boldsymbol{y} \in F(\Phi_F^\alpha, \mathcal{Q})$. Now, $L1, \boldsymbol{y} \to J1, \boldsymbol{y}$ is an axiom (the only one with $J1$, in the conclusion) so in a subsidiary derivation we may derive $J1, \boldsymbol{y}$ for exactly those $\boldsymbol{y} \in F(\Phi_F^\alpha, \mathcal{Q})$. By these subsidiary derivations we have shown

$$((\lambda 1, x)[\varphi_J^{M,I}]_{\alpha+1}(\mathcal{P}, x))(\mathcal{Q}) = F(\Phi_F^\alpha, \mathcal{Q})$$

for all $\mathcal{Q} \subseteq \mathcal{A}^n$, hence

$$(\lambda 1, x)[\varphi_J^{M,I}]_{\alpha+1}(\mathcal{P}, x) = \Phi_F^{\alpha+1}.$$

Thus the main derivation has advanced by the $\alpha + 1$st creation of a subsidiary operator, and it is, in fact, $\Phi_F^{\alpha+1}$.

Now, let λ be a limit ordinal, and suppose the main derivation has been carried out so that, for each $\alpha < \lambda$, an αth subsidiary operator has been created and we have determined that

$$(\lambda 1, x)[\varphi_J^{M,I}]_\alpha(\mathcal{P}, x) = \Phi_F^\alpha.$$

We say how to continue this pattern to include a λth creation.

After the steps of the main derivation just described, create the subsidiary operator yet another time; denote it

$$(\lambda 1, x)[\varphi_J^{M,I}]_\lambda(\mathcal{P}, x).$$

We will show that it equals Φ_F^λ.

Let $\mathcal{Q} \subseteq \mathcal{A}^n$, and run subsidiary derivations for

$$((\lambda 1, x)[\varphi_J^{M,I}]_\lambda(\mathcal{P}, x))(\mathcal{Q}).$$

In any subsidiary derivation we have $I1, \boldsymbol{x}$ for just the $\boldsymbol{x} \in \mathcal{Q}$, hence $K1, \boldsymbol{x}$ for just the $\boldsymbol{x} \in \mathcal{Q}$.

Rule 2 may be applied, using any previously created subsidiary operator (and $\mathcal{Q}$) as input. Our assumptions about the main derivation thus far then say we can put down $L1, \boldsymbol{y}$ for any $\boldsymbol{y} \in F(\Phi_F^\alpha, \mathcal{Q})$ for $\alpha < \lambda$. But $F(\Phi_F^\alpha, \mathcal{Q}) = \Phi_F^{\alpha+1}(\mathcal{Q})$, and $L1, \boldsymbol{y} \to J1, \boldsymbol{y}$ is an axiom. Hence we may derive $J1, \boldsymbol{y}$ for just those $\boldsymbol{y} \in \Phi_F^{\alpha+1}(\mathcal{Q})$ for some $\alpha < \lambda$. It follows that we can derive $J1, \boldsymbol{y}$ if and only if $\boldsymbol{y} \in \Phi_F^\lambda(\mathcal{Q}) = \bigcup_{\alpha<\lambda} \Phi_F^\alpha(\mathcal{Q})$.

Thus, since $\mathcal{Q}$ was arbitrary,

$$(\lambda 1, x)[\varphi_J^{M,I}]_\lambda(\mathcal{P}, x) = \Phi_F^\lambda$$

and we can return to the main derivation.

In this way we go through the "maximal" derivation, creating subsidiary operator after subsidiary operator, until we reach creation α_0 (notation still from the proof of Theorem 7.1). The above arguments tell us that

$$(\lambda 1, x)[\varphi_J^{M,I}]_{\alpha_0}(\mathcal{P}, x) = \Phi_F^{\alpha_0},$$

and the proof of Theorem 7.1 tells us that $\Phi_F^{\alpha_0}$ is the least fixed point of F.

Now we continue the main derivation, for $[\varphi](F, \mathcal{P})$ as follows.

We can derive $M\boldsymbol{x}$ for each $\boldsymbol{x} \in \mathcal{P}$; do so. $M\boldsymbol{x} \to K2, \boldsymbol{x}$ is an axiom (the only one with $K2$, in the conclusion) so we can derive $K2, \boldsymbol{x}$ for exactly those $\boldsymbol{x} \in \mathcal{P}$. Do so.

Create the subsidiary operator, with parameter 2 this time,

$$(\lambda 2, x)[\varphi_J^{M,I}](\mathcal{P}, x).$$

We determine its behavior. Let $\mathcal{Q} \subseteq \mathcal{A}^n$ and run a subsidiary derivation for

$$((\lambda 2, x)[\varphi_J^{M,I}](\mathcal{P}, x))(\mathcal{Q}).$$

In it we can get $I2, \boldsymbol{x}$ for just those $\boldsymbol{x} \in \mathcal{Q}$. $I2, \boldsymbol{x} \to K1, \boldsymbol{x}$ is an axiom so we can get $K1, \boldsymbol{x}$ for just those $\boldsymbol{x} \in \mathcal{Q}$ (we have no $I1, \boldsymbol{x}$ to work with so $I1, \boldsymbol{x} \to K1, \boldsymbol{x}$ can't be used). Also we have the previously created subsidiary operator

$$(\lambda 1, x)[\varphi_J^{M,I}]_{\alpha_0}(\mathcal{P}, x) = \Phi_F^{\alpha_0}.$$

Then by the rule 2 we can derive $L1, \mathbf{y}$ for just those $\mathbf{y} \in F(\Phi_F^{\alpha_0}, \mathcal{Q})$. But $L1, \mathbf{y} \to J2, \mathbf{y}$ is an axiom, so we can derive $J2, \mathbf{y}$ for just those $\mathbf{y} \in F(\Phi_F^{\alpha_0}, \mathcal{Q})$. Hence we have shown

$$(\lambda 2, x)[\varphi_J^{M,I}](\mathcal{P}, x) = (\lambda x)F(\Phi_F^{\alpha_0}, x) = \Phi_F^{\alpha_0}$$

since $\Phi_F^{\alpha_0}$ is a fixed point of F.

Now return to the main derivation again. We have derived $K2, \mathbf{x}$ for those $\mathbf{x} \in \mathcal{P}$. And we have created the above subsidiary operator, with parameter 2. Then by rule 2 yet again, we may derive $L2, \mathbf{y}$ for those $\mathbf{y} \in F(\Phi_F^{\alpha_0}, \mathcal{P}) = \Phi_F^{\alpha_0+1}(\mathcal{P}) = \Phi_F^{\alpha_0}(\mathcal{P})$.

Finally, $L2, \mathbf{y} \to N\mathbf{y}$ is an axiom. So, by proceeding in this manner we can derive $N\mathbf{y}$ for those $\mathbf{y} \in \Phi_F^{\alpha_0}(\mathcal{P})$.

We leave it to the reader to convince himself that no other derivation can yield anything more, and thus

$$[\varphi](F, \mathcal{P}) = \Phi_F^{\alpha_0}(\mathcal{P}).$$

This concludes the proof.

8. Higher types

The pattern we have begun can be continued to define *effective operators of types* 4, 5, 6, ... As a matter of fact, these all follow the style of type 3, and nothing greatly different is involved in the definitions. So, in the interests of simple formalism, we omit a full treatment of these higher type levels; indeed we even omit the definitions. We content ourselves with a single example of an effective type 4 operator, and leave the rest to the patient reader.

Let I, J be n-place, O, K be k-place, L, M be m-place, and N, P be p-place, both in variables and in axioms. Consider the set φ of elementary formal system axioms:

$$J\mathbf{x} \to I\mathbf{x}, \qquad O\mathbf{y} \to K\mathbf{y},$$

$$L\mathbf{z} \to M\mathbf{z}, \qquad N\mathbf{w} \to P\mathbf{w}.$$

Then

$$\left[\varphi_{P}^{\begin{bmatrix}\begin{bmatrix}\begin{bmatrix}I\\O\end{bmatrix}\\M\end{bmatrix}\\N\end{bmatrix},\begin{bmatrix}\begin{bmatrix}J\\K\end{bmatrix}\\L\end{bmatrix}}\right]$$

denotes an effective type 4 operator, whose behavior we wish to investigate.

Let F be a (monotone) type 3 operator of the order of

$$\begin{bmatrix}\begin{bmatrix}J\\K\end{bmatrix}\\L\end{bmatrix}$$

namely $\langle\langle n,k\rangle;m\rangle$. Let $\mathscr{F}$ be a (monotone) type 4 operator of the order of

$$\begin{bmatrix}\begin{bmatrix}\begin{bmatrix}I\\O\end{bmatrix}\\M\end{bmatrix}\\N\end{bmatrix}$$

namely $\langle\langle\langle n,k\rangle,m\rangle;p\rangle$. We start a main derivation for

$$\begin{bmatrix}\begin{bmatrix}\begin{bmatrix}\begin{bmatrix}I\\O\end{bmatrix}\\M\end{bmatrix}\\N\end{bmatrix}, \begin{bmatrix}\begin{bmatrix}J\\K\end{bmatrix}\\L\end{bmatrix}\\ \varphi \\ P\end{bmatrix}(\mathscr{F},F).$$

Right at the beginning of this main derivation we may *create the subsidiary type* 2 *operator* (with no parameters)

$$(\lambda x^1)\begin{bmatrix}\begin{bmatrix}\begin{bmatrix}J\\K\end{bmatrix}\\L\end{bmatrix}, \begin{bmatrix}I\\O\end{bmatrix}\\ \varphi \\ M\end{bmatrix}(F,x^1).$$

We then determine the behavior of this subsidiary operator by running subsidiary derivations. In fact, the work has already been done, in Example 2, of Section 6. The operator is just F itself. As a matter of fact, it is easy to see that, no matter when, in the main derivation, we create such a subsidiary operator, we always get F.

Now, returning to the main derivation, we may apply a rule analogous to rule 2 in Section 5, let us call it rule 3. In this case it says, having created a subsidiary operator, we may give it to $\mathscr{F}$ as input and add as a line in the main derivation, $N\boldsymbol{w}$ for any $\boldsymbol{w}$ we get as output. That is, in summary, we may add $N\boldsymbol{w}$ as a line in the main derivation for just those $\boldsymbol{w}\in\mathscr{F}(F)$. But

$\boldsymbol{Nw} \to \boldsymbol{Pw}$ is an axiom, the only one involving P. Hence there is a main derivation of $\boldsymbol{Pw}$ for just those $\boldsymbol{w} \in \mathscr{F}(F)$. Thus we say

$$\begin{bmatrix} & \begin{bmatrix}\begin{bmatrix}\begin{bmatrix} I \\ O \end{bmatrix} \\ M \end{bmatrix} \\ N \end{bmatrix}, \begin{bmatrix}\begin{bmatrix} J \\ K \end{bmatrix} \\ L \end{bmatrix} \\ \varphi & \\ \quad P & \end{bmatrix} (\mathscr{F}, F) = \mathscr{F}(F).$$

9. Indexing

In Chapter 6 we showed there is a "usable" indexing of enumeration operators in every recursion theory having an effective pairing function and in which equality is recursive (Corollary 6.14.5). The idea, stripped of all frills, was a simple one. An enumeration operator Φ is specified by an elementary formal system; an elementary formal system is a word; with equality and an effective pairing function available, such words can be coded by members of the domain, and these codes serve very nicely as indexes for Φ.

Now, effective higher type operators are also specified by giving elementary formal systems which, being words, can be coded by members of the domain under the above conditions. It is straightforward, but lengthy, to show that we get suitable universal operators at every type level, much as we did for type 1 in Chapter 6. We omit the proof, and simply state the result.

THEOREM 9.1. *Suppose* rec($\mathfrak{A}$) *is a theory with equality and an effective pairing function. For each type level, n, and for each order* $\langle s\,; k\rangle$ *at that type level, there is an effective operator* $\mathscr{U}$ *of type n and order* $\langle s\,; k+1\rangle$ *which is* universal *in the following sense. If* Φ *is any effective operator of type n and of order* $\langle s\,; k\rangle$, *then for some* $i \in \mathscr{A}$, *called an* index *for* Φ,

$$\Phi = S_i^{k+1}\mathscr{U}.$$

It also can be shown that an analog of the *output place-fixing property* holds at every type level, for each order. Then a proof like that of Theorem 8.3.2 can be given to show a *fixed point* result on indexes holds for every type and order. An analog of *Rice's Theorem* then follows.

The idea naturally suggests itself, then, to create an analog of indexed production system that incorporates higher types. We do not follow up on this here. We note that abstract treatments pertinent to higher types have been successfully developed in the literature, along different lines. See, for example, Moldestad [1977] and the references given there.

10. Conclusion

This is far enough to carry the development of higher type operators. We hope we have conveyed a general feeling for their basic behavior. The reader must be warned, however, that in the extensive literature on higher type recursion theory, operators as we defined them will not be found. It is *functionals* that are considered (partial functions from the domain to itself form the lowest level; partial functions from these functions to the domain, the next; and so on up). We have chosen the approach of this chapter because it is the most natural in the context of the rest of the book, not because it is standard in the field.

Actually, functionals can be introduced quite naturally into our development. Recall, in Ch. 2, §5 we noted that for a partial function f there are two meanings we might give to "computability"; (1) the graph of f is generated; and (2) f is generated pointwise. Now, the pointwise generated version can be extended naturally to higher type levels, making possible a notion of *effective functional of higher type* in the present framework.

We confess, we do not know the relationship between effective functional, developed as sketched above, and recursive functional developed in ways now standard in the literature. We will have to leave this for investigation elsewhere. It would surprise us, however, if the connection turned out to be other than simple and natural.

Whatever the exact nature of the relationship between higher type effective operators and higher type recursive functionals, we believe the work presented in this chapter gives the flavor of the elementary parts of higher type recursion theory. And it serves to introduce higher type operators, which are, we believe, natural objects which deserve further study for their own sakes.

Higher type recursion theory is an active area of research, with many profound results, and promise of much to come. But once the elementary parts of the subject have been passed, complexity grows exponentially. We conclude with a very brief sketch of its origins, and a few references for further reading, notably some survey articles which place the advanced work in the subject into its proper context.

Higher type recursion theory for functionals was introduced by Kleene in Kleene [1959] and Kleene [1963]. It was defined in a way quite unlike any of the then standard approaches to ordinary recursion theory. Briefly, the idea involved defining the recursive functionals, and an indexing for them simultaneously, allowing these indexes to themselves serve as items subject to computation. As Kleene put it, there should be "... means for reflecting upon computation procedures already set up as objects and computing further computation procedures from them." (Kleene [1959].)

In a series of papers (Kleene [1961], [1962], [1962A], [1962B]), Kleene argued for the naturalness of his concepts by showing that a number of quite disparate approaches lead to the same notions. Also see Kleene [1978] for a recent rethinking along different lines again.

But it has been the unusual approach of Kleene [1959] that has become the common one in the field. Actually, when restricted to the lowest type level, it provides an alternate way of developing ordinary recursion theory. And, unlike several other versions of ordinary recursion theory, this one was capable of a successful generalization to arbitrary structures. It is along these lines that Moschovakis developed *prime* and *search computability* in Moschovakis [1969].

Kleene's work has been the basis of a rich and intricate development. Even the initial ideas are quite complex, though. One might do well to consult Hinman [1977] for a survey of the basic, deep problems that are under investigation in the field. And Kechris, Moschovakis [1977] provides an account that makes full use of the years of investigation since the subject began in 1959.

Platek, in his dissertation, Platek [1966], approached the subject from quite a different point of view, partly as a result of feeling dissatisfaction with certain restrictions inherent in the Kleene approach. Basically, he felt that the intuitive idea of a recursive definition was captured in the process of finding the least fixed point of an operator. (As our proofs of Theorem 1.12.1 and Theorem 7.1 show, one produces the least fixed point by "building it up recursively.") So he defined the subject matter of higher type recursion theory to be: the smallest collection of functionals, containing certain "initial" and "combinatorial" ones, containing least-fixed-point functionals, and closed under composition. Note that this definition is index free. Also, as it happens, it applies to arbitrary structures, not just to arithmetic as in Kleene [1959].

Unfortunately Platek never published his work. Moldestad [1977] provides an available account. The Platek and the Kleene approaches do not coincide because Platek can "talk about" functionals that Kleene can not. But they agree on the common ground.

One of the earliest important results was Kleene's discovery of a connection between hyperarithmetic theory and higher type recursion theory, in Kleene [1959]. (See Example 4 of Section 4, above, for a version suitable to our development.) Hyperarithmetic theory yields a natural hierarchy for an interesting class of sets. Kleene's result suggested the possibility that higher type recursion theory could be used to produce other useful hierarchies. This area of research has developed rapidly. Hinman [1978] is a very extensive work on this and related subjects.

APPENDIX

Problems we'd like to see somebody work on

(1) Elementary formal systems can be broadened by allowing function symbols. For example, if s represents the successor *function* of arithmetic, the following generates the addition relation:

$$Px, 0, x,$$

$$Px, y, z \to Px, s(y), s(z).$$

Similarly, elementary formal systems can be restricted in various ways. One example: no variable may occur in an axiom unless it occurs in the conclusion (the above example meets this condition). Another example: in a derivation, only the most recent instance of an atomic formula may be used; thus if $P2, 3, 5$ is below $P2, 2, 4$, then $P2, 2, 4$ is not accessible. In effect, we can only have one thing in a "box" at a time. (This makes no change in what P represents using the axioms of the above example.)

In general, what is the effect of such broadenings and restrictions; do they allow us to characterize any "natural" classes of relations? In particular, can prime computability be characterized? (See Moschovakis [1969].)

(2) Each of the structures $\mathfrak{S}(\mathbb{N})$, $\mathfrak{S}(L_\omega)$ and $\mathfrak{S}(a_1, \ldots, a_n)$ has a notion of *primitive recursive*. (For $\mathfrak{S}(\mathbb{N})$, see Kleene [1952]. For $\mathfrak{S}(L_\omega)$ see Jensen and Karp [1971]. For $\mathfrak{S}(a_1, \ldots, a_n)$, see Asser [1960], or Machtey and Young [1978].) Also each has a notion of Δ_0 relation (all quantifiers bounded in an appropriate way). Are there any general relationships that obtain between these notions? Under embeddings between the recursion theories on these structures, do primitive recursive functions go over into primitive recursive functions? Similarly for Δ_0. If not, for what class of embeddings is this the case?

(3) For the structures of question 2, is there a natural way of characterizing either the primitive recursive functions or the Δ_0 relations using elementary formal systems? If so, does it suggest a generalization to other structures?

(4) For word and set extensions of a structure (Ch. 6, §2), what are the relationships between Δ_0^w and Δ_0^s, or Σ^w and Σ^s? Can similar notions be

developed for the pairing function extension, and related to the above?

(5) In Ch. 5, §4, *reflexive* production systems were defined. Conjecture: all reflexive first order theories are finite. Conjecture: all reflexive recursion theories are countable.

(6) For recursion theories, to what extent does reflexive mean the domain can be "effectively given?" In particular, in Ch. 1, §5, we defined *pure* elementary formal systems. What is the relationship between a recursion theory being reflexive, and a recursion theory being equivalent to its "pure part"?

(7) Let $\mathbb{A}$ be an admissible set, and let $a_1, \ldots, a_n \in \mathbb{A}$. By an $\mathbb{A}$-*word* over $\{a_1, \ldots, a_n\}$ we mean a function f with domain an ordinal, range $f \subseteq \{a_1, \ldots, a_n\}$, and $f \in \mathbb{A}$. The structure of $\mathbb{A}$-words under concatenation is a useful structure. Can such structures be characterized directly without first developing the "set structure" $\mathbb{A}$? Similarly for the generalization from a finite set $\{a_1, \ldots, a_n\}$ of "letters" to an $\mathbb{A}$-finite one.

This is really asking the following. In Chapter 6 we developed word and set structures over a structure $\mathfrak{A}$. The set version, $\mathfrak{A}^s$, has a successful generalization in admissible sets. Does the word version, $\mathfrak{A}^w$, have a similar *direct* generalization?

(8) Can an elementary formal system approach to admissible set recursion theory be designed to handle non-recursively listed admissible sets in a satisfactory way? Possibly one could drop the *chain* sup condition and replace it by a more general notion of limit.

(9) Δ_0 formulas as defined in Ch. 6, §4 do not allow negation, while as defined in Ch. 9, §2, they do. In standard literature on admissible sets, it is customary to allow negation. In effect, one assumes, of a structure, that the given relations *and their complements* are available for use in admissible sets. From our point of view this is less natural than an approach that does not automatically throw in complements of the given relations. Question: can a satisfactory generalization of admissible sets with urelements be developed in which the Δ_0 formulas are *not* assumed to be closed under negations?

(10) By the remarks in Ch. 9, §16, in α-recursion theory: If A is α-regular and α-hyperregular, then B is α-r.e. in A iff B r.e. in $\langle \mathrm{L}_\alpha, A\rangle$. But what is an appropriate generalization of enumeration reducibility along these lines? The trouble is, the definition of Δ_0, Σ, etc. for admissible sets allows negation (see question 9) and hence in the structure $\langle \mathrm{L}_\alpha, A\rangle$ we have, in effect, both A and $\bar{A}$ to work with. But to generalize enumeration reducibility this way, we would like to have A without necessarily having $\bar{A}$ too. Question: If a generalization of admissible sets along the lines of

question 9 is possible, can it be used to generalize enumeration reducibility as suggested above?

(11) A major topic in admissible set recursion theory is that of HYP (see Barwise [1975]). Briefly, for a structure $\mathfrak{A}$ (in which each relation and its complement is given, as well as $\neq_{\mathscr{A}}$), HYP($\mathfrak{A}$) is an admissible set and, if ω-rec($\mathfrak{A}$) has an effective pairing function, then: a relation $\mathscr{R}$ on $\mathfrak{A}$ is ω-r.e. in ω-rec($\mathfrak{A}$) iff $\mathscr{R}$ is Σ on HYP($\mathfrak{A}$), that is, iff $\mathscr{R}$ is r.e., in the admissible set sense, on HYP($\mathfrak{A}$). (See Barwise [1975], p. 230.)

It would be nice if this connection could be extended to operators as well. We conjecture that if admissible set recursion theory can be successfully generalized along the lines indicated in question 9, then the relationship between ω-rec($\mathfrak{A}$) and HYP($\mathfrak{A}$) (in the broadened sense) will extend to the operators in the compact core of ω-rec($\mathfrak{A}$).

(12) What is the exact relationship between higher type recursion theory, as in Platek [1966], and as in Chapter 10?

(13) What is the exact relationship between Indexed Production Systems, with the various finiteness assumptions, and Computation Theories? (See Moschovakis [1971], Fenstad [1974] and Stoltenberg-Hansen [1979].)

(14) What is an appropriate extension of Indexed Production Systems to axiomatize the basic features of effective operators of higher types, as defined in Chapter 10?

Note added in proof

This is a partial answer to item (5) above. All reflexive first order theories are finite, by the following argument.

Suppose $\boldsymbol{A}$ is reflexive (relative to $\dot{\cup}$ say) where $\boldsymbol{A} = \text{f.o.}(\mathfrak{A})$. Then $(\text{inj}_{\mathscr{A}})^{\dot{\cup}}$ is actually defined by a first order formula $\varphi(x, y)$ over the structure $\mathfrak{A} \dot{\cup} \mathfrak{A}$. Let $\mathfrak{B}$ be a structure elementarily equivalent to $\mathfrak{A}$. By the work in Fefferman, Vaught [1959], since $\mathfrak{A} \equiv \mathfrak{B}$ then $\mathfrak{A} \dot{\cup} \mathfrak{A} \equiv \mathfrak{A} \dot{\cup} \mathfrak{B}$. Now there is a formula that "expresses" the fact that φ defines a one–one correspondence. This formula is true over $\mathfrak{A} \dot{\cup} \mathfrak{A}$, hence it is true over $\mathfrak{A} \dot{\cup} \mathfrak{B}$. It follows that $\mathfrak{A}$ and $\mathfrak{B}$ are structures of the same cardinality. Since any structure elementarily equivalent to $\mathfrak{A}$ is the same size as $\mathfrak{A}$, the usual Skolem–Löwenheim arguments say $\mathfrak{A}$ must be a finite structure.

BIBLIOGRAPHY

Asser, G.
[1960] Rekursive Wortfunktionen, *Zeitschrift für mathematische Logik und Grundlagen der Mathematik* **6** (1960) 258–278.

Barwise, J.
[1969] Infinitary logic and admissible sets, *Journal of Symbolic Logic* **34** (1969) 226–252.
[1974] Admissible sets over models of set theory, *Generalized Recursion Theory*, edited by J.E. Fenstad and P.G. Hinman (North-Holland, Amsterdam, 1974) 97–122.
[1975] *Admissible Sets and Structures* (Springer-Verlag, Berlin, 1975).
[1975A] Review of "Elementary Induction on Abstract Structures" by Y.N. Moschovakis, *Bulletin of the American Mathematical Society* **81** (1975) 1031–1035.

Beth, E.W.
[1964] *The Foundations of Mathematics*, revised edition (North-Holland, Amsterdam, 1964); paperback version (Harper & Row, New York, 1966).

Cantor, G.
[1878] Ein Beitrag zur Mannigfaltigkeitslehre, *Journal für die reine und angewandte Mathematik* (Crelle's Journal) **84** (1878) 242–258. Reprinted in: *Gesammelte Abhandlungen mathematischen und philosophischen Inhalts*, edited by E. Zermelo (Teubner, Berlin, 1932; reprinted Olms, Hildesheim, 1962).
[1895] Beitrage zur Begrundung der transfiniten Mengenlehre, *Mathematische Annalen* **46** (1895) 481–512. Reprinted, in English translation, in: *Contributions to the Founding of the Theory of Transfinite Numbers*, translated and edited by P.E.B. Jourdain (Open Court Publ. Co., 1915; reissued by Dover, New York, 1955).

Church, A.
[1936] An unsolvable problem of elementary number theory, *American Journal of Mathematics* **58** (1936) 345–363. Reprinted in *The Undecidable*, edited by M. Davis (Raven Press, Hewlett, NY, 1965) 88–107.

Davis, M.
[1950] On the theory of recursive unsolvability, Ph.D. dissertation, Princeton University, 1950.

Di Paola, R.
[1978] The theory of partial α-recursive operators, *Notices of the American Mathematical Society* **25**, No. 78T–E59 (1978) A–497.
[1978A] The operator gap theorem in α-recursion theory, *Archiv für mathematische Logik und Grundlagenforschung* **19** (1978) 115–129.
[1979] The α-operator gap theorem, *Chinese Journal of Computers* **2** (1979) 163–173.
[1981] A lift of a theorem of Friedberg: a Banach–Mazur functional that coincides with no α-recursive functional on the class of α-recursive functions, *Journal of Symbolic Logic* **46** (1981) 216–232.

[198+] The elementary theory of partial α-recursive operators, to appear in *Annali di Matematica Pura ed Applicata.*

EILENBERG, S., ELGOT, C.
[1970] *Recursiveness* (Academic Press, New York, 1970).

EVEN, S., RODEH, M.
[1978] Economical encoding of commas between strings, *Communications of the A.C.M.* **21** (1978) 315–317.

FEFFERMAN, S., VAUGHT, R. L.
[1959] The first order properties of algebraic systems, *Fundamenta Mathematicae* **47** (1959) 57–103.

FENSTAD, J.E.
[1974] On axiomatizing recursion theory, *Generalized Recursion Theory*, edited by J.E. Fenstad and P.G. Hinman (North-Holland, Amsterdam, 1974) 385–404.
[1975] Computation theories: an axiomatic approach to recursion on general structures, *Logic Conference Kiel*, 1974, edited by Müller, Obershelp, and Potthoff, Springer Lecture Notes **499** (Springer-Verlag, Berlin, 1975), 143–168.
[1980] *Computation Theories: an Axiomatic Approach to Recursion on General Structures* (Springer-Verlag, Berlin, 1980).

FITTING, M.
[1969] *Intuitionistic Logic Model Theory and Forcing* (North-Holland, Amsterdam, 1969).
[1978] Elementary formal systems for hyperarithmetical relations, *Zeitschrift für mathematische Logik und Grundlagen der Mathematik* **24** (1978) 25–30.
[1979] An axiomatic approach to computers, *Theoria* **45** (1979) 97–113.

FRAÏSSE, R.
[1961] Une notion de récursivité relative, *Infinitistic methods, Proceedings of the Symposium on Foundations of Mathematics*, Warsaw, 1959 (Pergamon, Oxford; PWN, Warsaw, 1961) 323–328.

FRIEDBERG, R.M.
[1957] Two recursively enumerable sets of incomparable degrees of unsolvability, *Proceedings of the National Academy of Sciences* **43** (1957) 236–238.

FRIEDMAN, H.
[1971] Axiomatic recursive function theory, *Logic Colloquium '69*, edited by Gandy and Yates (North-Holland, Amsterdam, 1971) 113–137.

GÖDEL, K.
[1939] Consistency proof for the generalized continuum hypothesis, *Proceedings of the National Academy of Sciences* **25** (1939) 220–224.
[1940] *The Consistency of the Axiom of Choice and the Generalized Continuum Hypothesis With the Axioms of Set Theory* (Princeton University Press, Princeton, 1940).

GORDON, C.
[1970] Comparisons between some generalizations of recursion theory, *Composito Math.* **22** (1970) 333–346.

GRILLOT, T.J.

[1974] Dissecting abstract recursion, *Generalized Recursion Theory*, edited by J.E. Fenstad and P.G. Hinman (North-Holland, Amsterdam, 1974) 405–420.

HINMAN, P.G.

[1977] A survey of finite type recursion, *Set Theory and Hierarchy Theory V*, Proc. Third Conf., Bierutowice, 1976, Springer Lecture Notes **619** (Springer-Verlag, Berlin, 1977) 187–209.

[1978] *Recursion-Theoretic Hierarchies* (Springer-Verlag, Berlin, 1978).

HOROWITZ, B.M.

[1975] Theory of Effectively Non-arithmetical Sets, Ph.D. dissertation, City University of New York, 1975.

[1975A] Relative recursion theory via elementary formal systems (abstract), *Journal of Symbolic Logic* **40** (1975) 301–302.

[198+] Elementary formal systems as a framework for relative recursion theory, to appear in *Notre Dame Journal of Formal Logic*.

JENSEN, R. B., KARP, C.

[1971] Primitive recursive set functions, *Proceedings of Symposia in Pure Mathematics XIII*, Part I (American Mathematical Society, Providence, RI, 1971) 143–176.

KECHRIS, A., MOSCHOVAKIS, Y.

[1977] Recursion in higher types, *Handbook of Mathematical Logic*, edited by J. Barwise (North-Holland, Amsterdam, 1977) 681–737.

KLEENE, S.C.

[1938] On notation for ordinal numbers, *Journal of Symbolic Logic* **3** (1938) 150–155.

[1943] Recursive predicates and quantifiers, *Transactions of the American Mathematical Society* **53** (1943) 41–73.

[1952] *Introduction to Metamathematics* (D. Van Nostrand, North-Holland, P. Noordhoff, 1952).

[1955] Hierarchies of number-theoretic predicates, *Bulletin of the American Mathematical Society*, **61** (1955) 193–213.

[1959] Recursive functionals and quantifiers of finite type I, *Transactions of the American Mathematical Society* **91** (1959) 1–52.

[1961] Lambda definable functionals of finite type, *Fundamenta Mathematica* **50** (1961) 281–303.

[1962] Herbrand–Gödel style recursive functionals of finite types, *Proceedings Symp. Pure Math.* Vol. V (American Mathematical Society, Providence, RI, 1962) 49–75.

[1962A] Turing machine computable functionals of finite type I, *Logic, Methodology and Philosophy of Science*, Proc. of the 1960 International Congress, edited by E. Nagel, P. Suppes and A. Tarski (Stanford University Press, Stanford, 1962) 38–45.

[1962B] Turing machine computable functionals of finite type II. *Proceedings of the London Mathematical Society* (Ser. 3) **12** (1962) 245–258.

[1963] Recursive functionals and quantifiers of finite type II, *Transactions of the American Mathematical Society* **108** (1963) 106–142.

[1978] Recursive functionals and quantifiers of finite types revisited I, *Generalized Recursion Theory II*, edited by J.E. Fenstad, R.O. Gandy and G.E. Sacks (North-Holland, Amsterdam, 1978) 185–222.

KLEENE, S.C., POST, E.L.

[1954] The upper semi-lattice of degrees of recursive unsolvability, *Annals of Mathematics* (Ser. 2) **59** (1954) 379–407.

KREISEL, G., SACKS, G.

[1965] Metarecursive sets, *Journal of Symbolic Logic* **30** (1965) 318–338.

KRIPKE, S.

[1964] Transfinite recursion on admissible ordinals I & II (abstracts), *Journal of Symbolic Logic* **29** (1964) 161–162.

KURATOWSKI, C.

[1921] Sur la notion de l'ordre dans la theorie des ensembles, *Fundamenta Mathematica* **2** (1921) 161–171.

LACOMBE, D.

[1964] Deux généralisations de la notion de récursivité, *Comptes Rendus de l'Academie des Sciences de Paris* **258** (1964) 3141–3143.

[1964A] Deux généralisations de la notion de récursivité relative, *Comptes Rendus de l'Academie des Sciences de Paris* **258** (1964) 3410–3413.

LEVY, A.

[1965] *A Hierarchy of Formulas in Set Theory*, Memoir Amer. Math. Soc., Vol. 57 (1965).

MACHTEY, M., YOUNG, P.

[1978] *An Introduction to the General Theory of Algorithms* (North-Holland, New York, 1978).

MACLANE, S.

[1971] *Categories for the Working Mathematician* (Springer-Verlag, Berlin, 1971).

MARKOV, A.A.

[1946] On the impossibility of certain algorithms in the theory of associative systems, English translation in *Comptes Rendus (Doklady) de l'Academie des Sciences de l'URSS* (new series) **55** (1946) 583–586.

MOLDESTAD, J.

[1977] *Computations in Higher Types*, Springer Lecture Notes **574** (Springer-Verlag, Berlin, 1977).

MONTAGUE, R.

[1968] Recursion theory as a branch of model theory, *Logic, Methodology, and Philosophy of Science III*, Proc. of the 1967 Congress, B. van Rootselaar et al, editors (North-Holland, Amsterdam, 1968) 63–86.

MOSCHOVAKIS, Y.N.

[1969] Abstract first order computability I & II, *Transactions of the American Mathematical Society* **138** (1969) 427–464 and 465–504.

[1969A] Abstract computability and invariant definability, *Journal of Symbolic Logic* **34** (1969) 605–633.

[1971] Axioms for computation theories — first draft, *Logic Colloquium '69*, edited by R.O. Gandy and C.E.M. Yates (North-Holland, Amsterdam, 1971) 199–255.

[1974] *Elementary Induction on Abstract Structures* (North-Holland, Amsterdam, 1974).

MOSTOWSKI, A.
[1947] On definable sets of positive integers, *Fundamenta Mathematicae* **34** (1947) 81–112.
[1951] A classification of logical systems, *Studia Philosophica* **4** (1951) 237–274.

MUCHNIK, A.A.
[1956] On the unsolvability of the problem of reducibility in the theory of algorithms, *Doklady Akademii Nauk SSSR* (new series) **108** (1956) 194–197.

MYHILL, J.
[1955] A fixed-point theorem in recursion theory (abstract), *Journal of Symbolic Logic* **20** (1955) 205.
[1955A] Creative sets, *Zeitschrift für mathematische Logik und Grundlagen der Mathematik* **1** (1955) 97–108.

MYHILL, J., SHEPHERDSON, J.C.
[1955] Effective operative operations on partial recursive functions, *Zeitschrift für mathematische Logik und Grundlagen der Mathematik* **1** (1955) 310–317.

PEANO, G.
[1890] Sur une courbe, qui remplit tote une aire plane, *Mathematische Annalen* **36** (1890) 157–160. Reprinted in English translation in *Selected Works of Giuseppe Peano*, by Hubert C. Kennedy (George Allen & Unwin, London, 1973) 143–149.

PLATEK, R.
[1966] Foundations of Recursion Theory, Ph.D. dissertation (and supplement), Stanford University, 1966.

POST, E.L.
[1936] Finite combinatory processes. Formulation I, *Journal of Symbolic Logic* **1** (1936) 103–105. Reprinted in *The Undecidable*, edited by Martin Davis (Raven Press, Hewlett, NY, 1965) 289–291.
[1943] Formal reductions of the general combinatorial decision problem, *American Journal of Mathematics* **65** (1943) 197–215.
[1947] Recursive unsolvability of a problem of Thue, *Journal of Symbolic Logic* **12** (1947) 1–11. Reprinted in *The Undecidable*, edited by Martin Davis (Raven Press, Hewlett, NY, 1965) 292–303.

QUINE, W.V.
[1946] Concatenation as a basis for arithmetic, *Journal of Symbolic Logic* **11** (1946) 105–114.

RICE, G.H.
[1953] Classes of recursively enumerable sets and their decision problems, *Transactions of the American Mathematical Society* **74** (1953) 358–366.
[1956] On completely recursively enumerable classes and their key arrays, *Journal of Symbolic Logic* **21** (1956) 304–308.

ROGERS, H.
[1967] *Theory of Recursive Functions and Effective Computability* (McGraw-Hill, New York, 1967).

SACKS, G.E.
[1967] Metarecursion theory, *Sets, Models and Recursion Theory*, Proceedings of the Summer School in Mathematical Logic, Leicester, 1965, edited by J. Crossley (North-Holland, Amsterdam, 1967) 243–263.

SCOTT, D.
[1976] Data types as lattices, *SIAM Journal of Computing* **5** (1976) 522–587.

SIMPSON, S.G.
[1974] Degree theory on admissible ordinals, *Generalized Recursion Theory*, edited by J.E. Fenstad and P.G. Hinman (North-Holland, Amsterdam, 1974) 165–193.
[1978] Short course on admissible recursion theory, *Generalized Recursion Theory II*, edited by J.E. Fenstad, R.O. Gandy and G.E. Sacks (North-Holland, Amsterdam, 1978) 355–390.

SMULLYAN, R.M.
[1956] On definability by recursion (abstract), *Bulletin of the American Mathematical Society* **62** (1956) 601.
[1956A] Elementary formal systems (abstract), *Bulletin of the American Mathematical Society* **62** (1956) 600.
[1961] *Theory of Formal Systems*, revised edition (Princeton University Press, Princeton, 1961).

SPECTOR, C.
[1955] Recursive well-orderings, *Journal of Symbolic Logic* **20** (1955) 151–163.

STOLTENBERG–HANSEN, V.
[1979] Finite injury arguments in infinite computation theories, *Annals of Mathematical Logic* **16** (1979) 57–80.

STRONG, H.R.
[1968] Algebraically generalized recursive function theory, *IBM Journal of Research and Development* **12** (1968) 465–475.

TAKEUTI, G.
[1960] On the recursive functions of ordinal numbers, *Journal of the Mathematical Society of Japan* **12** (1960) 119–128.

TURING, A.M.
[1936] On computable numbers, with an application to the Entscheidungsproblem, *Proceedings of the London Mathematical Society* (Ser. 2) **42** (1936–37) 230–265. Corrections, ibid. **43** (1937) 544–546. Reprinted in *The Undecidable*, edited by Martin Davis (Raven Press, Hewlett, NY, 1965) 115–154.

WAGNER, E.G.
[1969] Uniform reflexive structures: on the nature of Gödelizations and relative computability, *Transactions of the American Mathematical Society* **144** (1969) 1–41.

WIENER, N.
[1912] A simplification of the logic of relations, *Proc. Camb. Phil. Soc.* **17** (1912–1914) 387–390.

INDEX